Ageing in a Nursing Home

Rosalie Hudson

Ageing in a Nursing Home

Foundations for Care

Rosalie Hudson
Thornbury, VIC, Australia

ISBN 978-3-030-98266-9   ISBN 978-3-030-98267-6 (eBook)
https://doi.org/10.1007/978-3-030-98267-6

© The Editor(s) (if applicable) and The Author(s), under exclusive license to Springer Nature Switzerland AG 2022
This work is subject to copyright. All rights are solely and exclusively licensed by the Publisher, whether the whole or part of the material is concerned, specifically the rights of translation, reprinting, reuse of illustrations, recitation, broadcasting, reproduction on microfilms or in any other physical way, and transmission or information storage and retrieval, electronic adaptation, computer software, or by similar or dissimilar methodology now known or hereafter developed.
The use of general descriptive names, registered names, trademarks, service marks, etc. in this publication does not imply, even in the absence of a specific statement, that such names are exempt from the relevant protective laws and regulations and therefore free for general use.
The publisher, the authors and the editors are safe to assume that the advice and information in this book are believed to be true and accurate at the date of publication. Neither the publisher nor the authors or the editors give a warranty, expressed or implied, with respect to the material contained herein or for any errors or omissions that may have been made. The publisher remains neutral with regard to jurisdictional claims in published maps and institutional affiliations.

This Springer imprint is published by the registered company Springer Nature Switzerland AG
The registered company address is: Gewerbestrasse 11, 6330 Cham, Switzerland

Preface

Spending the final chapter of your life in a nursing home is considered by many a fate worse than death, but this need not be so. The radical suggestion of this book is that with enlightened, imaginative care, even the frailest of lives can flourish. We all have a stake in building a society that values its most vulnerable citizens. The key is to replace the former custodial attitudes with contemporary, evidence-based care. Such a cultural/philosophical shift is perceived as an urgent need in a rapidly ageing population, described in Chap. 1, where welcoming diversity and creating hope are of the essence. In Chap. 2, communication is shown to be an important signaller of attitude and intent from the waiting list to the end of life, with a particular focus on pivotal points of staff 'handover'. Issues of non-discrimination are described in the chapter and reinforced throughout the whole book.

Government inquiries and COVID-19 pandemic revelations of sub-standard aged care call for systemic renewal. This book offers encouragement and practical guidance for those who strive for improvement: transformation springs from evidence-based practice. Chapter 3 highlights the building of confidence through partnerships with families and carers, promoting the use of knowledge that comes from residents' own experiences. The chapter describes the highest standards of clinical care by focussing on the social and gerontological aspects of the whole person.

Learning from the residents is key to responsive, enlightened dementia care, a notion permeating Chap. 4. Due to extended life expectancy, nursing home residents are now older and frailer than ever before, with over 50 per cent having a diagnosis of dementia. Most have other complex, chronic conditions requiring significant multidisciplinary team input from healthcare professionals with broad aged care expertise. Nursing home managers are encouraged to reflect on where their emphasis lies, and the type of workforce best equipped to respond to residents' increasingly complex cognitive care needs.

Chapter 5 describes the way a palliative approach to care brings comfort to the resident when cure is no longer the aim. Residents are older and suffering greater co-morbidity than ever before, justifying the need for clinical governance which includes provision of palliative end-of-life care. Every resident has the right to a second opinion from an appropriate specialist in response to their life-threatening medical conditions, and the benefits of palliation deserve greater attention.

Recognising that most residents admitted to a nursing home will die there, Chap. 6 brings the subjects of death and dying to the fore, rather than leaving the matter unaddressed or left until 'the end'. Compassionate, timely conversations inspire confidence that best practice end-of-life care is assured for each resident. Leadership is the focus for Chap. 7, where partnership with residents, carers and families is also highlighted. A key challenge is to ensure such care addresses the combined biological, psychological, spiritual and social factors that influence residents' health and well-being.

Many thousands of submissions to the 2019 Royal Commission into Aged Care Quality and Safety attest to the fact that, as a society, we do not value older people or aged care, prompting the title of the Interim Report: *Neglect,* and the Final Report (2021): *Care, Dignity and Respect.* Examples from these reports are cited in order to underscore the value of using evidence to guide practice.

Returning to the fears of those who would rather die than live in a nursing home, the last chapter (Chap. 8) describes how excellence may be achieved, while accepting the inevitable occasions when things go wrong. The subject of elder abuse is also brought into the open, with suggestions for ameliorating this scourge.

Vignettes in every chapter describe the unique lives of individual residents and their carers. This is, or should be, what nursing home care is about. Within the framework of optimal medical, nursing, spiritual, psychological, social, volunteer and allied health support we encounter the totality of the resident's life, prompting targeted, responsive care. If caring is about giving and receiving, the ultimate question may be: what can this person teach me about how to care?

In an ageist society that prizes independence and individualism, a nursing home can be a model of partnership where each resident is valued for their total life's experience, afforded the care they deserve and contributing to their own care where possible. Such care may help to reverse society's principles and aims wherever suboptimal care is accepted. A focus on best practice within vibrant communities helps to reverse this trend, particularly when at least some aspects of healthy ageing remain achievable for many residents. While it may not be a substitute for their former home, the nursing home may be a place which transforms the final chapter of their lives.

Thornbury, VIC, Australia   Rosalie Hudson

Acknowledgements

Many of the stories in this book are anecdotes drawn from the lived experience of staff and residents of a Melbourne nursing home, where I spent 12 challenging and satisfying years as director of nursing/manager. Decades after its closure, contemporary sources continue to teach me how to care for the frail older people who entrust us with their lives.

I am indebted to my family for their unfailing love and patience. My daughter, Dr Elizabeth Hudson, used her sharp writing skills to refine much of the detail, as well as ensuring I did not forget to eat while writing the book. My younger son, Professor Peter Hudson, contributed his unique knowledge of palliative care to hone the references to this emerging priority for residential aged care. My older son, Timothy Hudson, a constant source of wisdom and encouragement, together with my four adult grandsons, has motivated me with his unique comments and ready humour. I am grateful for my two daughters-in-law, both skilled senior nurses, from whom I have drawn much practical knowledge and inspiration.

My dear friend, Jennifer Richmond, co-author of two of my previous books, has added much needed advice from her editorial experience, and her intimate knowledge of nursing homes. This book would undoubtedly have been enhanced by her distinctive, creative writing.

My local federal MP Ged Kearney, whose special insight into the subject matter, both from nursing experience and a passionate desire to see justice for older Australians in residential care, has encouraged me to pursue this project. I thank her for her keen interest, albeit within her own very demanding political life.

The unfailing confidence of my loyal friends has sustained me, and they may now be relieved that my attention will no longer be single-focussed.

My late husband John provided more than 50 years of loving support and encouragement for every aspect and achievement of my nursing career. He would, no doubt, be proud to see this book.

Contents

1	**Ageing in a Changing Community**		1
	1.1 An Ageing Population		1
		1.1.1 Person or Individual?	3
		1.1.2 Ageism and Human Rights	4
		1.1.3 Ethics and Culture	6
		1.1.4 Funding	8
		1.1.5 For Profit?	9
		1.1.6 Trusting Our Aged Care Services?	10
		1.1.7 More or Less Homelike	10
		1.1.8 Ethno-Specific Nursing Homes	11
	1.2 Describing Aged Care Communities		12
		1.2.1 Home, Community or Facility	12
		1.2.2 Shared Rooms or Single	14
		1.2.3 Respite Care	15
	1.3 Welcoming Diversity		16
		1.3.1 Cultural Mix: The Nursing Home's Core	16
		1.3.2 Aboriginal and Torres Strait Islander People	18
		1.3.3 Lesbian, Gay, Bisexual, Transgender/Gender-Diverse, Intersex, and Queer (LGBTIQ) Residents	19
		1.3.4 Young People in Nursing Homes (YPINH)	20
	References		21
2	**Communication: The Key to Care**		25
	2.1 From Waiting List to Admission		25
		2.1.1 The Entrance Foyer	25
		2.1.2 Information	26
		2.1.3 Pre-Admission	26
		2.1.4 Choosing a Nursing Home	27
		2.1.5 The Waiting List	30
		2.1.6 Expectations	30
	2.2 Communication without Discrimination		31

		2.2.1	Communication Keys and Cues	31
		2.2.2	The Short-Term Stay	36
		2.2.3	Loved One: Unambiguous Communication?	37
	2.3	Between Agencies and Age Groups		38
		2.3.1	Relatives Caring	38
		2.3.2	Children Welcome	39
		2.3.3	Hospitalisation and Transfers	42
	2.4	Handover Conversations		44
		2.4.1	Depends Who's on	44
		2.4.2	Staff Reactions to Handover	44
		2.4.3	Hands-on Care	46
	References			47
3	**Social, Gerontological Care**			49
	3.1	Informal Carers		49
		3.1.1	Families and beyond	49
		3.1.2	Citizenship	50
		3.1.3	Volunteers	50
	3.2	Expanding Partnerships		52
		3.2.1	Planning Ahead	52
		3.2.2	Academic Communities Welcome?	54
		3.2.3	Engaging the Wider Community	56
	3.3	Clinical Care		57
		3.3.1	Learning from Nightingale	57
		3.3.2	Duty of Care	58
		3.3.3	General Practitioners (GPs)	59
		3.3.4	Telehealth	61
		3.3.5	Comprehensive Medical Assessment (CMA)	61
		3.3.6	Medication Management: Pharmacist's Role	62
		3.3.7	Staffing the Nursing Home	63
		3.3.8	Registered Nurse (RN)	63
		3.3.9	Enrolled Nurse (EN)	65
		3.3.10	Personal Care Assistant (PCA)	65
		3.3.11	Nurse Practitioner (NP)	66
		3.3.12	Skincare	67
		3.3.13	Nutrition and Hydration	67
		3.3.14	Mealtime Experience	69
		3.3.15	Enteral Feeding	70
		3.3.16	Alcoholic Beverages	70
		3.3.17	Dehydration	71
		3.3.18	Elimination: Bowel and Bladder Management	71
		3.3.19	Quality Continence Care	72
		3.3.20	Breathing Difficulties	72
	3.4	Infections		72
		3.4.1	Oral and Dental Care	74

		3.4.2	Pain Management	75
		3.4.3	Mental Health and Depression	75
		3.4.4	Cardiopulmonary Resuscitation (CPR) and Not for Resuscitation (NFR)	76
	3.5	Care for the Whole Person		77
		3.5.1	Who Makes the Decisions?	77
		3.5.2	Allied Health	78
		3.5.3	Music, Singing and Dance	79
		3.5.4	Pastoral, Spiritual and Religious Care	80
		3.5.5	Sexual Expression	83
		3.5.6	Psychosocial Care	84
		3.5.7	Loneliness	86
		3.5.8	Rituals, Celebrations and Laughter	86
		3.5.9	Laughter	88
		3.5.10	Podiatry	88
		3.5.11	Personal Clothing and Laundering	89
		3.5.12	Hairdressing	89
		3.5.13	Bus Trips and Excursions	90
		3.5.14	Dolls and Pets	90
		3.5.15	Sensory Loss	91
		3.5.16	Sleep	92
		3.5.17	Information Technology (IT)	92
		3.5.18	Robotic Care	93
		3.5.19	Independence, Privacy, Respect	93
	References			94
4	**Dementia Challenges**			99
	4.1	Dementia Misconceptions		99
		4.1.1	Dementia Definition, Diagnosis and Prognosis	100
		4.1.2	Types of Dementia	101
		4.1.3	Alzheimer's Disease (AD)	101
		4.1.4	Dementia with Lewy Bodies (DLB)	102
		4.1.5	Younger Onset Dementia	102
		4.1.6	Government Responsibility	103
	4.2	Who Is This Person?		103
		4.2.1	Who Am I and Where Am I?	103
		4.2.2	Loss of Self?	104
		4.2.3	Focus on the *Person*, Not the *Disease*	105
	4.3	Evidence-Based Care		107
		4.3.1	Leadership and Education in Dementia Care	107
		4.3.2	Qualified Carers	108
		4.3.3	Restraint Usage in Dementia	109
		4.3.4	Language and Dementia	110
		4.3.5	Pain and Behaviour	110
		4.3.6	Violent Behaviour	111

	4.3.7	Music for Pleasure	112
	4.3.8	Activities, Laughter and Play	113
	4.3.9	Segregation of People Living with Dementia	114
	4.3.10	The Dignity of Risk	114
	4.3.11	The *Person* Who Is 'At Risk'	115
	4.3.12	Nursing Home Culture	116
	4.3.13	Nursing Home Design	117
	4.3.14	Psychiatric Drug Prescribing	117
	4.3.15	Keeping Up with the News	118
	4.3.16	Aversion to Water?	119
	4.3.17	Communal Lounge	119
	4.3.18	Self-Care?	120
	4.3.19	Depression and Delirium	121
	4.3.20	Care at Night	121
	4.3.21	End-of-Life Care	122
	4.3.22	Bereavement Planning	124
	4.3.23	Limitation of Care Orders (LCOs)	124
	4.3.24	Social Worker Role	125
4.4	Families Flourishing or Floundering		125
	4.4.1	Information Deficit	125
	4.4.2	Family Assessment: Caring for the Carers	126
	4.4.3	Communication	127
	4.4.4	Wandering: A Management Problem?	128
	4.4.5	Visiting	129
	4.4.6	Spiritual Care	130
	4.4.7	Involving Families in Care Planning	131
	4.4.8	Touch	131
	4.4.9	Sexuality	132
	4.4.10	Mealtimes	132
	4.4.11	Beyond Stereotypes	133
References			134

5 A Palliative Approach ... 137
  5.1 What Is Palliative Care? ... 137
    5.1.1 Dispelling the Myths ... 137
    5.1.2 Who May Benefit from Palliative Care? ... 138
    5.1.3 Misunderstanding Palliative Care ... 140
    5.1.4 Comprehensive, Professional Reporting ... 141
    5.1.5 Multidisciplinary Team ... 141
    5.1.6 Specialist Palliative Care ... 143
    5.1.7 Managers' Knowledge ... 144
    5.1.8 Palliative Care 'Needs Rounds' ... 144
    5.1.9 New Technologies ... 145
    5.1.10 Food, Feeding and Fluids ... 145
    5.1.11 Dementia Needs a Palliative Approach ... 146

		5.1.12 Quality of Life Questioned	147
		5.1.13 Fear of Becoming a Burden	148
		5.1.14 Acknowledging Signs of Dying	148
		5.1.15 Terminal Stage	150
		5.1.16 Final Wishes	150
	5.2	Palliative Care in Partnership	151
		5.2.1 Liaison with Palliative Care Services	151
		5.2.2 Hospital in the Nursing Home	151
		5.2.3 Core Business or Optional Extra?	152
		5.2.4 GPs Delivering Palliative Care	153
		5.2.5 Pharmacists' Role	153
		5.2.6 Link Nurse	154
		5.2.7 Involving Families	154
		5.2.8 Palliative Care Education	156
		5.2.9 Palliative Care for Chronic Conditions	157
		5.2.10 Compassion, Sympathy, Empathy	158
		5.2.11 Compassion Fatigue	159
	5.3	Pain: A Vital Sign	159
		5.3.1 Learning from the Literature	162
		5.3.2 Learning from the Patient	162
	5.4	Palliation: A Cloak of Care	163
		5.4.1 Designing the Cloak	163
		5.4.2 Cloak in a Basket	164
		5.4.3 A Cultural Cloak	165
		5.4.4 A Legal Cloak	165
		5.4.5 A Comforting Cloak	166
		5.4.6 A Cloak for Suffering	167
		5.4.7 An Ill-Fitting Cloak	168
		5.4.8 A Psychiatric Cloak	169
		5.4.9 A Cloak of Sedation	169
		5.4.10 Frailty in Need of a Cloak	170
		5.4.11 Cloak or Cover-Up?	170
		5.4.12 A Cloak of Hope	171
	References		172
6	**Death and Dying**		177
	6.1	Preparing for Death	177
		6.1.1 Drawing on Ancient and Contemporary Literature	177
		6.1.2 The Conversation	178
		6.1.3 Euphemisms and Misunderstandings	179
		6.1.4 Preference and Choice	181
		6.1.5 Dying Naturally	181
		6.1.6 Premature Deaths	182
		6.1.7 What Is Known of the Dying Resident?	182
		6.1.8 Staff Preparedness	183

		6.1.9 Sudden Death	183
		6.1.10 Dying: A Time for Joyous Celebration	184
	6.2	Perceptions and Perspectives of Death	185
		6.2.1 Part of Life	185
		6.2.2 The Dying Process	185
		6.2.3 Anticipating Death	186
		6.2.4 Suicide	186
		6.2.5 Grief, Loss and Bereavement	188
		6.2.6 Grief Among Support Staff	189
		6.2.7 A Volunteer's Grief	190
		6.2.8 To Hospital Or?	190
		6.2.9 Impending Death	191
		6.2.10 Solitary Death	192
		6.2.11 Dignity, Euthanasia and Futile Treatment	192
	6.3	After the Death	194
		6.3.1 Death's Effect on Other Residents	194
		6.3.2 Documenting the Death	195
		6.3.3 Death Notification and Certification	196
		6.3.4 Verification of Death	196
		6.3.5 Preparation and Removal of the Body	197
		6.3.6 Death Review	198
	6.4	Community Reflections	198
		6.4.1 Death Denial	198
		6.4.2 Wish to Hasten Death	199
		6.4.3 In Touch with Death	200
		6.4.4 Death as Part of Life	201
		6.4.5 Humour and Death	201
		6.4.6 Acknowledging Family	202
		6.4.7 A Good Death?	202
		6.4.8 Cultural Differences	203
		6.4.9 Aboriginal and Torres Strait Islander People	203
		6.4.10 Notification of Dying	204
		6.4.11 Communicating the Fact of Death	205
		6.4.12 Reportable Deaths	206
		6.4.13 Death as Loss of Community	206
		6.4.14 Planning for Death	206
	References		207
7	**Leadership**		**211**
	7.1	Leadership and Governance	211
		7.1.1 What Makes a Good Leader?	211
		7.1.2 Emotional Intelligence (EI)	212
		7.1.3 Board Governance	213
		7.1.4 Clinical Governance	213
		7.1.5 Reforms Needed	214

		7.1.6	Choosing the Right Staff	215
		7.1.7	Caring for Staff	217
		7.1.8	Acknowledging Night Staff	218
	7.2	Systems Guiding the Community		219
		7.2.1	Accreditation and Quality Monitoring	219
		7.2.2	Policies and Procedures	220
		7.2.3	Communication: The Management Culture	221
		7.2.4	Staff Induction and Orientation	222
		7.2.5	Acknowledging Long Service	222
		7.2.6	Staff Departures	223
		7.2.7	Exit Interview	223
		7.2.8	Accepting Gifts?	224
	7.3	Education and Students		224
		7.3.1	Routine Education Sessions	224
		7.3.2	Innovative Education Role Play	225
		7.3.3	Students	226
	7.4	Involving Families		227
		7.4.1	Who Is the Resident's 'Family'?	227
		7.4.2	Family Meetings	229
		7.4.3	Families' Needs	232
		7.4.4	When Problems Arise	235
		7.4.5	Which Family Member to Ask?	236
		7.4.6	Family Carer's Ageing	238
		7.4.7	Recognising Family Suffering	238
		7.4.8	False Comfort	239
		7.4.9	The Absent Family	239
		7.4.10	Happy Families?	240
	References			240
8	**Community Expectations**			243
	8.1	Meeting the Standards		243
		8.1.1	Required Standards	243
		8.1.2	Mission Statements	245
		8.1.3	Transparency	245
		8.1.4	Workplace Health and Safety (WHS)	246
		8.1.5	Prospective Residents	246
		8.1.6	Resident Records: Beyond the Clinical	248
		8.1.7	Quality: A Slippery, Subjective Notion	249
	8.2	When Things Go Wrong		251
		8.2.1	Staffing Inadequacies	251
		8.2.2	Making and Managing Complaints	252
		8.2.3	Why Staff Don't Speak Out	255
		8.2.4	The Serial Complainant	256
		8.2.5	Complaints About the Complaint's Process	257
		8.2.6	Managing and Responding to Risk	257

	8.2.7	Physical Restraint.	258
	8.2.8	Chemical Restraint.	259
	8.2.9	Alternatives to Restraint.	261
8.3	Negligence, Neglect and Abuse		261
	8.3.1	Negligence and Neglect	261
	8.3.2	Elder Abuse Defined	263
	8.3.3	Deliberate Killing: The Ultimate Abuse	265
	8.3.4	Resident-Resident Aggression	265
	8.3.5	Reporting Elder Abuse	266
	8.3.6	What Is the Planned Response to Elder Abuse?	267
	8.3.7	What Else Can Be Done About Elder Abuse?	268
	8.3.8	Preventing Elder Abuse and Sexual Assault	269
	8.3.9	Investigative Reporting	270
8.4	Home Sweet Home: Expectations Met?		270
	8.4.1	Residents/Relatives Committees	270
	8.4.2	Liaison with Other Agencies	271
	8.4.3	A Homely Community.	271
	8.4.4	Size Matters	273
	8.4.5	Design Matters.	274
	8.4.6	Home or Hospice	275
	8.4.7	Home Sweet Nursing Home	276
References.			276

Index. 279

About the Author

Rosalie Hudson, RN, Dip Arts, B App Sci adv.nsg., BTheol, MTheol, Grad Dip Geront., PhD, is associate professor and honorary senior fellow at the School of Nursing and Social Work, University of Melbourne, Australia.

Rosalie is a registered nurse, former nursing home manager with 12 years' experience and several years as community nursing supervisor. Her doctoral research focused on death and dying in nursing homes.

As an 'expert witness', Rosalie assists with legal investigations into cases of negligence in aged care. She is an academic supervisor and examiner, and reviewer of academic journals. Rosalie is also an experienced keynote speaker at conferences in Australia and overseas on topics including aged care, ethics, dementia, palliative care and spiritual care.

Rosalie has been a member of several academic boards, currently for Meaningful Ageing Australia, for whom she is honorary research consultant. She chaired a major hospital's ethics committee for 10 years.

Rosalie's teaching experience includes university undergraduate and postgraduate lectures on aged care, and regular sessions at other teaching institutions.

In 2018, she was awarded Victorian Senior of the Year.

Her major publications include four authored and co-authored books on aged care.

Chapter 1
Ageing in a Changing Community

1.1 An Ageing Population

Changing population trends indicate that nursing home residents are living longer. In addition to the sheer increase in demand, we can also expect changes in residents' needs, with differing patterns of disease, personal preferences and expectations of care, wide-ranging wealth levels, developments in technology and the implications of the COVID-19 pandemic.

Broad questions of accessibility, affordability and sustainability have been raised over several decades, asking whether the current aged care system remains fit for purpose. The aged care sector has faced many scandals about the quality of its services, particularly in nursing homes, prompting reviews and changes to improve the safety and health care of long-term residents. All of these reviews have concluded that substantial reform is needed; significantly, the Royal Commission into Aged Care Quality and Safety Interim Report: *Neglect* states as follows:

> We have uncovered an aged care system that is characterised by an absence of innovation and by rigid conformity. The system lacks transparency in communication, reporting and accountability. It is not built around the people it is supposed to help and support, but around funding mechanisms, processes and procedures. (Royal Commission into Aged Care Quality and Safety, 2019, p. 1)

'For the first half of the twentieth century, state institutions such as asylums and hospitals were the main source of accommodation for frail older people' (Smith, 2019, p. 2). Smith describes the Australian government's changes in 1954 to assist those in need of hostel accommodation through the provision of capital grants to religious and charitable organisations. In 1997, a package of changes was introduced, based on the unification of the former hostel and nursing home sectors.

While (some of) the stories are based on factual situations, real names and other details have been altered to protect the identity of the persons concerned. Resemblance to any particular person is therefore purely coincidental.

Funding was restructured, accommodation and administration of residential care united into one system, allowing services to offer the full continuum of care. Accreditation measured against new care standards was linked to funding, and certification arrangements were put in place to improve building quality (Smith, 2019, pp. 2–3).

While there are many calls for reform, the aged care system remains underpinned by the *Aged Care Act 1997* and the Aged Care Laws in Australia (Australian Government Department of Health, 2020). Changes in demographics will continue to impact residential aged care in the future. 'That is, the share of Australians aged 85 years or older is expected to increase by 83% in the next 40 years compared with the last 40 years' (Commonwealth of Australia, 2019, p. 85). It is clear that a new priority is needed, focusing on the residents' needs rather than on the owners'/providers' goals.

Deficiencies within the *Aged Care Act 1997* include cursory comments on elder abuse, failure to address infection control issues, lack of attention to residents' rights or their lived experience, inadequate oversight and accountability, negligible staff training and insufficient funding. In order to increase confidence in the provision of residential aged care, major reforms are urgently needed. Such reforms would emphasise the uniqueness of every resident, allowing their voice to be heard, with funding applied to their individually assessed needs and preferences.

Nursing homes provide support and accommodation on either a permanent or a temporary (respite) basis for people who have been assessed as needing higher levels of care than can be provided in their own homes, for example:

- Help with day-to-day tasks (such as cleaning, cooking, laundry)
- Personal care (such as bathing, dressing, grooming, toileting)
- Clinical care (such as wound care and medication administration) under the supervision of a registered nurse
- Other care services

Given the increasing frailty, comorbidities and dementia diagnosis of those entering nursing homes, care requirements have become more complex, with no staffing adjustments to compensate. Elements of reform suggested by the Royal Commission into Aged Care Quality and Safety (Commonwealth of Australia, 2021, p. 90) include

- 'Older people being treated with respect and dignity'
- 'Aged care staff having the skills and training needed to provide appropriate care and support'
- 'The provision of services and supports for daily living that assist older people's health and well-being'
- 'Older people feeling safe and comfortable'

Major cultural reform is also needed which focuses on residents' rights, transparency and financial accountability, particularly for the tax payers who help to fund aged care. To focus on the ageing population is to look, first, at the individual needs of each person in the context of their community.

1.1.1 Person or Individual?

The Universal Declaration of Human Rights avoids the term *individual* in favour of *person*. The link is then clearly made from *person* to *community*, implying that persons are not individuals alone, but persons engaged with other persons, within a community.

> A cat or a dog, or even a tree, can be an individual, but only a human being (or God and the angels) can be a person. *Person* is far more specific to the human race; it is a far more humanistic term… Persons are social beings before they are aware of having their own distinctive personalities. Persons come to fulfilment only in community, and communities have as their end and purpose the raising of persons worthy of their inherent dignity. (Novak, 1999, p. 41)

In sharp contrast to Novak's description of *person* is the contemporary propensity to speak of the *individual*. The 'cult of the individual' is given supremacy, posing a challenge for the development of communities in which frail, dependent older people not merely survive, but thrive. The changing nature of families is also apt; when bereft of full-time care or other support, the older person may be perceived as an individual needing residential care rather than a person with a continuing role in the community.

Person-centred care (PCC) provides a welcome emphasis away from individuals surviving in isolation. However, as Nay contends, 'for PCC to be more than rhetoric it must be embedded from bathroom to boardroom' (Nay, 2016). This means, in practice, that staff engaged in the deeply personal challenges of responding to a resident's 'bathroom needs' require appropriate preparation and continuous education. For managers and board members, it means acknowledging each resident as a person with distinctive needs and responding accordingly. PCC assumes a cultural, philosophical understanding of personhood, defined some decades ago:

> A standing or status that is bestowed upon one human being by others, in the context of relationship and social being. It implies recognition, respect and trust. Both the according of personhood, and the failure to do so have consequences that are empirically testable. (Kitwood, 1997, p.8)

When leaders fail to acknowledge residents, staff, visiting professionals, volunteers and families as 'persons', in Kitwood's view, the care will be inadequate. Staff need to know they are recognised, respected and trusted; they need to be treated as persons, not merely nameless individuals or 'one of the staff'. The same applies to residents and all who visit the nursing home; their inherent, personhood is to be recognised, together with their *preferred name*.

To focus on the *person* is to raise the subject of nomenclature. It ought not to be assumed, for example, that every nursing home resident prefers to be called by their first name. Some may prefer the formality of Mr, Mrs, Miss, Ms, followed by their surname, or alternate gendered or non-gendered appellations; others prefer the informality of first names or 'nicknames'. Terms of endearment are welcomed by some residents, whereas others may be offended by 'pop', 'darl' or 'sweetie'. Use of collective pronouns such as 'we' may also cause offence or be regarded as infantilising: 'what are *we* doing today?' or 'how are *we* feeling this morning?' An articulate resident may well respond to such an inane question: 'I don't know about *you*, but I'm feeling a little better thank you.'

Each resident's preferred name should be clearly documented and respected. Focus on the person can also be enhanced by the use of name tags for carers.

> *The suggestion came from a meeting with relatives and carers. 'Why don't the staff wear easily readable nametags?' Steps were taken to print name tags not only for staff, but for residents, relatives and visitors. Name tags were of uniformly large size with easy-to-read print and kept in one readily accessible location, with 'blanks' and felt-tipped pens for extras.*

This simple innovation, used to effect in many nursing homes, produced increased personalised attention and heightened communication.

1.1.2 Ageism and Human Rights

Older people have much to fear from their inability to swim against the strong current of ageist attitudes in whose waters the ebb and flow of their lives is measured only by disease, dependency and designated categories. In the face of an economic and ageing tidal wave, they must withstand the mainstream emphasis on youth, which leaves those who are frail and weak stranded as the waters sweep past. Ageism is manifested in prejudice, stereotypes, myths – as well as discriminatory practices in areas such as housing and employment. It is also evident in epithets, cartoons and jokes at older people's expense. A World Health Organization (WHO) report finds that ageism is ubiquitous, 'with every second person holding ageist attitudes' (Burke, 2021). The WHO report also reveals ageism in research where older persons are systematically excluded from clinical trials and denied certain medical treatments. (Undertaking research for this book, I found a dearth of contemporary evidence supporting best practice care for Australian nursing homes.)

Rather than merely a quaint phenomenon, ageism and its effects can lead to depression, as well as a 'decreased quality of life and premature death' (Burke, 2021). Ageism divides the world into 'us' and 'them' as though 'they' belong to a society apart from 'us'. Hitchcock opposes this distinction:

> The elderly, the frail *are* our society: they gave birth to us, nourished us, protected us, paid their taxes diligently, went to war, ate bread with sugar when there was no butter. They worked and loved and lived — and can continue to do so. They are our parents and grandparents, our carers and neighbours, and they are every one of us in the not-too-distant future. (Hitchcock, 2015, p. 69)

We may ask Hitchcock, and in the context of the COVID-19 pandemic, why it has taken such a crisis to bring to government and public awareness the shocking effects of ageism in keeping matters of substandard care (such as lack of infection control) hidden from the wider community? Lack of adequate protection resulted in hundreds of preventable deaths. In a younger population the same statistics would, no doubt, have been considered a greater scandal.

'As a conceptual tool, ageism suffers from the same intellectual parochialism that plagues social gerontology generally. It is neither informed by broader social or psychological theory nor grounded in historical specificity' (Cole, 1983, p.35). Similarly, those who propound the opposite view – that 'old people are (or should be) healthy, sexually active, engaged, productive, and self-reliant' – represent a 'positive mythology' which 'fails to acknowledge the intractable vicissitudes of aging' (Cole, 1983, p.35).

> Unable to infuse decay, dependency, and death with moral and spiritual significance, our culture dreams of abolishing biological aging… Unless we grapple more openly with the profound failure of meaning that currently surrounds the end of life, our most enlightened view of old age will amount to perpetual middle age. (Cole, 1983, p.39)

Countering the 'positive mythology' view, ageism and stigma are evident when older people are regarded as useless and their lives less valuable than others. The consequences of such an ageist view for those in need of care are, according to one writer, to 'warehouse' them, or to minimise their needs, leaving them devoid of stimulation in nursing homes. Ageism, in such contexts, may well be equated with human rights abuse.

Ageism is not merely a benign attitude towards older people. '… ageism can lead to undertreatment, particularly for complaints of pain, fatigue, depression, sexual disorders, and cognitive impairment, as well as overtreatment, with needless tests, procedures, medications, and admissions to the intensive care unit' (Brauser, 2018). Examples of benign attitudes or 'benevolent ageism' include using pet names, slow or sing-song speech, or plural pronouns such as 'we're just going to have a look at your tummy' or 'we're going to have our breakfast now'. Careful scrutiny of such language (also discussed in the section 'Person or Individual') reveals not only the nonsensical grammatical inference; it fails to give due emphasis to the *person* being addressed, ignoring their human right to be treated accordingly. Residents also have the right to feel offended by other descriptors such as 'she's so *cute*' (a term more appropriately used of a doll or pet animal).

Regardless of their cognition level, each resident deserves due respect, including being addressed by name, spoken to as an adult, the right to fair and equitable treatment, the right to voice their opinions and to make a complaint when necessary.

It seems that, at least in some contexts, age discrimination is perpetuated, if not intensified, adding weight to the cultural and philosophical emphasis in this chapter, and indeed the entire book, where the *person in relation to their community* is the primary reference point. This person-centred philosophy is, of course, readily apparent in many nursing homes – exemplars of older persons' generational knowledge and wisdom, together with the upholding of their human rights. Philosophical change is described in the Royal Commission Report (Commonwealth of Australia, 2021) which recognises the need for Australians to have access to an aged care system which is not rationed but available to every person who needs it. The system should also be *visible,* first and foremost by reversing the *invisibility* of the people it is meant to serve, enshrining their rights in a new Aged Care Act which places the older person at the centre.

In a media release in May 2021 the Prime Minister Scott Morrison foreshadowed a new Aged Care Act for 2023 which will improve service suitability 'that ensures individual care needs and preferences are met' and improve access to and quality of residential care, a 'bigger, more highly skilled, caring and values based workforce', new legislation and 'stronger governance' (Prime Minister of Australia, 2021). This statement signifies an important benchmark to test how well residents' 'individual care needs and preferences' will be met.

Much-needed reforms aiming to improve outcomes for older people are identified in the Grattan Institute's report (Duckett et al., 2021) which decries the lack of leadership in aged care, graphically portrayed in horrific stories of substandard care, weak accountability, perverse funding models and an 'overstretched, under-trained, and underpaid workforce' (Duckett et al., 2021, p. 3). The report's authors recommend a new Aged Care Act which would enshrine older persons' rights through transparent governance, minimum care hours and dramatically increased funding.

A global initiative by the WHO (2021) describes ageism as 'how we think (stereotypes), feel (prejudice) and act (discrimination) towards others or ourselves based on age'. The website includes a toolkit and presentation with suggestions for initiating a conversation on the subject and suggestions for overcoming ageism in our communities.

1.1.3 Ethics and Culture

Somerville reminds her readers: 'The "ethical tone" of a society is not determined by how it treats its most powerful, strongest, most affluent members, but by how it treats its weakest, most in need, most vulnerable members'(Somerville, 2020). Using the example of nursing homes in the COVID-19 pandemic, Somerville says we need to have 'respect for the dignity of the person; balancing aged care institutions' obligations to each resident and to all `residents' (Somerville, 2020). Noting also the government's responsibility for aged care funding, Somerville links ethics and culture to society's ageism:

> The collision of two pandemics – Covid-19 and ageism – is indeed lethal. We do not respect elderly people or their human dignity. Our Aboriginal brothers and sisters are an exception and we other Australians have much to learn from them in this regard. (Somerville, 2020)

Somerville criticises the impersonal language used by some journalists describing residents being 'decanted' to hospitals, saying that this type of language 'allows us to disidentify from the elderly person: it reassures us that we are not like them and never will be', adding the descriptive comment: 'Elderly people are "ethical canaries" in the coalmine during the Covid-19 pandemic' (Somerville, 2020).

Beyond the crisis of a pandemic, when applied to the ordinary daily care of nursing home residents, staff are reminded that ethics is not a remote concept beyond their responsibilities. Ethics includes

1.1 An Ageing Population

- Choices, particularly at the end of life
- Informed consent when required
- Comfort and dignity issues for every resident in every situation
- Identifying who is the decision-maker when the resident loses capacity

One way of checking the ethics of a given situation is to weigh the *burden* versus *the benefit*. Focusing on these 'two big b's', staff and families may be prompted to consider the most appropriate action at the time – reviewing and reconsidering when circumstances change. Without promising perfect solutions, the burden/benefit analysis provides an instructive means for problem solving and weighing the ethical consequences of a particular situation.

Similarly, a broad ethical view of ageing is not confined to matters of money. Hitchcock, describing the 'black heart of a health system based solely on utilitarian economics' (Hitchcock, 2020, p. 27) where the care of unproductive citizens is no longer cost effective, says: 'The young rule the world; we stomp around doling out mean rations to the old, the machinery of our secure, able bodies purring to us the myth that we will live forever' (Hitchcock, 2020, p. 28).

'Poppa' may well have been regarded as one of society's 'unproductive citizens'.

> *He lies curled up in his bed, making no response other than to look intently at each person who comes to attend him. His poorly shaven chin is often stained with the food that dribbles out of his mouth. None of the staff have learned to pronounce his long name, so they called him 'Poppa' for short. No one knows what language he prefers. He has no family, arriving at the nursing home with a small battered suitcase containing soiled underwear and an old electric toaster. When an inquisitive nurse discovered he had money deposited with the State Trustees, she sought permission to buy him some clothing and toiletries. Within weeks his appearance was transformed with attractive clothing and a professional haircut. An electric shaver made it easier to remove stray whiskers and a portable radio brightened his days. Staff were instructed to call him by his full name. Mr. Papandreou expressed his gratitude with a wan smile.*

The action and response (described above) came from a carer who dared to enter the unfamiliar territory of a resident with little speech and no common cultural background. Other carers are sometimes heard to complain when in a similar situation: 'I can't understand a word he's saying.' It follows that no attempt is made to communicate in any meaningful way.

By way of contrast, a positive attitude is prompted by the services of a culturally competent health professional, particularly when assessing a resident's needs. Rather than making judgements or presumptions, assuming 'No English' is a legitimate rationale for lack of dialogue, it is far preferable to seek advice from an interpreter service or community health worker. Appropriate cultural response prompts exploration of the resident's and family's attitudes and values, including

- The meaning of illness and ageing in each situation
- Truth telling in relation to diagnosis and prognosis
- Communicating about death and dying
- Alternative healing practices versus Western health care

Residents from a variety of cultural backgrounds will have beliefs which may differ from what is considered 'the norm'; so, it may be appropriate to ask each person what is most important to them. Culturally appropriate care is fostered by determining the historical and political context of the resident's life, for example:

- Refugee or immigrant status
- Socio-economic status
- Language/s spoken
- Degree of assimilation into Western culture
- Degree of support from and integration with their cultural community

Resources with the potential to assist staff in unfamiliar cultural contexts include healthcare professionals, community groups, religious leaders and traditional healers. Understanding the general decision-making style of the resident's cultural group is fostered by questions such as

- How are decisions about health care made in your family?
- Who is the head of the family?
- Can you tell me what you need?
- Is there anyone else I should talk to in your community about your care?

Ultimately, undertaking a thorough cultural assessment will help nursing home staff and volunteers deliver ethically safe and person-centred care to all residents regardless of their race, ethnicity, culture or language.

1.1.4 Funding

The funding and regulation of aged care services is predominantly the role of the Australian government, although all three levels of jurisdiction – local, state and federal – are involved. Funding is also complicated by the variation in ownership and purpose. For example, some for-profit nursing homes may have a different rationale from those in the private non-profit sector, the latter varying according to their religious and/or charitable status. Financial performance varies according to the size of the nursing home and the level of care required. Victoria is the largest public provider of nursing homes in Australia, with over 180 Public Service Residential Aged Care Services (PSRACS) (Victorian State Government, 2019).

The government pays subsidies and supplements to approved providers for each resident receiving care under the Act. The basic care subsidy for each permanent resident is calculated using the Aged Care Funding Instrument (ACFI) – a tool the provider uses to assess the resident's care needs. The ACFI consists of 12 questions and 2 diagnostic sections. Needs are classified under three funding domains: (a) activities of daily living, (b) behaviour management and (c) complex health care. Funding for each domain is calculated as 'nil, low, medium or high'. 'The subsidy consists of the combined amounts payable for the level for each domain' (Australian Government, 2020a).

Additional supplements are provided for residents who need enteral (tube) feeding or oxygen supplies and for others with special needs, such as homeless veterans and people in remote areas. Fees are determined according to the resident's income and assets: extra services and a higher standard of accommodation attract additional costs (Australian Department of Health, 2021). Nursing home care is provided at no cost to veterans with a 'gold card', who may also receive a veterans' supplement.

Funding includes (a) operational, which supports day-to-day services such as nursing and personal care, living and accommodation expenses; and (b) capital financing, which supports the construction of new nursing homes and the refurbishment of existing facilities. To place the funding in perspective: 'Spending on aged care is expected to be the fastest growing budget item after the National Disability Insurance Scheme. The aged care sector is one of the fastest growing industries in Australia' (Smith, 2019, p. 30). From another perspective, 'Australia spends less on aged care than similar countries ... Netherlands, Japan, Denmark, and Sweden spend between 3 per cent and 5 per cent of their GDP on long-term care. The Australian government spends only 1.2 per cent' (Duckett et al., 2021, p. 21), with an associated lack of accountability. While increased transparency is needed for the amount currently spent on staffing and other care components, it is clear that more public money is needed to effect major transformation.

The Australian Aged Care Classification (AN-ACC), designed to replace the ACFI in October 2022, is the new tool for documenting a resident's functional, cognitive and physical capability. It is designed to improve aged care funding, emphasising consistency, reducing time spent on paperwork and using external assessors.

The document itemising these funding and accommodation details is aptly named 'Navigating the maze' (Smith, 2019), and in this chapter, no attempt is made to find an easy pathway through it.

1.1.5 For Profit?

If the main aim in providing residential care for older people is profit, how is this reflected in the nursing home's philosophy and quality of care? Considerable evidence from observational studies shows that care delivered in for-profit facilities is inferior to care delivered in public or non-profit facilities (Ronald et al., 2016, p. 2). This causal link is not a new phenomenon according to these researchers, citing evidence from the 1980s. One example of profit's relationship to quality lies in staffing: higher staffing levels are associated with lower staff turnover, which in turn impacts favourably on residents' continuity of care. Concomitantly, when profit motives dictate employment of fewer staff, and those with minimum qualifications (or none), the quality of care will be poorer, compromising residents' care.

Staffing issues have come to the fore in the COVID-19 pandemic, resulting in many residents either being transferred to acute care or qualified personnel seconded to the nursing home to provide urgently needed clinical care. Although this phenomenon awaits comprehensive analysis, emerging stories show the marked degree of satisfaction enjoyed by residents when the number of registered nurses (RNs) was increased – albeit for a short period.

Some nursing homes enjoy financial profits while providing excellent care. For others, a profit motive may run counter to a philosophical approach which prioritises residents' quality care. Attaching another dimension to 'profitable' the assumed aim would be to benefit (or 'profit') the consumer, in this case the nursing home resident.

1.1.6 Trusting Our Aged Care Services?

A sampling of over 1700 members of the general public found that only 18 per cent trust the aged care industry and only 13 per cent consider it open and transparent. Very few respondents rated aged care services as providing a high level of residential care, showing empathy, or taking the time to properly assess the residents' individual needs (Bastian, 2019). This survey shows a deficit in levels of trust and transparency not only in the service delivery but in funding arrangements impacting care.

However, these findings indicate that the problems lie much deeper than systems and funding issues. Caregivers reported the emotional and physical stresses arising from lack of preparation for their role, claiming the experience had a negative impact on their lives. Those interviewed agreed that the burden would have lessened with greater levels of support and recognition. For many older people and their families, the move from their community and/or their own home is literally a life-changing event. When treated merely as a 'location change' or when the emphasis is mainly on financial arrangements, the personal, familial and cultural implications remain unaddressed.

Measures to increase transparency would include readily accessible funding details so that consumers know, for example, how much is spent on food, staffing and other relevant services. A greater understanding of older peoples' attitudes to aged care services may also prompt more innovation and the delivery of care tailored to meet their needs; hence, increasing their trust. However, transformation and lasting change will only come about through political will, matched by the necessary dollars.

1.1.7 More or Less Homelike

A decade ago, it was not uncommon for a resident to live in a nursing home for up to and over five years; now, it is much less, indicating the need for greater concentration on the residents' needs during this final phase of their life. The emergence of large corporate players buying up many of the smaller, independent operators impacts the capacity to respond to residents' idiosyncratic needs. Inherent in such changing patterns are issues which run deeper than dollars, often touching on profound philosophical, psychological and spiritual factors affecting an older person's life and death. Commentators have also questioned how other community changes will impact residents' acuity; for example, whether Consumer-Directed Care (CDC) and the National Disability Insurance Scheme (NDIS) will result in older people remaining at home for longer, resulting in shorter periods of nursing home residence (Herinton, 2016).

Together with residents' increasingly complex needs and shorter nursing home stays, the changes (referred to above) indicate a depersonalising of many aged care services. For some, the admission process itself may be technologically more proficient (e.g. with arrangements made online) while the *person* about to enter the nursing home has little attention paid to their unique needs and preferences. The deeply personal implications are readily apparent, leaving one prospective resident to opine: 'I'm just a number.'

Other analysis (Hampson, 2018) found that for accommodation in a small homelike facility (such as 15 beds for residents with dementia) the cost was comparable with (and in some cases lower than) larger institutionalised nursing homes. Further scrutiny is needed to compare funding arrangements, bed numbers and other financial factors influencing the care of nursing home residents across Australia. For example, in some smaller homes, residents may have the opportunity to share in domestic activities as part of their occupational therapy. Others have a greater choice of mealtimes, saving wastage. Greater emphasis on sharing the costs and the care would bring welcome relief, not only financially, to many current and future residents and their families, particularly those who favour a more homelike environment.

1.1.8 Ethno-Specific Nursing Homes

Contemporary research is lacking on specific advantages and disadvantages of ethno-specific nursing homes or data pertaining to the need for additional nursing home beds. While some research shows that the availability of preferred food, bilingual staff and other cultural preferences may be catered for in ethno-specific homes (such as Italian and Greek), details of other ethnic groups are hard to find.

It has been proven, however, by at least one notorious case in Melbourne, Victoria, that a specific religious, ethno-specific nursing home does not necessarily provide exemplary care or care in line with its ethos. On the contrary, this tragic example attests to the fact that, in spite of adequate funding, this nursing home failed to keep residents safe during the COVID-19 pandemic, resulting in many hospitalisations, deaths and the eventual closure of the home.

This brief discussion highlights the need for 'mainstream' nursing homes to accommodate older people from any background, according them the dignity and care commensurate with their meticulously assessed needs.

1.2 Describing Aged Care Communities

1.2.1 Home, Community or Facility

When the title 'aged care facility' is preferred to 'nursing home', it suggests that 'nurse' and 'nursing' have been eradicated from the very place where older people in need of continuous hands-on nursing will spend the remainder of their lives. Is it possible to reverse this not-so-subtle linguistic trend when the workforce has been deskilled to the point of having only a token number of registered nurses, if any, and employing others with no mandated health education or even rudimentary understanding of gerontological nursing?

Analogously, a preference for either 'facility' or 'nursing home' suggests a choice between an institution and a home. The term 'nursing home' is, however, a reflection of what residents need: nursing care. And this is no mere (domestic) 'home'. Its residents have profoundly disabling, complex medical conditions requiring full-time care; most having more than one life-threatening, incurable disease. It is not unrealistic to expect they should be cared for by skilled nurses and care attendants who understand the ageing process and the concept of persons as whole, even in their physical frailty. When such care is provided, they may come to regard their new and final residence as 'home'.

The term 'facility' can signify a policy shift. Its use in the *Aged Care Act 1997* (Commonwealth of Australia) demonstrates what some believe to be a carefully calculated change in terminology consistent with shrinking government funding for care provision. Facility can also mean 'being easy, absence of difficulty, equipment or physical means for doing something, aptitude' – descriptions derived from the word 'facile', which is often used pejoratively. Furthermore, when 'facility' indicates 'that which offers unimpeded opportunity' the irony is doubly scored, considering that for the funding tool residents are categorised according to their *inabilities.*

Even the term 'resident' can be fraught with difficulty, particularly when the residence lacks any sense of community, or family, or homelike attributes, where each person is regarded merely as an isolated individual. In contradistinction to a community, many residents have no meaningful engagement with others and may not even know their neighbour's name. However, for want of a better alternative, 'resident' is used in this book in terms of its positive connotations.

A nursing home resident is also known as an 'aged care recipient'– a term connoting an inert parcel waiting to be identified, in the dependent, impersonal, one-way path of consumerism. Coupled with the term 'placement' the image is of something being picked up from one location to be deposited in another. This notion of 'being put' or 'being placed' belongs only to objects, not people. When people are regarded merely as recipients of care for a specific period according to their level of dependency, they are depersonalised and cast adrift from their community. Their full personhood is denied, and holistic care remains in the realm of rhetoric.

1.2 Describing Aged Care Communities

Some nursing home managers place great emphasis on the 'community' aspect of their residence. Others see nursing home admission as moving *from the community*. What makes a residential aged care facility (RACF) a community? What makes a 'nursing home' a home? While the dictionary definition of 'facility' implies a place for a particular purpose it also connotes the ability to perform or function. Many residents would fail to meet the criteria attached to such a definition. Similarly, many residents and families regard the RACF as an *institution* rather than a *home*. It appears that nomenclature, design and size have a significant impact on residents' perceptions, not to mention their well-being.

'Clustered domestic models of residential care are associated with better quality of life and fewer hospitalisations for residents, without increasing whole of system costs' (Dyer et al., 2018, p. 433). This research found that, apart from the size, residents in smaller homes enjoyed independent accessible outdoor areas, meals cooked on site and increased participation in daily routines. Flying in the face of such research is the propensity for some corporations to develop more expansive, multiple-level accommodation, sometimes including several large buildings with no ready outdoor access. Providing a 'homelike environment' in such a context creates a challenge, if not an impossible aim. However, some manage to achieve a homely atmosphere even when accommodating a large number of residents. Size does matter, but only when matched with high standard care.

What defines a 'home' when compared with an 'institution'? Families' opinions of the latter, sought by way of a research project, included the following factors:

- Loss of autonomy (leading to passivity)
- Impersonal attitudes
- Rigid mealtimes
- Fixed seating in the dining room (with no flexibility)
- Residents' lack of personal control over space, noise, diet (Russell, 2017, p. 3)

A nursing home wishing to improve its homelike atmosphere may learn from distributing a questionnaire to elicit a variety of views from those who live, work and visit there. Responses such as those recorded above can be readily maintained by means of regular surveys. Such engagement reassures residents, families and staff that their opinions are welcomed and suggestions noted. Given residents' increasing physical frailty, the 'home' needs also to be supported by accessible, responsive health services.

One of the implications of the COVID-19 pandemic has been the urgent need to define and describe what medical care is available to nursing home residents in order to avoid their unnecessary transfer to hospital. One criterion is to ask whether the resident's condition is treatable and whether or not acute care would improve their quality of life. Such decisions are best made on an individual basis, respecting the resident's and family's wishes and preferences wherever possible, guided by expert opinions from a geriatrician and/or gerontologist where necessary (Colenda et al., 2020). A comprehensive, up-to-date advance care directive (ACD) is an important reference in such situations. However, Colenda and colleagues acknowledge an ageist bias, exemplified by the derogatory term 'boomer removers', noting the response to the pandemic would be different when describing its effects on younger people.

1.2.2 Shared Rooms or Single

Advantages of a shared room are noted here, particularly when the life of one resident draws to a close, exemplified by the following.

> 'Sarah' was discussing her mother's long-term accommodation in a two-bed room, where she had, for the previous year, lost her capacity to communicate in words, resulting in minimal verbal exchange with her daughter. Sharing a room with another person posed no problems either for resident or daughter. 'The wonderful thing about the shared room is that mum's room-mate has taken on the role of advocate. She's an outspoken member of the residents' committee and seems to speak out for all the others, and especially for my mum. She always knows what's going on and she keeps me informed. She doesn't seem to be disturbed by mum's inarticulate groanings and other idiosyncrasies. She sometimes interprets what she believes mum is trying to say. It's wonderful to receive all the information she gives me, not only about my mum, but about the rest of the nursing home. I'd find it so isolating to have mum in a single room'.

Sarah's story demonstrates the capacity of older people with distinctively different characteristics, families and care needs, not only to adapt to living in a shared room, but to benefit from the experience. In other circumstances, however, the problems might outweigh the benefits. One of the many complexities relates to visitors, who may be welcomed by one resident in a shared room, and shunned by others. While it may not be possible to satisfy the preferences of every resident and their family, reassurance and trust are gained by open discussion. Choice is of the essence here, enabling each prospective resident to have their preference met wherever possible. Open discussion is also important where one option may be to share a room with a person of the opposite (or different) sex: a scenario abhorred by some and accepted by others. One undeniable factor in such discussions is the voice of the resident (wherever possible) in choosing the room which will become their home.

Advantages of single rooms include decreased risk of cross-infection, greater privacy, undisturbed sleep and expanded visiting options for a resident close to death. Some claim that residents in single rooms show fewer signs of aggression and less agitation than those in shared rooms, although the opposite may also be the case. Another advantage is that a resident who needs increased attention, particularly throughout the night, does not disturb others, especially when increased lighting is needed. Issues of family preferences and residents' satisfaction also need to be considered, with many favouring single rooms. These factors need to be weighed against the capital costs which may, for some proprietors, be prohibitive.

Definitive research is lacking regarding the advantages of single rooms with en suites in nursing homes. While residents' and families' views and preferences need to be considered, the main factor has to do with residents' care, a subject of particular interest to the staff. Do the residents receive better care in single rooms rather

1.2 Describing Aged Care Communities

than shared? How may each resident's care be optimised regardless of whether they are in a single or shared room? Here, as elsewhere, more evidence from research would provide a helpful guide.

1.2.3 Respite Care

> Respite care enables older Australians to live in their own homes and still receive the care they need from their families, friends, and others. Far fewer people access respite care, as opposed to permanent residential care, but it is growing at a far greater rate; approximately four times faster. (Coredata Research, 2019, p. 11)

Respite services offer some relief for family members caring for an elderly, dependent person. As of June 2018, one in five people using residential care services were in respite care (Coredata Research, 2019, p. 15). Increasing preference for home care is also impacting nursing home occupancy, together with the significant issues revealed in the government enquiry, leading to a major distrust in the system.

Considerable increase in the use of respite care in recent years may also reflect a 'try before you buy' model, allowing potential residents an option before committing to permanent care. As more people are being cared for at home, albeit with high level of needs, further increases in respite care are predicted. These findings emphasise the support needed for carers, as well as a timely reminder for nursing home managers providing respite care. A measure of quality may well be their capacity to welcome residents on a short-term stay, with the assurance of excellent care, especially for residents with dementia. Respite care requires a comprehensive clinical focus on all of the resident's needs, together with a clear plan and opportunity to evaluate the care.

> It may be only for a couple of days or weeks but it is a significant change in the often fine balancing act of looking after a frail older person. Therefore, residential respite care should be considered more than just a short-term 'sitting' service. It is a service where the outcomes should be beneficial not only for the carer but to the ongoing care and life of the older person. (Victorian Institute of Forensic Medicine, 2016, p. 6)

Whether or not some caregivers regard respite as a 'sitting' service, for others it is associated with the grief of separation. 'My mother knew that no one would recognise her grief at the commencement of Dad's respite stay.' On further reflection and in her 'daughterly heart' trying to understand this experience, Sutton says: 'If we expect carers to provide the back-breaking, year in year out, care of dependent loved ones, we can't expect them to be able to leap out of it and utilise a respite care option whenever it's available' (Sutton, 1997, p. 6).

Sutton (1997) describes her mother's ability to 'build sacred personal space' and the tragedy of those who trample roughshod over private terrain. Such personal insights prompt increased recognition of the loss involved when a long-time carer is separated from their partner. A key challenge is to ask whether or not the resident and their family seeking respite are welcomed into the nursing home community and provided with specific resources according to their unique needs.

Investigations by the National Ageing Research Institute (NARI) found complaints regarding the respite services related to fees and staffing in particular. Respite care residents reported poor food quality as well as being 'lonely, bored, call bells not answered' (National Ageing Research Institute Ltd, 2020, p. 87). If this report indicates a 'try before you buy' exercise it does not augur well for those wishing to access full-time nursing home care.

1.3 Welcoming Diversity

1.3.1 Cultural Mix: The Nursing Home's Core

Australia is one of the most culturally and linguistically diverse nations in the world. 'Almost 25 per cent of the population were born overseas, over 200 languages are spoken, and 116 religions practiced' (Australian Government, 2016). Although all Australians have the right to equitable health care, people from culturally and linguistically diverse (CALD) backgrounds, including Aboriginal and Torres Strait Islander peoples, suffer significantly more adverse events than others. They are more likely to experience medication errors, misdiagnosis, incorrect treatment and poorer pain management. Nursing home staff need to be alert to the specific needs of a resident whose first language is not English or one whose cultural background is atypical. It seems unclear in contemporary research whether the number of CALD nursing home residents is an accurate representation of the broader CALD diversity within the community.

One way of defining culture is to describe ways of thinking and behaving that are socially accepted among a particular group or society. Culture is not static. It is influenced by political and economic conditions, varying with factors such as age, gender, class, religion, education and personality. However, the aspects of culture we can easily see (e.g. food, dress and language) can be likened to the tip of an iceberg: the majority is hidden from view, often undiscovered, not revealed and never coming to the surface.

What is being communicated, culturally, within nursing homes whose residents come from a variety of backgrounds? What can be seen and what is hidden from view? Research has uncovered misunderstandings, miscommunication and culturally unsafe care by healthcare professionals with CALD patients often describing feelings of powerlessness, vulnerability, loneliness and fear when undergoing treatment (Johnstone & Kanitsaki, 2006). While this research focused on the acute hospital setting, it is a cogent reminder of the cultural vulnerability of nursing home residents. Rather than a single cause of the inequalities in health care experienced by CALD hospital patients, researchers found that clinical encounters failing to address cultural factors lead to adverse outcomes and health inequality (Johnstone & Kanitsaki, 2008). Applying these findings to residential aged care, it is clear that practical strategies are needed for improving culturally competent care, including the following:

- Adapting each resident's care for consistency with their cultural values and beliefs
- Developing open communication including the capacity for self-awareness and identification of personal biases and prejudices among staff
- Accessing qualified interpreters when required (rather than relying on the resident's family, except in an emergency)
- Including cultural assessments as a part of routine resident care planning

It is universally acknowledged that comprehensive assessment underpins safe practice. Aged care nurses are all too familiar with the need to routinely conduct falls risks assessments, pressure care assessments, nutrition assessments and measurement of vital signs. Although cultural assessments are equally important, their use in nursing homes is not always given high priority or considered a vital part of routine care. Some, however, have developed their own comprehensive cultural assessment forms, and others seek advice from the wider community when confronted by unfamiliar customs. Such a broad perspective helps the nursing home to achieve 'cultural competence'.

> Cultural competence is the ability to participate ethically and effectively in personal and professional intercultural settings. It requires being aware of one's own cultural values and world view and their implications for making respectful, reflective and reasoned choices, including the capacity to imagine and collaborate across cultural boundaries. (Sherwood, 2019)

Sherwood's definition includes the key concepts for staff education. As a microcosm of society, a nursing home may showcase to the wider community how older people of diverse backgrounds are welcomed and cared for, particularly at the end of life. However, such reassurance is not always apparent. Many older people from CALD backgrounds are reluctant to enter a nursing home, expecting to find the experience alienating and disempowering. Language barrier is not the only reason; failure of staff to acknowledge their unique cultural mores adds to their negative perceptions.

Creating a positive, empowering environment for CALD residents includes a heightened awareness of their unique communication needs. Researchers ranked companionship and individual recognition by staff as important issues cited by the residents. Others fear their beliefs will not be understood, nor their diverse values or their language respected. While some of these issues are addressed in culturally specific nursing homes, and others place a high priority on cultural awareness, all residents from diverse cultures deserve care that respects their differences (Rawson, 2019). Policies should be inclusive, avoiding inequities and impositions arising within a system which is, for some, an alien culture.

> *'Beware the culture folder!'* This comment was made by a nurse seeking to find some culture-specific information relevant to a newly admitted resident, disappointed if not greatly perturbed to find scant, unhelpful, out-dated 'information'. Taking a proactive stance, the nurse appealed to the manager to obtain up-to-date 'fact sheets' pertaining to current and prospective CALD residents.

As in all areas of care, individual assessment is the key. Rather than 'Greeks like this type of food' or 'Italians don't like…' fact sheets can only cover broad cultural issues.

1.3.2 Aboriginal and Torres Strait Islander People

'Aboriginal and Torres Strait Islander people's care needs are intimately connected to family, community and culture, to Country and to language, including sign language' (Commonwealth of Australia, 2019, p. 174). When a person from this cultural milieu enters a nursing home, their preferred connections need to be acknowledged, together with at least a brief understanding of their spirituality.

> Aboriginality is not about skin tone, curly hair or living in the desert. It is about a connection to each other and the dirt of the land. It is about a shared heritage, a shared pride—and a shared pain. It is about being passionate that our children inherit those things and our history. It is about what is in our hearts. It is an inside, not outside concept. It is about pride. (Burney, 1997, p. 3)

A brief foray into their unique culture indicates that initial assessments are best facilitated by the use of an interpreter or Aboriginal and Torres Strait Islander worker. It is considered most unwise to use a family member as an interpreter for any resident with limited English or from an unfamiliar culture; the discussion needs to be free of any bias or misinterpretation. A formal interpreter who understands the person's language preference, food taboos, gender roles, consent issues, family factors and priorities around death and dying enables the formulation of a culturally appropriate care plan. Employing a qualified interpreter takes time and incurs costs, which may or may not be met by the resident, further emphasising the need for a comprehensive assessment and sensitive planning.

Cultural recognition may also be assisted more broadly by displaying a plaque depicting the land on which the nursing home is built and acknowledging important cultural ceremonies. Many older Aboriginal and Torres Strait Islander people live away from their original lands or country, a loss exacerbated during times of stress, particularly at the end of life. Many older people have a strong desire to go back to their country at these times, regarding it as more important than receiving health treatment. Each resident's priorities need to be respected and reflected in a thoughtful, accurate, culturally specific care plan, which incorporates the unique circumstances associated with their end-of-life care (Mcgraph, 2010).

'The importance of older people remaining in their community close to family and country is well documented', and the nearest nursing home could be in a town or city a great distance from their familiar environment (LoGiudice et al., 2014, p. 185). A holistic approach to care acknowledges the many cultural, language, traditional and spiritual differences among Aboriginal and Torres Strait Islander people – depending on their location. It is important therefore not to reduce such assessments to a single (or simple) 'aboriginal culture'. While it is preferable for people to be cared for in their own community, where language and healthcare practices are familiar to them, admission to a 'mainstream' nursing home is sometimes necessary. Such was the case for 'Joy'.

> *'Joy' had been flown from a rural/remote area for urgent treatment in a major city hospital. When her death did not occur as expected, she was transferred to a suburban nursing home. Fortunately, one staff member understood Joy's specific cultural needs. When it became*

clear, after some weeks, that Joy, aged 60, was dying, this nurse learned that Joy's greatest wish was to have the smoking ceremony she would have been given in her home community. The nurse sought advice from management and was told such a ceremony would be contrary to fire regulations. On further investigation the nurse found that a small section of the nursing home where Joy's bed was located could be isolated for the very few minutes of the smoking ceremony, with little chance of adverse effects elsewhere. The following day, the nurse was able to secure the services of an aboriginal elder, with the necessary equipment for the very short and simple smoking ceremony. The facility's safety officer was notified and with management approval the section surrounding Joy's bed was isolated from the smoke alarm system. Within a few minutes, the ceremony was completed, and Joy died peacefully several hours later.

This story illustrates the importance of providing culturally specific care, even when the nursing home is not the person's first preference. It is also important to note the life expectancy of Aboriginal and Torres Strait Islander people compared with the non-Indigenous population: '8.6 years lower for males and 7.8 years for females' (Australian Government, 2020b, p. 2), indicating their timely need for palliative and end-of-life care.

Guarding against inappropriate or inaccurate assumptions means communicating with the individual, their family and community, as well as Aboriginal health workers, in a sensitive way that values their cultural safety. Careful communication ensures that the right information is shared with the right people, identifying the correct nationhood of the Aboriginal person. Particular sensitivity is needed when preparing for a resident's death to ensure accurate information is obtained regarding their spiritual needs, ceremonies or practices. It is also important to check expectations after the death, recognising that Aboriginal and Torres Strait Islander cultures are not all the same and assumptions should be avoided. A person-centred approach would take into account the central importance of connection to family, kinship networks, communities and country. Appointing Indigenous staff, where possible, to care for Indigenous residents would be a step in the right direction.

1.3.3 *Lesbian, Gay, Bisexual, Transgender/Gender-Diverse, Intersex, and Queer (LGBTIQ) Residents*

Addressing the diverse sexual orientation, gender diversity or non-gendered identity of persons seeking residential aged care involves a respectful, non-discriminatory attitude, avoiding bias and judgement. This means that older LGBTIQ people should be provided with the same opportunities and options in aged care that are available to all Australians.

A first-hand report conveys the type of responsive care enjoyed by at least one such resident.

Mr B, who describes himself as 'the G part of the LGBTIQ', said the nursing home is a real home to him. 'I've been very comfortable here. People don't discriminate against us,' he said. 'Everyone's accepted…. Everyone calls each other by their first names, providing each resident with reassurance of their identity. It makes me feel that I belong,' Mr. B explained. (Egan, 2020)

However, survey results showed that 49 per cent of respondents do not believe that aged care services meet the needs of the LGBTIQ community. Most of those surveyed emphasised the importance of 'inclusive language and behaviour, treating disclosures with sensitivity, and support for partners' (Egan, 2020). Nursing homes should enact policies ensuring no discrimination for minority groups, including, for example, LGBTIQ married partners.

Clearly, there is a need for more research in this area, together with information and advice for nursing home managers and staff regarding inclusivity and tolerance of differences. Drawing on external resources helps to raise awareness, ensuring all staff receive appropriate, regular education on this topic.

1.3.4 Young People in Nursing Homes (YPINH)

> Advocacy for the plight of young people in residential aged care is not new. Over the past two decades neither two investigations by the Australian Senate Community Affairs Reference Committee, nor the roll out of the National Disability Insurance Scheme, have been able to reduce the number of young people living in nursing homes. (Ibrahim, 2019)

The Royal Commission Report (Commonwealth of Australia, 2019, pp. 234–235) provides an overall picture of the approximately 6000 people under the age of 65 years living in nursing homes, aptly naming the section 'Falling through the gaps'. The report acknowledges that these arrangements are 'singularly inappropriate for young people who wish to live on their own terms'. Four categories of young people in nursing homes are noted: (a) those with a disability, (b) those requiring palliative care, (c) those with age-related conditions, such as early-onset dementia, and (d) those assessed as needing early aged care services, such as Aboriginal and Torres Strait Islander people.

It is clear that nursing homes were never designed to meet the needs of younger people, particularly those requiring long-term care; however, in many instances there is no other option. The report cited above provides many examples of younger people experiencing frustration and dissatisfaction with the level of care, as well as the inherent loneliness, grief and isolation of living in an environment clearly unsuited to their needs. Writers of the report argue that there is an urgent need for reform, clearly identifying the prevalence of young people in nursing homes as a 'human rights issue' demanding immediate resolution (Commonwealth of Australia, 2019, p. 241).

There are some notable exceptions of young people in nursing homes being well cared for. In some instances, this is only made possible with additional funding from external sources, together with a team of committed family and friends. One example is given a full, detailed description in the first comprehensive nursing home model for care of a young person with acquired brain injury (ABI), which includes descriptions of a support group advocating for a young man with no speech and almost no independent movement (What Does Chris Want, 2007).

There was no 'road map' for the care of this 26-year-old man following his tragic, catastrophic brain injury. He was transferred after 6 months in a major acute care hospital to a community nursing home caring exclusively for older people. He had almost no independent movement and no speech. Demonstrating what can be achieved through 'partnership in practice' Chris (his name used with permission) has enjoyed comprehensive care in a nursing home for almost three decades, communicating his appreciation with broad smiles and wide eyes. The partnership of carers includes nursing home staff, supported by a team of attendant carers and therapists largely funded by Chris's family and support network.

The model of care developed for Chris and nine others over a 10-year period in one nursing home demonstrates 'the *highly specialised* nature' of care required for this population: a model 'based on a philosophy of *"partnership"* (italics in the original) with a socio-medical focus' (What Does Chris Want, 2007, p. 72). Other young people with severe brain impairment also receive excellent care in many nursing homes. 'These stories ... demonstrate that appropriate care is not merely a nice option. It is a matter of life and death' (What Does Chris Want, 2007, p.73). Stigma and negative attitudes are, however, by no means absent, reflected in the following comment by one nurse.

> *I didn't come to this nursing home to care for people in their twenties! My skills set and my preference is directed to aged care. These guys need to be transferred elsewhere.*

Regardless of the instances of high-quality care, together with the many negative stories, the case remains that young people with complex needs require an age-appropriate environment, staffed by those with specialised skills.

References

Australian Department of Health. (2021). *Schedule of fees and charges for residential and home care.* https://www.health.gov.au/resources/publications/schedule-of-fees-and-charges-for-residential-and-home-care

Australian Government. (2016). *2016 Census.*

Australian Government. (2020a). *Aged care funding instrument.* Australian Government. https://www.health.gov.au/initiatives-and-programs/residential-aged-care/funding-for-residential-aged-care/the-aged-care-funding-instrument-acfi#what-is-the-acfi

Australian Government. (2020b). *Deaths in Australia.* https://www.aihw.gov.au/reports/life-expectancy-death/deaths-in-australia/contents/life-expectancy

Australian Government Department of Health. (2020). *Aged care laws in Australia.*

Bastian, D. (2019). Few Australians trust aged care: Report. *Aged Care Insite.*

Brauser, D. (2018). Ageism in medicine must stop, experts say. *Medscape.*

Burke, C. (2021). Every second person in the world has ageist attitudes: WHO report. *Aged Care Insite.*

Burney, L. (1997). *Education and social justice* 1996 Frank Archibald Memorial Lecture, University of New England.

Cole, T. (1983, June). The 'enlightened' view of aging: Victorian morality in a new key. *The Hastings Center Report*, 34–40.

Colenda, C., Reynolds, C., Applegate, W., Sloane, P., & Zimmerman, S. (2020). COVID-19 pandemic and ageism: A call for humanitarian care. *Journal of American Geriatrics Society, 68*(8), 1627–1628.

Commonwealth of Australia. (2021). *Royal Commission into aged care quality and safety final report: Care, dignity and respect, volume 3A the new system.*

Commonwealth of Australia. (2019). *Royal Commission into aged care quality and safety interim report: Neglect volume 1.*

Coredata Research. (2019). *Aged care State of the Industry report July 2019.*

Duckett, S., Stobart, A., & Swerissen, H. (2021). *The next steps for aged care: Forging a clear path after the Royal Commission.* The Grattan Institute.

Dyer, S., Liu, E., & Gnanamanickam, E. (2018). Clustered domestic residential aged care in Australia: Fewer hospitalisations and better quality of life. *The Medical Journal of Australia, 208*(10), 433–438.

Egan, C. (2020). The nursing home meeting the needs of the LGBTI community. *Hellocare*.

Hampson, R. (2018). Australia's residential aged care facilities are getting bigger and less homelike. *The Conversation*.

Herinton, J. (2016). Ten years in aged care. *Residential Aged Care Communique, 11*(4).

Hitchcock, K. (2015). Dear life: On caring for the elderly. *Quarterly Essay*, (57), 1–78.

Hitchcock, K. (2020). *The medicine: A doctor's notes.* Black Ink.

Ibrahim, J. (2019). The aged care royal commission's 3 areas of immediate action are worthy, but won't fix a broken system. *The Conversation*.

Johnstone, M., & Kanitsaki, O. (2006). Culture, language and patient safety: Making the link. *International Journal for Quality in Health Care, 18*(5), 383–388.

Johnstone, M., & Kanitsaki, O. (2008). Cultural racism, language prejudice and discrimination in hospital contexts: An Australian study. *Diversity and Equality in Health and Care, 5*, 19–30.

Kitwood, T. (1997). Dementia reconsidered: the person comes first. Open University Press.

LoGiudice, D., Flicker, L., & Smith, K. (2014). Health and care of older aboriginal and Torres Strait islander peoples. In R. Nay, S. Garratt, & D. Fetherstonhaugh (Eds.), *Older people: Issues and innovations in care*: Churchill Livingstone Elsevier.

Mcgraph, P. (2010). The living model: An Australian model for aboriginal palliative care service delivery with international implications. *Journal of Palliative Care, 26*(1), 59–64.

National Ageing Research Institute Ltd. (2020). *Inside the system: Home and respite care clients' perspectives.*

Nay, R. (2016). The good, the bad and the downright ugly: Reflections on 10 years. *Residential Aged Care Communique, 11*(4), 6.

Novak, M. (1999). Human dignity, human rights. *First Things, 97*(November), 39–42.

Prime Minister of Australia. (2021). *$17.7 billion to deliver once in a generation change to aged care in Australia.*

Rawson, H. (2019, October 7). Nursing homes for all: Why aged care needs to reflect multicultural Australia. *The Conversation*.

Ronald, L., McGregor, M., Harrington, C., Pollock, A., & Lexchin, J. (2016). Observational evidence of for-profit delivery and inferior nursing home care: When is there enough evidence for policy change? *PLoS Medicine, 13*(4), 1–12.

Royal Commission into Aged Care Quality and Safety. (2019). *Royal Commission into aged care quality and safety volume 1, Interim report: Neglect.* Commonwealth of Australia.

Russell, S. (2017). *Living well in an aged care home.* Research report, Issue. https://www.aged-carematters.net.au/living-well-in-an-aged-care-home/

Sherwood, J. (2019). *Participate ethically and effectively in intercultural settings.* National Centre for Cultural Competence.

References

Smith, C. (2019). *Navigating the maze: An overview of Australia's current aged care system.* Background Paper, Issue. //agedcare.royalcommission.gov.au/news-and-media/navigating-maze-overview-australias-current-aged-care-system

Somerville, M. (2020). The lethal collision of two pandemics: Covid-19 and ageism. *Mercatornet.* https://mercatornet.com/the-lethal-collision-of-two-pandemics-covid-19-and-ageism/65804/

Sutton, J. (1997). Why I'd rather be a citizen than a consumer. *Health Issues, 52*(September 1997), 6–8.

Victorian Institute of Forensic Medicine. (2016). *Residential aged care communique.*

Victorian State Government. (2019). *Public sector residential aged care services.*

What Does Chris Want. (2007). *A socio-medical model of partnership for the care of young people with severe acquired brain injury and related conditions.* WDCW Carlton.

World Health Organisation. (2021). Elder abuse. https://www.who.int/news-room/fact-sheets/detail/elder-abuse.

Chapter 2
Communication: The Key to Care

2.1 From Waiting List to Admission

Meaningful, responsive communication can be measured from the first step; when prospective residents and relatives make a tentative move towards nursing home admission. From this decision-making time, through the waiting period, to the day of admission, clear communication is vital for inspiring confidence and laying the foundations of a trusting relationship. Disharmony arises from confusion over waiting list times and changes. The tone (either harmonious or discordant) is influential in creating first (and often lasting) impressions.

2.1.1 The Entrance Foyer

What is conveyed when a prospective resident or their representative enters the foyer or entrance lobby? Does it seem welcoming? Is there ready access to a person who can answer initial queries? Is it clean and free from unpleasant odours? Are there signs of life and activity, or is it devoid of personal interaction? In a large nursing home, are the directions to specific sections clearly marked? Is the signage visible in other languages, depending on the mix of non-English-speaking residents and families? In other words, first impressions count. However, if judgement is made solely on the entrance foyer, other benchmarks could be missed.

First impressions may elicit the proverbial 'five stars' rating, while on closer examination the nursing home deserves far fewer. One survey seeking evidence for

---

While (some of) the stories are based on factual situations, real names and other details have been altered to protect the identity of the persons concerned. Resemblance to any particular person is therefore purely coincidental.

© The Author(s), under exclusive license to Springer Nature Switzerland AG 2022
R. Hudson, *Ageing in a Nursing Home*, https://doi.org/10.1007/978-3-030-98267-6_2

the Royal Commission into Aged Care Quality and Safety found that of the many homes inspected only 1.3 per cent attracted the top rating (Eagar, 2020).

2.1.2 Information

One vitally important area of communication relates to costs, charges and fees, indicating the need for clear information and transparent financial management systems. How is the money spent? What proportion of residents' fees is dedicated to their care? Is the information available in languages other than English? Other questions include medical and nursing oversight, visiting, pets, activities, meals, privacy and ready access to the resident's records. What is the staff/resident ratio, and what are the qualifications of the former? Are families included in regular meetings about the residents' care? Are changes in each resident's health condition promptly communicated? What opportunities are there for relatives to contribute to the nursing home life, if desired? When formal and informal information systems are evident, the scene is set for clear communication. (For details regarding standards of care and complaints' procedures, see Chap. 8.)

Residents are entering nursing homes with increasingly complex healthcare needs, including (a) dementia and (b) frailty as the most common. Dementia and frailty are often combined with a list of several other comorbidities, including incurable, life-threatening diagnoses. Prospective residents and families need to be assured that appropriate medical care, including referral to specialists, is readily available. Information is needed about the options for end-of-life care to ensure that residents are able to remain in the nursing home until their death, avoiding hospitalisation if that is their wish.

For those whose condition allows a more active life in the nursing home, information is needed about the options for enjoyable activities and leisure pursuits (described in more detail in Chap. 3).

2.1.3 Pre-Admission

Topics such as those listed in the 'Information' section (above) are covered in some formal pre-admission sessions for prospective residents and families where discussion may give rise to questions not necessarily raised voluntarily. In one such information session a very articulate, although physically frail, current resident was asked to give her impressions of nursing home life to the group.

> *'Patricia' was younger than the average nursing home resident. Aged 62, she was diagnosed with a rare disease, leaving her with profound physical disabilities while her mental acuity was retained. With the manager's encouragement Patricia developed a twenty-minute description of life in the nursing home, drawn from her own experience. She was delighted to present this overview at each of the regular information sessions for*

prospective residents and families. This firsthand account proved very popular with all who attended and her sense of humour was warmly welcomed. Prior to her illness, Patricia had enjoyed full time employment in a high-ranking government bureaucracy as head of communications. After settling in to the nursing home, and following a period of feeling useless, she enquired of management what contribution she could make, suggesting involvement in the information sessions. Taking on this role, Patricia found her life was no longer meaningless; her excellent communication skills proved an invaluable asset for the entire nursing home.

First-hand experience can be an appropriate addition to a resident's and relative's information session, offering a compelling glimpse into the everyday realities of nursing home life. Regular meetings provide opportunity for newsworthy items to be effectively communicated, together with answers to questions posed by the residents and/or families, both current and prospective. Meetings which are well planned, efficiently timed and skilfully chaired can prove an excellent means of communication. Disharmony arises from 'mixed messages' or inconsistent messages (or *no* messages), particularly in the absence of senior personnel from these important meetings.

A prospective resident found communication and information to be sadly lacking on her visit to one nursing home, described in her own words.

My daughter took me to this large posh-looking place but we couldn't find one human being to answer our queries. We tried to make an appointment for a formal interview but we were told: 'Oh, we don't bother with those formalities! Just pop in any time and have a look around.' This attempt at reassurance failed to satisfy us: we had loads of questions we wanted to ask.

For some people, a laissez-faire attitude may suit their temperament: the preferred option to a more formal approach involving appointments and meetings. However, discernment is needed to ascertain whether such an offhand attitude is also reflected in the care.

2.1.4 Choosing a Nursing Home

What are the key factors when a person is seeking nursing home admission for themselves or (as in most cases) for a relative or other person? Is location important, ensuring easy access for visiting, especially for a person without transport or with poor mobility? Is size important for those who are reassured by a large home with maximum infrastructure or for others who prefer a smaller, homelike environment?

The following first-hand account reflects some tension between the prospective resident's preferences and the nursing home's regulatory requirements.

I remember someone telling me to ignore the chandeliers and smell the couch cushions. In my tour of various nursing homes, I found a million things you couldn't do for 'health and safety' reasons. These included bringing my father's favourite Indonesian food, and his pet dog to visit. It seemed more like a prison than a home, so that was scratched off the list. (Woodlock, 2018)

On another occasion when this family member (cited above) was given an appointment for a tour of a place that seemed 'very posh' she decided to make an unannounced visit. She found 'a mattress in the hallway and a woman, who seemed to be in some distress, calling out without being answered'. Lending weight to the point about first impressions, this nursing home was also promptly removed from the list. Finally, after visiting several others, when she looked at the local council-run home it was like 'we had stumbled on the promised land'. This family member explained that her father didn't demand too much: just good care, good food, kind and gentle interactions with appropriately trained staff and a 'can-do attitude from the higher-ups'. She described her father's death two weeks later. 'The night before he died, the nurses asked if I'd like to stay over, and arranged for a camp-bed for me to sleep in next to his bed. They were angels, the staff at that home' (Woodlock, 2018).

The quality of a nursing home may be measured from a person's first visit. What/who is the 'face' of the nursing home? What is the first impression given to prospective residents and families? For many, the visit comes after much soul searching and decision-making about admission to full-time care. Such was the case for 'Annabel'.

> *Annabel was distraught, sleepless with anxiety over the decision to have her dad admitted to residential care, having exhausted all avenues for home care. Although increasingly frail due to a series of strokes, Annabel's dad remained mentally alert, eager to be involved in every step of the decision-making process. On their visit to one nursing home, he asked to see the lounge room, remarking to his daughter: 'I don't want to come here. Listen! They treat them like children!' Obviously the first impressions were not at all persuasive. 'And where will I put my computer, and my books?'*

While a significant number of nursing home residents have some form of cognitive decline, it ought also to be recognised that others, although physically incapacitated, may wish to pursue academic interests and/or keep up to date with current affairs. While direct and important questions relate to the prospective resident's physical capacity, what information is gleaned about the person's likes and dislikes, preferences and peculiarities? What is conveyed in the nursing home's advertising material which would be relevant for a person like 'Annabel's dad' who, albeit with serious physical limitations, remained mentally alert, involved in the world outside the nursing home and wishing to pursue his intellectual interests?

For others, some of the implications of living in Australia's multicultural society will be a significant factor in choosing a nursing home. Where wide choice is readily available, a nursing home appropriate for the person's distinctive cultural background and interests may be found. In every situation the need for sensitivity to specific cultural aspects is paramount. While issues of language are extremely important, including ready access to interpreter services and multilingual written material, the importance of exploring other cultural factors should not be overlooked. For example, how may this particular person's previous lifestyle be adapted to living in an unfamiliar environment? What cultural sensitivities need to be drawn to the attention of staff who may not be familiar with each resident's unique background? (For expanded discussion on multicultural perspectives related to death and dying, refer to Chap. 6.)

2.1 From Waiting List to Admission

Another important factor when choosing a nursing home is its accreditation status.

> ... a declining proportion of nursing homes are being re-accredited in full.... As of June 30, 2019, 92.3 per cent of re-accredited nursing homes had been given three years' accreditation – the longest re-accreditation period possible. This figure is down from nearly 97 per cent at 30 June 2018. (Ireland, 2020)

Accreditation involves a nursing home being reviewed or audited against the Aged Care Quality Standards by the Aged Care Quality and Safety Commission (ACQSC). Accreditation status is a guide for making comparisons: however, figures and statistics do not constitute quality communication when compared to personal engagement such as information sessions. Questions about the nursing home's *purpose* are seldom explicitly stated (Eagar, 2020), nor are their *goals*, with outcomes regularly analysed and published. Is healthy ageing promoted for those whose prognosis allows for physical, spiritual or psychological growth? Are the nursing home's goals clearly enunciated, together with quantitative data related to benchmarks?

For some families or other prospective visitors, the nursing home's proximity to their private home is an important factor, depending on their mode of transport and visiting patterns. For some with independent transport, distance will not be a problem; others prefer a nursing home within walking distance.

Given some of the suggestions outlined above, it may be advisable for prospective residents and families to devise a list of questions, covering what they consider the most important aspects of nursing home admission. Assuming adequate time and attention is given to the preliminary meeting, discussion of these issues will prove pivotal in making comparisons and informing decision-making. The following questions may serve as a guide, rather than constituting an exhaustive list.

- How many qualified nurses are employed and how many staff per shift?
- How many residents and how many single and shared rooms?
- Are the meals prepared on site, and is a sample menu available?
- Is there provision for residents to access drinks and snacks 24/7?
- What is their infection control policy?
- What access to medical personnel and other specialists, including allied health, is available?
- What activities are available, including options according to residents' choice?
- Is there a garden accessible to residents and families?
- Are pets allowed?
- Are visitors permitted any time day or night?
- Is there a volunteer programme?
- Do family members have access to the resident's care plan if desired?
- Are family meetings held regularly?
- Is the complaints policy readily available?
- What pastoral care/chaplaincy services are available?

2.1.5 The Waiting List

Some prospective residents and/or relatives may wish to secure the person's name on a waiting list in order to be certain of a place if or when needed. The following time frame is a helpful guide:

> The time it takes for older Australians to enter a nursing home after being assessed as needing residential care has blown out almost 50 per cent in two years... The Productivity Commission reports that the median 'elapsed time' between getting approval from an aged care assessment team (ACAT) and going to a nursing home was 152 days in 2018-19. (Ireland, 2020)

Communication about the waiting list is another important factor to consider. What are the criteria for including (or removing) a prospective resident's name on the list? Processes and procedures vary across nursing homes, so it is a legitimate question to ask about the means of notification when the list changes.

2.1.6 Expectations

The advertising material may boast 'a home away from home', but, for many, it is a residence whose primary aim is keeping older people 'safe'. Taken at face value, safety is an important issue; but when it involves strict rules and restrictions, the atmosphere is hardly reminiscent of a home. Fixed routines for bedtime, meals and toileting do not constitute what 'Molly' had expected.

> *Molly was pleased to be accommodated in a nursing home, relieved to lift the burden from her daughter, but she had never experienced a 'home' like this. Time tables supplanted spontaneity and freedom of choice; she was unable to vary her diet or her meal times, she felt she was flouting the rules if she left her TV on after 11pm. Nobody seemed interested that she had worked as a nurse in at least two other countries. She thought they'd appreciate hearing of her comparisons, not to mention some of her escapades! Expressing surprise at seeing her take an avid interest in the Olympic Games, the care attendant was even more surprised to hear that, many years previously, Molly had competed in the Olympic trials as a long-distance runner. Her legs could now carry her no further than the bathroom, but she did like to recall her more active days. She had no wish to boast; merely to expand the conversation topics beyond meals and toileting.*

Molly's experience epitomises a key difference between an institution and a home. The former may aptly describe an aged care facility built for purpose and providing safe custodial care, where residents must conform to strict timetables; the latter is a place where one can not only receive care but also 'be oneself'.

Apart from these comparisons, it is evident that a twenty-first-century nursing home accommodates older people with far more complex healthcare needs than the homes described in the *Aged Care Act 1997*. Far fewer residents are independently mobile, with more than 50 per cent requiring significant assistance, and a greater percentage needing dementia support. 'This is a highly dependent population'

(Eagar, 2020). Taking account of their increased physical and psychological dependence, residents' need for social engagement is equally crucial.

How may residents and families be supported, particularly when nursing home accommodation is alien to their experience? Are there any procedures for welcoming newcomers? With increasing numbers of residents living in single rooms, is there any opportunity for meeting other residents? Are visitors made to feel welcome, for example, by introductions to other residents and their families? The emphasis here, as elsewhere, is on the nursing home as a community of persons rather than a collection of individuals.

Prior to nursing home admission, most residents would have enjoyed access to a telephone or other personal communication system. This may now be denied to them, leaving some residents dependent on others for their personal, private contact with others. In addition to mobile phones, other emerging forms of communication (such as IT) are readily available and enthusiastically promoted in some nursing homes, particularly for those who wish to keep in touch with 'the outside world'.

2.2 Communication without Discrimination

2.2.1 *Communication Keys and Cues*

Communication, referred to as *'therapeutic'*, is derived from the Latin 'to heal'. In other words, communication from the outset is intended to cause no harm and may indeed be healing. Every effort should therefore be made by management to ensure staff have appropriate education on the subject. Improvement in communication skills does not necessarily lead to increased length or frequency of clinical time with residents: in other words, good communication may save time. The simple initial inquiry 'How are you?' or 'How are you feeling today?' may elicit responses about the resident's mood, speech pattern, pain experience and general condition. In sharp contrast to a carer merely stating 'I've come to give you a shower' an open-ended question invites dialogue, not to mention maintaining the resident's unilateral decision-making capacity regarding their own care.

Timely reminders may be needed for those staff who claim (of a resident): 'She can't communicate!' To which one manager responded: 'If she can't communicate, she must have died!' This exchange is intended to reinforce the meaning of communication, which includes non-verbal language such as eye movements, nodding or shaking the head, or hand movements. Staff may need to be reminded that all non-verbal behaviour is communication. As further evidence of the paucity of research into nursing home life, there is little available guidance from the literature on this important matter. However, in the broader literature examples abound of the many facets of non-verbal communication, for example:

> Poets and pundits, sages and songwriters have all waxed lyrical about
> the powers of nonverbal communication, their sentiments captured in

> Edward Sapir's (1949) now famous quote: 'We respond to gestures with an extreme alertness and, one might almost say, in accordance with an elaborate and secret code that is written nowhere, known to none, and understood by all.' (Hoobler, 2002, p. 240)

Creative, practical examples of an 'elaborate and secret code' show that placing keywords in large print on the walls of a resident's room may prompt communication where there is a language barrier or where a resident has little or no speech. Families and friends are often more than willing to contribute to varied means of communication; another instance of staff learning skills from others intimately involved in residents' lives.

Formal assistance from a speech pathologist may be warranted in some instances; however, whether the service is publicly funded or available only through the private system varies. It remains beyond question that, as with other therapies, ready access to a full-time speech professional should be mandatory for all nursing homes. The benefits, not only for communication but for residents' swallowing problems, are well attested in the literature, referred to in many places throughout this book. Apart from comprehensive assessment, the advantages of speech pathology include enhancing residents' capacity to express themselves, as well as preventing abuse and neglect, for example, by enabling them to call for help if needed. As for other 'extra services', this important area of communication enhancement depends on government funding priorities; all too often deficient in this context.

Examples drawn from the COVID-19 lockdown include inventiveness and other creative ideas for enhancing communication: families have shown ingenuity and residents have learned new skills via various electronic means, particularly where face-to-face contact was disallowed. For others, in situations where only one visitor was permitted for one hour per day, careful consideration was given to timing the visit during a mealtime, so that the visitor could provide optimal assistance. The advantages of communal dining for the purpose of enhancing communication have been one of the many casualties of the pandemic.

Unfortunately, there are many other examples of poor communication or no meaningful communication in the day-to-day life of the nursing home; again, brought to the fore through a pandemic with associated high mortality rates. In some instances, managers have taken the view that no news is preferable to bad news. For example, when a number of residents contracted COVID-19, resulting in hospitalisation or death, one manager either deliberately chose not to communicate this news or lacked the confidence to convey the facts with competence and compassion. Others appeared to have no established communication protocol to guide staff on how to respond to such a crisis. According to many news reports, families have been devastated by the lack of communication, sometimes learning of a resident's death or hospitalisation some days after it happened. In one other instance a family member was advised of an available place in the nursing home for her mother, four months after the resident had died (in the nursing home) from the coronavirus.

Other managers have been very proactive and innovative in their communication: devising a daily news bulletin (with relevant translation where needed) for the interest of the whole nursing home community. While it is evident that the COVID-19 pandemic constitutes a world-first experience, many have risen to the challenge through expert crisis management and clear communication skills. These characteristics would seem to be important factors in the selection and recruitment of managers. It is also crucial for 'front of house' staff to be good communicators, creating positive first impressions for all who enter the nursing home.

For newly admitted residents and their families, learning which staff to approach with specific queries is an important component of orientation and communication, thus avoiding the all-too-common scenarios below.

'Clement' was utterly bewildered, having spent his first night in the nursing home. He was not sure why he was in this strange place. Where was his wife? How could he find the toilet? It seemed a long time since he'd had a cup of tea and he didn't know whether he was going to get any breakfast. Who were all these strangers passing his door?

'Miss Francis', aged 96, was mentally alert although physically frail. She was aware of her need for assistance but not sure whom to ask. She'd met several different staff members on her first day in the nursing home, although she didn't know what their various responsibilities were. When she asked one nurse assistant for help with her walking frame, she received the reply: 'Oh, you'll have to ask the physio about that.' Miss Francis had to wait until the following week to find out who the physio was and when they could be contacted.

'Hilda', the wife of another resident, had lain awake most of the night at home; anxious, fretful and grieving since being separated from her husband of 62 years. Reluctantly, she conceded that her husband's care needs were now beyond her capacity but she lacked confidence that nursing home staff would provide anywhere near the amount of loving care she'd been giving him. On arriving at the nursing home to visit, she spent what seemed a very long time wandering around the corridors trying to find someone to ask about her husband, as he was not in his room. It seemed to Hilda that many of the residents were seeking attention but receiving no answers to their calls. She wished she could reverse her decision, and take her husband home again.

Some of the details in the scenarios above may of course have perfectly reasonable explanations, or they may constitute initial impressions which later prove no cause for concern. However, much bewilderment, not to mention harm, insecurity and fear, can be prevented by an intentional discussion with the resident and family as soon as possible after admission. Much can be gleaned from a short, carefully planned meeting, potentially putting the resident and family at ease from the outset. Such discussions can transform a negative decision-making process resonating with pessimism, nihilism, guilt and shame into a positive outcome at a time of great need, both for resident and family.

Without comprehensive communication procedures, the settling in processes for some residents are fraught with difficulties, as in this example.

Discussion at handover turned towards the resident admitted three days previously. 'We haven't seen any family. No-one has been to visit him and he doesn't seem to have many belongings.' Admission details referred to the resident's wife as next of kin although no other information was available. One nurse commented rather scathingly, 'You'd think she'd visit him at least once or twice a week!' Another said, 'I think we should contact the social worker to see if there's any family issues we should be aware of'. On further investigation it was found that the resident's wife was being investigated in hospital for rapidly advancing symptoms of metastatic cancer, and the couple had no children.

This brief scenario merely skims the surface of a complicated social history, with evidence of an abusive relationship within a long marriage, and with little attention given to the resident's cognitive decline over many years. Judgemental, condemnatory comments by various members of staff did little to increase the understanding of the situation until a proactive nurse initiated a social work referral. Rather than a source for gossip, or assumptions being drawn from hearsay, clear factual, non-judgemental documentation provided a basis for a holistic approach, leading to thoughtful, responsive care.

Communication is enhanced when information is obtained on referral, supported by clear policies and procedures for referring agencies to provide details relevant to the resident's needs. Clear communication prevents scenarios such as the following, occurring at afternoon handover.

> *We have a new resident admitted this morning. As usual, there's absolutely no information about ongoing specialist involvement for his severe arthritis, or what his treatment options are. He seems to have a number of complications from other comorbidities but few details about them. His diagnosis also says (as usual!) 'dementia' with no information as to when it was diagnosed or what type of dementia it is.*

Rather than lamenting the lack of referral details, the nursing home would do well to develop comprehensive communications systems alerting referring agencies to the kind of information required to care for the resident being transferred. For example, in the chapter on 'dementia' (Chap. 4), full weight is given to the discussion of this issue, illustrating the importance of relevant data conveyed between agencies. Sensitive dementia care is based on knowledge of the different types of dementia, foreshadowing the unique responses needed. Regarding dementia diagnosis, it is important also to understand what has been conveyed to the resident and/or family, thereby avoiding the following unfortunate scenario.

> *Nurse one: "The referral sheet only has 'dementia'" and when I asked the family, they said they'd been told: 'Your mum has a bit of old age memory loss so we think she'd be better off in a nursing home.'*
>
> *Nurse two: 'I asked the GP but she has no other information either, so we're completely in the dark.'*

Rather than remaining 'in the dark' these nurses would provide better care by returning to the referral source, requesting more details, including diagnostic procedures. Or, in the absence of other information, to arrange a referral to a geriatrician or psychogeriatrician in order to obtain an accurate picture. Such details are essential to best-practice clinical care, particularly when the prognosis is uncertain. The following scenario is, regrettably, all too common.

> *The nurse arranging a patient's admission from hospital to nursing home was concerned about the lack of diagnosis details. When asking the family, the daughter responded: 'I asked them in the hospital what was wrong with mum. I said that our family, and the nursing home, need more details, especially seeing mum's condition seems to be deteriorating. The doctor who was at the desk replied: "Don't worry. We decided your mum would do better in long term care. They don't need to know all the medical details"'. On further enquiry, the daughter was told: 'Stop worrying! She'll probably thrive in the nursing home!'*

2.2 Communication without Discrimination

Perhaps the doctor's response (above) was due to ignorance or unrealistic expectations about nursing homes' expertise in end-of-life care. On receipt of detailed diagnostic details and recent test results the experienced nurse developed a care plan which took account of the resident's serious comorbidities and obvious clinical deterioration, and, accordingly, assisted the family to plan for her death.

Clear communication lies at the heart of relationships between hospitals and nursing homes. When staff in the former understand that care in the latter is dependent on having relevant information about a resident's diagnosis, prognosis, medications and other associated details, a relationship of mutual trust develops. Rather than nursing home staff lamenting: 'They've given us no information', a phone call requesting relevant pathology reports, diagnostic procedures, specialist medical advice, would be considered by both parties as essential for optimising the resident's care. In many situations, a clear understanding of both agencies ensures the communication is prompt, detailed and includes contact details for further follow up when needed.

Communication, of course, works both ways. Hospital staff have been heard to make derisive comments about the poor quality of referral information from nursing homes. On the other hand, when time is taken to ensure all relevant details are available, the groundwork is laid for a professional relationship which, in turn, enhances the resident's care. As one senior staff member commented: 'It's always good to receive a referral from this nursing home; the communication is excellent and the paper work is always up to date.'

Communication between agencies involves more than admission, discharge and referral documentation. Personal contact attests to the care and concern for the resident's condition and treatment, shown in this example at hospital handover.

> *The nursing home's senior RN rang this morning to make sure we had all the information we needed for caring for Charles. It was a short, but very helpful phone call, explaining why we won't be seeing any visitors for him. His wife is too frail to travel alone, and their children are all interstate.*

Similarly, when a resident returns to the nursing home from hospital, it is important to learn of any follow up and to have a contact number when/if more details are required. A trustworthy liaison system ensures a smooth transition between agencies. On the other hand, poor communication breeds dissatisfaction, disappointment and, in some cases, deep distress, particularly regarding end-of-life care.

The method for assessing each resident's communication needs varies across nursing homes. A pivotal point to all care planning is to ask what is the communication goal and whether it can reasonably be met? Generic systems do not necessarily meet each resident's idiosyncratic needs. Documenting this aspect of care may be of equal, if not greater, importance than their physical needs, although not always commanding comparable attention. As with many other aspects of care, families who may have far more intimate knowledge and experience can provide valuable assessment data, particularly for residents whose first language is not English.

Communication is a vast subject: this section has described some of the 'keys and cues' to prompt carers in developing the 'art'. While some may argue 'We don't have time to collect all this information', others demonstrate every day how clear documentation actually saves time. Clear, comprehensive communication systems are considered by many nursing homes to be of paramount importance: setting the tone for exemplary care. Systems for communication in a language other than English are also vitally important in an increasingly multicultural environment.

Multicultural Assessment

Cultural assessment is discussed in Sect. 1.3; additional comments here relate specifically to communication. Within the nursing home, multicultural communication involves more than linguistic diversity. While issues of language are extremely important, other cultural factors should not be overlooked. For example, how may this particular person's previous lifestyle be adapted to living in an unfamiliar environment? What cultural sensitivities need to be drawn to the attention of staff who may not be familiar with the resident's background? How might families be supported, particularly when nursing home accommodation is alien to their experience? What form of communication suits the disparate needs of residents and families, so that all of these questions can be raised and discussed appropriately?

One community with unique communication needs is the broad but diverse population of Aboriginal and Torres Strait Islander people who 'are not being well served by the current aged care system, which, in many respects, fails to grapple with the realities of the barriers this part of our community faces'. This evidence before the Royal Commission showed that they prefer to receive aged care services 'from people and organisations they know and trust, and, where possible, that are Aboriginal and Torres Strait Islander controlled and staffed' (Commonwealth of Australia, 2019, p. 166). Where this is not possible, nursing homes with responsibility for the care of older persons from within these communities need access to culturally relevant aged care experts. For example, for some people from Aboriginal and Torres Strait Islander background body language is important: it is a sign of respect to divert one's eyes when addressing an older person. As there are many different communities within this particular cultural group, to honour a specific community's rites or etiquette, appropriate referral and specific person-centred advice are needed.

2.2.2 The Short-Term Stay

The following story illustrates several important factors regarding communication in a short episode of care; namely, how a resident's life can be enhanced even at the end, the importance of teamwork and the impression gained by the doctor involved.

> *'Zena's' short stay of less than six weeks had quite an impact on her carers. It was very pleasing to see Zena gradually 'blossom' from her early signs of isolation and withdrawal and to observe the positive impact this had on her family. Her complex medical condition tested the skills and resources of the care team. The manager communicated his thanks to*

the night staff who used their many therapeutic skills to ensure Zena was relaxed and peaceful prior to her death. He also acknowledged the great team effort, including a quick after-hours response by the pharmacist to a request for urgent medication. The GP communicated her praise for the staff, telling the manager: 'Staff in so many nursing homes panic at the slightest thing and continually call the doctor unnecessarily, but your staff are very professional and skilled. Zena could not have received better care.'

In an environment where good communication is not always given high priority, these comments relayed to staff resulted in increased work satisfaction as well as raising morale.

2.2.3 Loved One: Unambiguous Communication?

What is conveyed by the ubiquitous term 'loved one'? In many contexts it infers a genuine relationship between two or more people. When, however, it is used carelessly and loosely to describe *every* relationship, a distorted message is conveyed. In an ideal world every older person would be a 'loved one', cared for and cherished by a person or persons close to them; in such circumstances the term conveys a positive message. In other instances, the person being referred to is, sadly, not a 'loved one' at all; or may be loved by some and not by others. Clear communication requires an understanding of each resident's circumstances. That is not to say an in-depth psychological analysis is required into their personal relationships. It serves as a reminder, however, that assumptions should not be made about those relationships or about the nature of feelings between two or more people. A sensitive assessment provides essential details about the resident's connection (or lack thereof) to family and friends, paving the way for clear communication. The following scenario indicates the potential harm when such assessment is lacking.

The night nurse phoned the resident's wife to report that her husband had died, unexpectedly: 'Mrs. Stanton, I'm so very very sorry to be ringing you with some very sad news. I'm afraid it's the worst kind of news you will want to hear. Jack has taken a very unexpected "turn" and I'm sorry to tell you he died before we had time to call an ambulance.' There was a pause of several seconds before Mrs. Stanton replied. 'Well, thanks for letting me know, but you may be surprised to know I'm not at all sad as we'd been living separate lives for many years but never got around to divorcing. There was no love lost between us and, quite frankly, I'm relieved to hear of the old bugger's death! Now, I can take myself off overseas and be free at last!'

The nurse was clearly surprised, if not shocked, to receive such a frank response. While it may not be the nurse's role to have detailed knowledge of each resident's personal relationships, careful assessment provides an honest description of 'next of kin', which is often subject to misunderstanding.

Definitions of 'next of kin' and 'substitute decision-maker' vary across states and territories in Australia, so nursing home staff would benefit from up-to-date information, routinely made available to staff within their location. For example, 'next of kin' may not necessarily be recognised in the relevant law.

The question at issue here relates to the information needed in order to provide appropriate, sensitive care to every resident and their family, respecting their privacy while being aware of circumstances which may impact on their care. Communication (written and/or verbal) with night staff is also of critical importance; they often lack the detailed information readily available to the day staff.

On the subject of relationships and kinship, what lies behind the statistics revealing that a significant number of nursing home residents receive no visitors, including pre-pandemic? Is this due to a lack of 'loved ones', an attitude of denial, physical constraints preventing visiting, perceived lack of communication with residents who suffer from dementia or a seemingly callous attitude: 'out of sight, out of mind'? Or is it, in some circumstances, reflective of families being 'burnt out' after years of caring for the older person, and now feeling relieved to have some time of their own? It may also be the case that the (former) main carer has full-time employment, together with other family responsibilities, and lacks any spare time for nursing home visits. Nursing home staff have neither the time nor the skills necessary for such analysis; however, it reinforces the fact that 'loved one' is an anomalous term in some circumstances. Clear and honest communication is the key, drawn from accurate admission assessment and social worker involvement when indicated.

2.3 Between Agencies and Age Groups

This section provides some commentary on the kind of communication needed for effective relationships between nursing homes and other agencies, demonstrating that true partnership is more than a management device. In this context, partnership is considered as a narrative, entering the story of each resident and their family, as well as participating in the wider story of the nursing home community and beyond. Contrasted with the short-term nature of most hospital admissions, the unique environment of the nursing home is considered a place where older persons can become part of a larger narrative, each one leaving a distinctive legacy.

2.3.1 Relatives Caring

Further to the section on 'loved ones' (above) other aspects are discussed here to indicate some of the innumerable facets of relatives' caring. For some relatives, the older person's nursing home admission can bring welcome relief from long-term full-time caring, providing them with opportunity for travel, other activities or much-needed respite. Visits to the nursing home may then be few and far between or ceased altogether. Others may wish to continue their caring role, visiting for several hours each day, assisting with many personal care tasks, as in the following example.

2.3 Between Agencies and Age Groups 39

> '*Mrs. B*', *married to* '*Mr. B*' *for over sixty years, had cared for him at home until his advancing dementia and other serious comorbidities became* '*too much*'*. She told the manager:* '*I don't want to stop looking after him, but the doctor persuaded me my own health will be affected if I don't relinquish the full-time carer role. But I'd like to come in every day from about 9am to 6pm with a break for lunch if that's alright with you? I'd like to help with his meals, play music for him, and generally reassure him that I haven't abandoned him.*' *Mrs. B found she was now able to sleep without interruption and wake refreshed. With no other family responsibilities, she wanted to devote her days to her husband's care.* '*I want to be with Bert for as much time as possible. I know I won't have him for much longer.*'

Various staff members aired their own (in most cases unfounded) criticisms of Mrs. B's visiting. 'She's *so* dependent on this relationship.' 'What's she going to do when he dies?' 'I think she'd be better off to visit less often, in preparation for when she's completely alone.' Others, who had taken the time to speak to Mrs. B about her visiting patterns, saw no need to make judgements or to criticise.

As for many others in similar situations, Mrs. B soon became a well-loved, regular presence in the nursing home, and staff appreciated the care she gave Mr. B, particularly at mealtimes when he could become resistive, requiring more patience and time than they could provide. Effective communication is exemplified by staff showing their concern, asking from time to time: 'Are you okay, Mrs. B? How is your own health? Are you still finding it all right coming in every day?' There is also a place for clear, non-judgemental documentation so that the facts are clearly communicated to all carers, leaving little or no room for baseless innuendo or false assumptions.

Further commentary on the impact of COVID-19 is apt. What would be the repercussions for Mrs. B and her husband if the former was (a) no longer permitted to visit or (b) have her visiting reduced to one hour, provided she wore personal protective equipment (PPE)? Descriptions have emerged of some managers devising creative, safe options in such carefully and individually assessed circumstances, rather than adhering to strict, risk-averse, inflexible rules. Examples include arranging outside garden areas where social distancing can be maintained during visits, encouraging families to send letters, parcels and pictures to the resident, and glass 'windows' which provide protection while allowing face-to-face communication. Relatives have shown their willingness to adapt, using these and other alternatives for communication with residents. The rapidly developing world of IT offers increasingly creative ideas for bridging the 'social distancing' gap.

2.3.2 Children Welcome

In some nursing homes children's presence is a rare phenomenon, noted with dismay by this relative.

> *What* '*Brenda*' *noticed on the day her father was admitted to the nursing home was the complete absence of children. She wondered whether her own grandchildren (her father's great-grandchildren) would be welcome. On making enquiries at the office, she was told:* '*You're quite right. We have very few children come in here. I think they'd find it too depress-*

ing.' Brenda thought about this comment and after some time, ventured to speak to the manager, offering some suggestions about changing the culture, to welcome children of all ages.

Following discussion with several staff members, including the activities coordinator, music therapist and workplace safety representative, a plan was drawn up to incorporate children in a variety of ways, including the following:

- Brightly coloured sign emphasising 'children welcome'.
- Liaison with kitchen staff to have children's drinks and snacks readily available.
- Meeting with the principal of the local primary school to arrange regular visits.
- Regular intergenerational playgroup established with local community pre-school.
- Musical instruments for children and residents to accompany the weekly singalong.
- Children learning how to 'dance' with a person using a walking frame and how to gently move a resident in a wheelchair, in time to the music.
- Advice from a music therapist on additional ways to incorporate children.
- Children with musical ability invited to 'perform' at an informal 'concert'.

Imagination and creativity come to the fore when welcoming children into the nursing home, particularly if their visit is coordinated by another agency, such as a school. What would be their first impression? An attractive welcome sign with photos of other visiting children? In one nursing home, an appropriately skilled staff member designated to organise and supervise such visits resulted in the following.

It was the first Monday in the month, and the music therapist had prepared to welcome a group of students from the local primary school, to be involved in the regular one-hour music session with selected residents. She had brought a variety of musical instruments, ensuring one for each student. Following morning tea and general introductions, the music therapist took the students to the lounge to meet the residents. Some of the regular students had by now learned some residents' names and greeted them spontaneously. Others wanted to give a resident a gift, such as a picture they had made; others enjoyed sharing their morning tea.

Examples abound, of intergenerational playgroups, where a mix of spontaneity and well-designed programmes meets the needs of both age groups. Meaningful relationships are formed, particularly when the same group visit the same residents regularly. Name tags are useful adjuncts to clear communication, helping both residents and children relate on a more intimate level. One of the benefits of regular visits from the same group of children is conveyed by the following story.

'Scarlett', aged 10, had been visiting 'Herb' for several months, developing a close, endearing relationship. On the regular monthly visit Herb was nowhere to be seen and Scarlett became quite upset. 'Where's Herb?' (He had invited her to use his first name.) The music therapist had not realised how upset Scarlett would be to find him missing. Gently, she told Scarlett that Herb was very ill in hospital. After sensitive negotiations between Scarlett, her parents, and Herb's relatives, a visit was arranged for Scarlett to see Herb in hospital. Although very ill, and close to death, Herb brightened considerably to see Scarlett. 'She's the granddaughter he never had,' explained Herb's wife.

Unsurprisingly, news of this visit spread quickly through the nursing home, with various opinions being stated loudly: 'Fancy taking that poor young child to see that old man in hospital, surrounded by tubes and near to death!' Others were more sympathetic, being sensitive to the special relationship developed over several months. 'It must lift his spirits to see his special young friend.'

As with similar episodes of this nature, communication is the key. In the capable hands of the music therapist, Scarlett's visit to Herb proved deeply meaningful to both, not to mention having a profound impact on some of the staff. 'It moved me to tears', confided one nurse.

A popular Australian television programme captured the introduction of young children to residents in a retirement village. It was not a 'one-off' visit but a timetable of activities covering several weeks. After three weeks, the halfway point, there were noticeable improvements in the residents' physical abilities, including their ability to bend down to the children's level. Most markedly, at the beginning of the programme nearly all of the residents suffered depression, two of them quite profoundly. 'After six weeks, none were registered as depressed. They had completely changed their outlook on life and in their hope for the future' (Stewart & Johnson, 2017). As a result of this 'experiment' plans were made to continue the programme on a permanent basis and with a long-term view of incorporating a permanent nursery in the village. The sessions also captivated the interest of some nursing homes, inspiring them to plan similar intergenerational programmes.

In another nursing home, children from a neighbouring primary school were invited to join morning tea in the garden with some of the residents. After the pleasure of specially prepared drinks and food, the children were prompted to meet their new friends, pick some flowers, then to come close and even to *touch*, as in this heart-warming account.

> 'Robert', aged 8, was absolutely fascinated when invited to meet 'George', the 95-year-old in his wheelchair. 'Can I touch him?', Robert asked. Reassured by George's response, the nurse helped Robert raise George's trouser leg to his knee. 'That's what a 95-year-old leg looks like', he told Robert. Again, prompted to touch, Robert felt George's leg tenderly, assuring himself that Robert was, indeed, alive.

Activities involving children, well planned and regular (rather than 'one-off'), contribute to lessening society's inherent ageism: in some contexts, older people are regarded as the 'untouchables'. 'Touch' (or the lack thereof) has particular implications during the COVID-19 pandemic; no doubt future research will show the impact of such enforced isolation.

Family relationships are changing, particularly when so many grandparents and great grandparents, not to mention step-grandparents, are caring for young children while coping with the vicissitudes of their own increasing age and frailty. There is potential for grief when the older person, who has had a significant place in the child's life, now enters residential care and subsequently dies. How might children be involved in a resident's dying process? What information do they need, or request? How may they be encouraged to participate in the resident's care if/when appropriate? How is their grief acknowledged following the death, and how and by whom would it be addressed? These questions serve as a reminder of the need for a family meeting in such circumstances: a meeting to which children are invited when appropriate or desired. In such instances, engaging a social worker or school teacher may prompt further discussion, insights and responses.

As for many other areas discussed in this book, formal research into the benefits (or otherwise) of children's presence in the nursing home would add further reinforcement for the development of meaningful programmes.

2.3.3 Hospitalisation and Transfers

One of the more common transfers of older people is from acute hospital to nursing home, and some of the issues are described in the section 'Communication Keys and Cues'. 'They haven't sent us any notes at all' is a frequently heard comment. The assumption is made that in the latter a resident's test results, amended diagnosis or changed prognosis are of no importance to 'the home'. Such lack of communication puts at risk any continuity of care, particularly between medical staff and GP, the latter not necessarily being informed of important changes in a resident's condition. What is communicated in the transfer notes, whether comprehensive or sketchy? Is there any expectation of continuity of care, and if so, what information is provided? While countless examples would convey the wide variation in communication systems, the following scenario is not atypical.

> *At handover from morning shift to afternoon, notification was given of a patient being transferred from the major city hospital where he'd been for seven weeks, with multiple comorbidities and major surgical procedures. His cognition was poor and he spoke little English. No definitive diagnosis of dementia was provided. It was evident that his surgical wounds were slow to heal but there was no documentation from the wound care nurse. Neither were any pathology or haematology results provided. The RN conveying the information also noted the absence of comprehensive pain management, although 'morphine prn' (morphine whenever necessary) was mentioned.*

What were the expectations for continuity of care in this brief, all-too-familiar story? What was communicated in the notes provided? What follow-up specialist advice would be available for this resident and his complex medical problems? The GP also expressed her frustration. 'I'm not confident about managing his complex wounds, and I'd be reluctant to prescribe morphine for a ninety-two-year-old.'

Fortunately, the care manager was familiar with the referral procedures required. He immediately arranged for a wound care specialist, dietician, pharmacist, interpreter and palliative care nurse to visit. In consultation with the family, he also ensured appropriate chaplaincy services were available. All of this care planning of course took more than one day and was not finalised until the nursing home manager requested more details from the hospital. 'How can we provide continuity of care', he prompted the hospital liaison officer, 'when we don't have all the patient's current details including his communication needs?'

This scenario is indicative of some contemporary medical care which concentrates exclusively on physical issues, ignoring the social, spiritual, psychological, sexual, economic, not to mention any particularly idiosyncratic interests the resident may have. In other words, and considering care within a holistic framework, who is this person in the totality of their being? The answer invites a personal response, rather than reducing the resident to their diagnostic details or 'vital signs'. Specific information about cultural, religious, language and sexual preferences is needed, together with the person's CALD status, or their relationship within an LGBTIQ community. Referring to the 'vital signs' mentioned (above), signs of equal (vital) importance are not necessarily confined to heart and blood pressure rate: they include signs of distress, sadness, bewilderment or pain.

Fostering relationships between nursing homes and (large) hospitals is not necessarily easy. Making personal contact, reinforced by comprehensive documentation, are means by which such partnerships are transformed, with both agencies emphasising they belong to the one community. However, tensions may also arise, often due to poor communication.

When inter-agency communication is taken seriously, comprehensive care planning is optimised and the resident assured of ongoing care. 'I rang the hospital', said the RN at handover, 'and they seemed surprised we would want to receive copies of radiology reports and updated medication advice.' There are, in many places, positive examples of professional, productive working relationships between agencies, particularly when reinforced by a liaison nurse. Continuity is then fostered, giving rise to a tangible community of care.

Lack of continuity, not to mention clear communication, is evident in the following distressing account of a frail non-English-speaking resident transferred to hospital for emergency treatment.

> 'Miss C' had lived in the nursing home long enough for all the staff to know she must not be showered; a bath or bed sponge was to be preferred. On her first morning in the hospital Miss C was wheeled to the shower and assisted to undress. When the shower hose was directed to her, from behind her back, she began screaming, fighting, gesticulating, unable to be pacified; lack of English preventing her explaining her terror. A senior nurse left her desk to investigate the cause of this commotion. Within a second or two, standing at the door of the shower room, the nurse immediately 'diagnosed' the problem. Miss C had a number tattooed on her forearm. She was one of the few surviving victims of Auschwitz concentration camp at the end of World War II, where 'going to the shower' meant extermination in the gas chamber.

The serious import of this scenario required an explanation from the senior nurse to the inexperienced attendant unfamiliar with the war's atrocities. It was also a vitally important reminder of the need for clear communication between agencies. All the nursing home staff knew why Miss C must not be showered because of the memories it would evoke of the gas chambers from which she had, miraculously, survived. However, they failed to communicate that matter to the hospital: a 'vital' sign equally as important as her pathology results.

Another factor concerning transfers from hospital to nursing homes is the predicted survival rate, with many patients dying before discharge (Gerber et al., 2021,

p. 431). These authors suggest that there is a need for sub-acute palliative options for those no longer needing acute care, too frail to return home, yet not suitable for long-term nursing home accommodation.

2.4 Handover Conversations

2.4.1 Depends Who's on

One of the most important means of communication in the nursing home is the handover; when vital information by those at the end of their work period is conveyed to their colleagues commencing a new shift. However, research on the subject is sparse, mostly concentrated on communication in acute care (Australian Commission on Quality and Safety in Health Care, 2018). Handover in nursing homes deserves more focused attention, including regular education drawn from policies and procedures to ensure consistency.

Depending on rostered hours, nursing home handover generally occurs early morning, mid-afternoon and late evening. When shifts vary in lengths and personnel, handover becomes an even more important means of communication. Residents benefit from the continuity of care, and staff achieve satisfaction through up-to-date advice. By contrast, poor communication is exemplified by comments such as 'Depends who's on!' Or, 'Nobody told me he was going to hospital.'

Handover is the time for important information that may affect residents' lives, such as isolating a resident or residents as an infection control procedure, or conveying accurate information about a resident close to death. Communication is not necessarily confined to information about residents' physiological status. Other important details relate to visitors, or absence thereof, phone calls made or awaiting attention, GP visits, communication with allied health staff or a volunteer. Information about a recently admitted resident needs to be conveyed with consistency, optimising the resident's introduction to a new environment and reinforcing their confidence that staff will know how to care for them.

2.4.2 Staff Reactions to Handover

An important litmus test for a successful handover relates to how staff members feel at the end of the session: confident to commence the shift, overwhelmed by too much information, frustrated by lack of details, buoyed by positive examples of care and/or an open invitation to ask questions? Handover also provides an opportunity to welcome new staff, agency staff or those returning from leave. These personal touches foster a community of care, as well as acknowledging each member's unique place in the team.

2.4 Handover Conversations

Following handover, some staff will have ready access to residents' file notes, with time to check the written/computerised record; others will rely solely on the handover conversation. While one person may begrudge the time spent at handover, another will appreciate it as an important means of communication. However, time is of the essence, indicated by the following:

> Oh, my god, Suzie's on today. She'll keep us at handover for over half an hour, chatting on about irrelevant stuff, not realising we have time constraints. It turns into a gossip session rather than giving us important details about the residents' care.

One way of ensuring timely, well-organised handover sessions is to use a pro forma document or agenda listing the time allocated, important topics to be covered, including which issues require urgent attention for staff on the next shift. An appropriate template ensures consistency, encouraging staff to 'stick to the agenda' rather than using valuable time for idle talk. When time permits, handover may include a brief (5–10 minutes) teaching session; for example, regarding a safety issue arising from an incident, a clinical crisis which may have been avoided or useful information about an uncommon diagnosis.

Notwithstanding all of the above, in 'real life' handover seems to prioritise a discussion between care managers, in many cases excluding the enrolled nurses (ENs) and personal care assistants (PCAs) who have the greatest contact with the residents. As one PCA commented:

> Although I have access to the electronic documentation, I rarely have time to enter any notes, or to read the notes of others. Documentation is not a requirement for PCAs, so we rely on verbal reports to and from the care manager if and when they are available. When I do attend handover, I'm often called away to answer call bells or other resident issues. However, I believe it's important for us to receive a good handover, as residents' needs change so quickly, and when I'm only here two or three days each week, I need to know what's different about the residents I care for.

Another PCA commented: 'We're not allowed to attend handover. We're told it's much more important to care for the residents.' It seems anomalous, to say the least, that those involved in the most intimate contact with residents have the least knowledge and information to guide their care. The statement also (falsely) assumes that handover has nothing to do with 'care'. Yet another anecdote confirms that lack of relevant information compromises the very nature of that care.

> The care attendant was familiar with the resident's routine. Although he attended part of the handover, he was not provided with the full details, so that when he attempted to assist the resident with the shower, he was not aware that emergency medical support had been needed through the previous night following the resident's fall and the resident had been administered strong pain medication. Cognitive deficits prevented the resident relating the incident. The care attendant was, therefore, unprepared for the resident's collapse in the shower, causing injuries to both of them.

It may not be necessary to convey all residents' details to PCAs at handover; neither is the PCA expected to know the resident's detailed medication regimen. However, this incident shows the importance of the person in charge discerning what information is relevant to the PCA's role; namely, the side effects of a powerful drug and the propensity for further falls. As for other 'procedures' a clear policy document should guide the handover session: prompting consistency while allowing for variations according to need.

2.4.3 Hands-on Care

A review of handover procedures paves the way to improved communication, streamlining the agenda to achieve optimum time management and providing opportunity for feedback from staff at all levels. Such a review, when carefully crafted, can be achieved by written survey or staff meeting discussion where questions may include the following:

- What is being handed over?
- Is it time well spent or a waste of resources?
- What does it convey about 'hands-on' care?
- What information is the most helpful for those involved in individual resident care?
- Is there opportunity for staff to ask questions?
- Is all the information appropriate for the immediate shift?

The first principle of handover is accountability and transfer of information about residents' needs, linked to the standards of care, particularly in preventing serious, harmful events (Australian Commission on Quality and Safety in Health Care, 2018). Handover need not be excessively formal or devoid of humour. Open communication may invite comments and tips such as the following from a PCA:

Discussion focussed on the difficulty staff were experiencing in persuading a particular resident to have a shower. The PCA offered her own wisdom.
'I find that when George refuses to go to the shower I invite him to have a dance, and we waltz our way to the bathroom'.

This brief scenario reinforces the benefits of knowing the resident's 'story' which is far more enlightening than the constant negative focus on his 'behaviour'.

Further analysis and discussion about handover give rise to other questions about 'hands'. What is communicated or impeded by gloved hands? What is the resident's response to having a hand laid on them? Is it a soothing, gentle hand? Or is it rough and uncaring? What is conveyed through a carer's hands when the resident has no sight or speech or common language? What are the implications of a 'hands-off' protocol? What does it mean to 'hand over' a resident's care?

Whatever the answer to these questions, there is no doubt that hands are a powerful means of communication. A welcoming hand is a personal invitation to join with others. The handover session presumes staff belong to a team, where every hand is important, benefitting all who belong to the nursing home community.

References

Australian Commission on Quality and Safety in Health Care. (2018). *Communication at clinical handover*.

Commonwealth of Australia. (2019). Royal commission into aged care quality and safety interim report: neglect Volume 1.

Eagar, K. (2020). Australian aged care is understaffed. *Medical Journal of Australia, 212*(10), 507–508.e1.

Gerber, K., Tuer, Z., & Yates, P. (2021). Who makes it out alive? – predicting survival to discharge of hospital patients referred to residential aged care. *Collegian, 28*, 431–437.

Hoobler, G. (2002). Judee K. Burgoon. *Handbook of Interpersonal Communication*, 240.

Ireland, J. (2020, January 23). Older Australians' increasing wait before nursing home care: Report. *The Sydney Morning Herald*.

Stewart, M., & Johnson, M. (2017). What happened when we introduced four-year-olds to an Old People's Home. *The Conversation*.

Woodlock, R. (2018). The trials of finding a good nursing home. *Eureka Street, 28*(18), 2. https://www.eurekastreet.com.au/article/the-trials-of-finding-a-good-nursing-home

Chapter 3
Social, Gerontological Care

3.1 Informal Carers

3.1.1 Families and beyond

'Family, friends and community are a crucial part of the aged care system' is a claim made by the Royal Commission into Aged Care Quality and Safety (Commonwealth of Australia, 2021, p. 201) who describe their value as 'difficult to overstate, but their work is largely invisible'.

Some family members become informal carers when the older person is admitted to a nursing home, generally welcoming the opportunity for daily involvement in the resident's care. A comprehensive plan includes the family's role: expected visiting patterns, aspects of care they wish to provide, together with the family's preferred response and contact details in a crisis. Families who wish to be involved should have free access to residents' data: care plans, medication charts and other relevant documentation.

Not all families are able or wish to be involved in the resident's care. This personal decision requires honesty, understanding and compassion. Rather than staff assuming the family will be involved, expectations need to be clarified on admission, with allowances for changing circumstances. A resident's 'family' ('kith and kin') may also involve others beyond immediate blood relationships. A pertinent question to ask the resident on admission: 'Who do you regard as your family?' (For other issues related to families, see Sect. 7.4.)

While (some of) the stories are based on factual situations, real names and other details have been altered to protect the identity of the persons concerned. Resemblance to any particular person is therefore purely coincidental.

© The Author(s), under exclusive license to Springer Nature Switzerland AG 2022
R. Hudson, *Ageing in a Nursing Home*, https://doi.org/10.1007/978-3-030-98267-6_3

3.1.2 Citizenship

One important aspect of the resident's care is their continued life as a citizen: retaining their rights (to the extent they are able and interested) to enjoy the benefits, responsibilities and possibilities pertaining to their role as members of society. A protective, cautious, reactionary culture is likely to erode their citizenship status. Whenever a resident is denied freedom of choice, for example, in what they wear, where they sit or walk to, what they eat or drink, when they get up, when they go to bed, what activities they engage in, including sexual activity, their basic citizenship is denied. When decisions are made *for* them rather than *with* them, their basic rights are jeopardised. More broadly, citizenship rights include engaging in political action, voting and expressing their views. Families and volunteers play an important part in residents retaining their citizenship status. Where physical or mental frailty precludes a resident's active participation, families may act on their behalf.

The right to vote is not annulled by residence in a nursing home, and mobile polling stations can be arranged through the Australian Electoral Commission when an election is due. For residents incapable of understanding the process, a letter from their GP facilitates the removal of their name from the roll. Lifestyle coordinators play an important role in advertising and making necessary arrangements for this important activity which is included in the life of many nursing homes.

3.1.3 Volunteers

A well-structured volunteer programme has immeasurable benefits for residents, particularly those without family or other visitors, shown in this account:

> A new volunteer offered her services: a jolly, wise, kind woman with a winning smile and light-hearted sense of humour. It seemed 'just right' to match her with 'Gertrude' who was totally isolated, having no relatives and no visitors. In her weekly visits, the volunteer would greet Gertrude warmly, offering to paint her nails, feed her whatever meal and drink was on offer and chatter away as though Gertrude could understand every word. Mute, bedbound and with few discernible responses, Gertrude's merry eyes seemed to convey her love for this stranger who had befriended her. Asked by a matter-of-fact, practical nurse why she responded so positively and warmly to the seemingly unresponsive Gertrude, the volunteer replied: 'She just crept into my heart'. (Commonwealth of Australia, 2006, p. 157).

Busy nurses may wish to engage residents at the 'level of the heart', experiencing frustration when time constraints dictate other priorities. In these situations, the service of volunteers is an invaluable asset. Even if the resident is incapable of verbal response, the volunteer's presence may provide a welcome relief from the day's routine, if not tedium. The scenarios above and below testify to the fact that communication can be richly rewarding, albeit without speech.

> The volunteer wanted to visit someone who received no other visitors, so she was assigned to 'Flo'. It was explained to the volunteer that Flo was dying of dementia; she had no

3.1 Informal Carers

speech, no independent movement and was socially isolated. On her first visit the volunteer was heard to say, seated close by Flo's bedside: 'Well, Flo, is it all right if I call you Flo? I'm Patsy. I wonder what the cook has made today? I'd say by the colour that it's pumpkin and something else. I hope you like pumpkin. . . . You should see what a terrible day it is outside, Flo. I thought I'd get drenched coming here. . . . Now, here's something I'm sure you will like – jellied fruit. Look how pretty it is, the golden fruit in the red jelly. This is one of my favourites.' The lack of verbal response meant little to Patsy, who continued a similar ritual every Wednesday until Flo died, still bereft of other visitors (paraphrased from. (Hudson & Richmond, 2000, p.142).

This insightful volunteer had the wisdom, knowledge, experience and intuition to be confident that her constant chatter was acceptable to Flo. Accustomed to Flo's non-verbal responses, Patsy knew that the intimate presence of another human person was far more preferable than remaining isolated with close contact largely confined to impersonal clinical tasks. She also knew instinctively that Flo's status as a human being was not dependent on her capacity for speech.

Assigning a volunteer to a resident requires careful thought, ensuring as far as possible they are well matched. Merely employing volunteers who express a willingness to 'help' the residents does not guarantee meaningful engagement. Patsy (in the story above) made some wise assumptions and was careful not to 'talk down' to Flo. Volunteers need to be carefully screened, counselled not to treat residents like infants or compound their dependency.

Another important issue for volunteering is a focus on dementia care. It may be appropriate in some circumstances to include a person who has early stages of dementia into the volunteer programme, providing added insight and increasing their self-esteem as a person capable of contributing to the care of others.

An external, formally constituted volunteer service is available through such programmes as the Community Visitors Scheme which supports regular visits to people who are at risk of social isolation or loneliness. The scheme provides friendship and companionship by matching residents with volunteer visitors. Visits can be one-on-one as well as in groups (Australian Government, 2019, p. 48).

Whether through a formal, funded scheme or by local arrangement, the presence of volunteers in nursing homes adds incalculable value to resident care, deserving of both casual and formal acknowledgement by management. 'Along with informal carers, volunteers are an integral part of the aged care system... Volunteers can provide important connections for older people from diverse backgrounds' (Commonwealth of Australia, 2021, p. 212), particularly those from LGBTI communities and those who have lost the ability to communicate in the English language.

Recruitment and supervision of volunteers include checking their motivation, providing training, establishing procedures for review and feedback, maintaining standards and including policies and procedures in the relevant nursing home manuals. Volunteers are also provided for veterans, through ex-service organisations, who provide them with relevant training.

In a caring environment, volunteers also deserve care, including regular acknowledgement of their service. A qualified coordinator of volunteers is key to the appointment, support and follow-up of suitable applicants.

3.2 Expanding Partnerships

3.2.1 Planning Ahead

Partnerships with schemes such as advance care planning (ACP) assist in optimising residents' holistic care. While other references throughout the book are made to care planning, this section provides a comprehensive account of formal procedures, beginning with definitions and descriptions.

> ACP promotes choice and control over future medical treatment decision-making, for a time when an individual may lack decision-making capacity… personal choice and agency in medical treatment is fundamental to their right to live autonomously and with dignity. An Advance Care Directive (ACD) is '… completed and signed by a competent adult, but only comes into effect when the person loses capacity to make decisions'. (Austin Health, 2019).

An Advance Care Directive (ACD) is a formal, legally binding document incorporating specific medical decisions, personal values and goals. During the COVID-19 pandemic, advance care planning became 'more important than ever', including details from a 'bio-psycho social health perspective' (Sinclair et al., 2020, p.922). Extra attention was needed for planning in consultation with families, particularly where visits were restricted or denied. Future research will undoubtedly uncover the unacceptable outcomes for residents and families where such lack of planning was evident. Reactive, unplanned, hasty hospitalisations resulted in many residents being separated from families to die lonely, isolated deaths which may have been avoided through proper planning (Parks & Howard, 2021). Being deprived of social contact may, for some, have proved more deadly than the virus itself.

While ACP procedures vary between states and territories, each jurisdiction allows for a substitute decision-maker for financial, personal and medical matters. Despite the importance of these documents, giving the older person control over their future, including end-of-life decision-making, ACPs remain relatively uncommon (Detering et al., 2010). Without a comprehensive plan, many residents are hospitalised unnecessarily, an unknown number against their will. In some instances, futile medical treatments are prescribed, resulting in suboptimal outcomes. Poor understanding of the process is exemplified by the following:

> *Oh, yes, we have ACPs in our nursing home – one for every resident, but the manager (who is not a health professional) fills them in. Residents and families are not involved, and so their wishes are not respected. It's all done to satisfy accreditation.*

Digitalised, generic ACP documents leave little room for an individual resident's preferences to be recorded or for changes to be made at any time. It is not surprising, therefore, that they are readily ignored. When ACP is well understood and robust policies and procedures enacted, residents will also have a current, clear, comprehensive ACD, ensuring their wishes are respected, especially at the end of life, including *where* they would prefer to die. The ACD should be completed, wherever possible, with the full cooperation of the resident and family, stored safely and accessibly, with a copy for the GP, and reviewed when circumstances change. An

3.2 Expanding Partnerships

important aspect of ACP is the opportunity for residents' choice; for example, to forego acute interventions in favour of comfort care, dignity and support. This information is vitally important if/when the resident is assumed to require hospitalisation; careful checking may prevent unnecessary transfers.

For CALD residents, comprehensive ACP policies highlight the 'promotion of resources and research to support the unique needs of culturally and linguistically diverse groups… ' (Austin Health, 2019). Spending time discussing residents' cultural differences is a key aspect of ACP, noting any differences in the way the resident and/or family make decisions, and emphasising their right to choose.

Where additional advice is needed regarding ACP procedures and documentation, assistance is readily available. Organisations which focus on planning include End of Life Directions and Aged Care (ELDAC) which 'seeks to promote changes enabling a reduction of avoidable hospital admissions, with shortened stays, and improved quality of care for people supported in residential and community aged care programs' (End of Life Directions and Aged Care, 2018). ELDAC also advises on substitute decision-makers for those lacking independence in this matter.

The optimum effect of ACP documents would be that every nursing home resident has a clear, up-to-date, regularly reviewed, individualised, *readily accessible* plan, developed in partnership with family where appropriate, ensuring their wishes are heard and respected. The resident's own words may offer a powerful stimulus towards that end, such as the following:

> I've had a good life – when my time comes, I'm ready to go. Don't stick me full of needles and tubes. Don't hang on to me – I think I'd hate that. Let me go quick, in my own bed, with you holding my hand. (Waird & Crisp, 2016, p.29).

ACP discussions do not always 'go to plan' due to barriers such as the following:

- Families' failure to accept a resident's poor prognosis.
- Lack of understanding the complex issues around life-sustaining interventions.
- Absence of discussion and documentation of decisions about ACP.
- Discrepancy between the documentation and the resident's own wishes (Siu et al., 2020).

The need for comprehensive planning comes to the fore particularly in the context of dementia. 'Advance care planning aims to ensure that care received during serious and chronic illness is consistent with the person's values, preferences and goals. However, less than 40% of people with dementia undertake advance care planning internationally' (Sellars et al., 2018). ACP discussions in this context depend on an understanding of dementia as a terminal illness, indicating the need for early planning while the person has maximum cognitive capacity. It is also not widely recognised amongst carers that health events which would be non-life-threatening in others may prompt early death in a person with dementia. ACP in this context requires a detailed, frank discussion with families and education for all staff, provided by health practitioners familiar with the disease.

The evidence is clear, that time needs to be given to ACP discussions for those with dementia, early in the disease process (Bryant et al., 2021): focusing on the *person* will be rewarded by clearly articulated and documented end-of-life decisions. However, as other researchers emphasise, there is a need for education to support staff in having these conversations, particularly for those without formal training (Spacey et al., 2021).

One researcher questions the validity of ACP, especially when it does not achieve the desired outcomes. Criticisms include the lack of sophisticated knowledge on which decisions are made and the failure to regularly review documentation about goals and preferences (Sean Morrison, 2020). This article serves as a reminder that for ACP to be effective, specialised skills are required and the system needs to be regularly reviewed.

The key to increased take-up of ACP is a concerted effort by management to ensure that all staff receive regular education, and residents and families are well informed. It is evident from the discussion (above) that a pro forma plan, completed at a distance by a third party, is the antithesis of effective planning for the end of a resident's life.

3.2.2 Academic Communities Welcome?

The proportion of undergraduate education specifically directed to the care of older persons varies, and 'placement' in a nursing home is not always high on a student's agenda, nor reflected in the curriculum. Clinical teachers' qualifications, direct knowledge and experience influence the quality of the placement (Abbey et al., 2006). For many, the idea of gaining direct clinical experience in a nursing home is fraught with anxiety and foreboding, indicated in the following:

> Ten university nursing students had arrived at the nursing home on the first day of a two-week clinical placement. 'What immediately comes to mind when you think of a nursing home?' asked the manager. 'Smells.' 'Old people who can't communicate.' 'Depressing sights.' 'Death.' Other responses indicated these students were there under sufferance and with little expectation of gaining valuable clinical knowledge and experience, let alone deriving any pleasure from their placement.

One way of breaking down perceived barriers between the nursing home and the broader community is to offer a positive experience for students and academic supervisors responsible for this part of the academic curriculum. The first challenge is to correct the negative stereotypical views described by some of the students. Creating a positive impression begins by showcasing the nursing home as a thriving community rather than a depressing, confining, custodial institution.

Such was the case in one nursing home when the undergraduate students' apprehension about their aged care placement was soon reversed by a warm welcome, careful assignment to residents and thoughtful pairing with selected mentors. At the end of this particular placement students recorded their evaluations.

> 'Fantastic! I never thought I'd enjoy working here.'
> 'The resident I was caring for had a great sense of humour. I will miss her!'

3.2 Expanding Partnerships

'I thought I'd be overwhelmed by horrible sights, sounds and smells. I've found it surprisingly pleasant.'
'I've made a firm friend in George and plan to continue visiting him.'
'It's great to see excellent clinical care combined with everyday enjoyment of life.'

One student had confided to the manager that she had dreaded coming to the nursing home. Her friend had recently been on a similar placement, finding it disorganised, if not chaotic, with no obvious preparation for the students' visit, no mentoring and no opportunity to write evaluations at the end. In contrast, this student told the manager: 'You made us feel so welcome here, and I've learned so much!'

The following scenario shows the possibilities for reversing students' negative perceptions:

> At the end of their two-week placement the students were asked to evaluate their experience both for academic requirements and for an internal nursing home audit. Many were surprised to witness only one resident close to death; they had assumed that of the 120 residents most would be relatively lifeless. Some were also surprised to learn of the productive lives many of the residents had previously enjoyed. 'Did you know that Mr. Brown was once the mayor of a large provincial city?' 'Have you seen the photos of Milly dancing at the Tivoli?' One by one, the students' perceptions were changing. They learned that these residents had a wealth of experience as productive citizens and an accumulation of life's wisdom previously unimagined by the students, who had regarded all nursing home residents as 'past their use-by date'. During their placement they witnessed residents' participation in bus outings, counter lunches at the local pub, pets being welcomed, school children visiting, concerts and a lively movement-to-music session. They also witnessed excellent clinical care in response to serious manifestations of chronic disease. Most importantly, they learned they could actually enjoy interacting with many of the residents. One student summed it up well: 'There is life in a nursing home!' (Hudson & Richmond, 2000, pp. 148, paraphrased).

While the students witnessed the care of a typical mix of residents – those with dementia, physical incapacity, emotional instability, mental frailty, some close to death, others relatively independent and mentally alert – they also observed laughter and enjoyment. As a way of removing barriers, this group of students had the potential to take with them into their wider nursing curriculum some positive experiences of aged care. At least for this small cohort from the academic world nursing homes need no longer be regarded as alien places, far removed from the wider community.

Another means of removing barriers is for a nursing home representative to provide a 'guest lecture' within health professionals' academic curriculum. Ensuing benefits are exemplified in the following:

> The nursing home director/manager was invited to give a lecture on 'aged care' to final year undergraduate medical students. While lacking confidence that the subject matter would evoke enthusiasm, she was delighted and surprised by the responses. 'Thanks for providing these insights. I had no idea what went on in nursing homes.' 'I now understand how seriously ill some residents are. I thought they were there just because they were "old".'

The prevailing perception of aged care is 'the bottom rung of the ladder' for graduate nurses' and other health professionals' career choices. On the other hand, many benefits derive from a positive experience for undergraduate students of medicine, physiotherapy, occupational therapy, music therapy, dieticians and other disciplines. Such lectures and/or 'placements' have the potential to change negative assumptions about aged care, which in turn may give rise to a more enlightened, multidisciplinary workforce in the future. The development of national standards would also accentuate the value of such relationships (Laugaland et al., 2021).

3.2.3 Engaging the Wider Community

Another way of changing negative attitudes is by fostering residents' engagement within their local geographic community, as in this account:

> *The nursing home happened to be in a street with a small, friendly pub on the corner. The manager introduced herself to the publican and discussions focussed on accommodating a small group of residents for a fortnightly 'pub lunch', supported by a nursing home staff member. Residents enjoyed choosing from a menu, paying with their own money for their food and accompanying drinks and, most of all, they appreciated a welcome change from the nursing home meals. Family members were also invited to share a meal with their own relative and/or others. Residents from the private homes in the street were agreeably surprised to see some very frail residents being ably assisted on these outings. 'It's wonderful to see them enjoying some normal life. I thought they would all be confined to bed all day.' The publican soon learned the names of the 'regulars', enjoying a beer with some of them. Staff were delighted and surprised to see signs of independence in some of the residents, thought previously to be totally reliant on others. 'Did you see George take such pride in producing his wallet to pay for his meal?' Students who happened to be on placement also expressed surprise and pleasure at this regular routine in the life of the nursing home.*

When the nursing home is regarded as separate from 'the community' misunderstandings may develop, not to mention residents being deprived of their own community. For 'Joyce', maintaining links with her cherished social milieu was vitally important.

> *'Jeremy', 'Joyce's' son, visited his widowed mother two or three times each day. He had taken early retirement, so he could dedicate more time to her care, taking her out for a drive each day, and often calling in at 'odd times' of the day or evening, as he lived quite close. Staff were inclined to 'Tut, tut' about an 'unhealthy relationship'. 'What will he do when she dies?' 'I think we should caution him not to come so often.' 'I don't think he realises she won't live forever; he doesn't acknowledge she's deteriorating.' Other staff who knew Jeremy better, had a different view. 'He knows full well how she's failing; it's becoming very difficult for her to get in and out of the car, but he knows how much she enjoys these short drives, reciting all the streets as though proving she still knows where she is.' These staff were not so pessimistic about Jeremy's future. 'I think he'll get on with his life, resuming his other interests. I don't think it's an unhealthy attachment at all.' It seemed that some well-meaning staff were more intent on speaking **about** Jeremy's relationship with his mother rather than having a frank discussion **with** him.*

For Joyce and her son (her only child), it was important to maintain visible relationships with the wider community, so well known to both of them. Moving to the nursing home community meant, for Joyce, a move *within* her well-known surroundings; nonetheless she appreciated actually *seeing* the houses, shops and streets to which she had a lifelong connection.

Another means of community partnership is through the involvement of churches, senior citizens' clubs and others, including those catering for particular ethnic groups. While comparatively 'a world apart' from nursing homes, these members of community groups may have their views expanded, prompting some to reverse their negative views, after a well-planned visit to a local nursing home.

3.3 Clinical Care

3.3.1 *Learning from Nightingale*

An apt reference point for a discussion on clinical care is the wisdom of Florence Nightingale. Far more than 'the lady with the lamp', Nightingale provided counsel for the government of the day, recommending that St Thomas's Hospital should have a maximum of 32 patients per ward to minimise cross-infection. She was a fierce advocate for regular, thorough handwashing for the same reason. Nightingale also pioneered the use of the pie chart for statistical records and evidence-based nursing. In addition to her nursing and scientific skills she excelled in Latin and mathematics; she published broadly, travelled widely and shunned publicity. She died in 1910 at the age of 90 years. Marking the 200th anniversary of her birth, Preston (American history professor at the University of Cambridge) reminded his audience of Nightingale's achievements which '… resonate with us more than ever. As our doctors and nurses fight the coronavirus pandemic, it's wholly appropriate that new field hospitals around the country have been named NHS Nightingale Hospitals' (Preston, 2020).

What can nursing home proprietors and staff learn from Nightingale, whose extensive scientific credentials, medical knowledge and research skills are not always widely acknowledged? Nightingale realised the necessity of nurse training: neither she nor any of her colleagues had been taught how to nurse. It was universally assumed that the only qualification needed for taking care of the sick was to be a woman. Such ignorance had disastrous consequences. Nightingale wrote in 1845: 'I saw a poor woman die before my eyes because there was nothing but fools to sit up with her, who poisoned her as much as if they had given her arsenic' (Cook, 1913, p. 45).

More than a century later, people are dying in some nursing homes from inadequate care, if not direct abuse and neglect. Not only are they left unwashed and unfed, they are sent to hospital (many times needlessly) where they die in emergency or intensive care units. Nightingale may well describe such contemporary 'care' as being in the hands of 'fools' whose methods are 'equivalent to poisoning them with arsenic'(Cook, 1913, p. 45). Noting Nightingale's words of advice, it can be stated with confidence that the art of caring is not (or should not be) divorced from relevant scientific protocols and procedures designed to keep older people safe: the epitome of a carer's duty and the hallmark of socio-gerontic care.

In contrast to the inadequate care (described above) it is important to acknowledge and showcase the nursing homes that provide exemplary care, demonstrating every day and in practical terms their understanding of 'duty of care'.

3.3.2 Duty of Care

Duty of care is defined as 'an obligation owed by one party to another ... to avoid causing harm or injury by negligence' (Mosby, 1983, p. 353). Guided by this definition it is important to note that 'duty of care' does not require practitioners to give futile or inappropriate treatment. 'The ... aim is to reach the best compromise plan that accommodates all parties... There is no place here for ideological purity where theoretical principles trump compassionate pragmatism' (Ashby & Mendelson, 2011, p.5). This emphasis provides a helpful reminder to health professionals who are swayed, albeit against their professional knowledge and judgement, as well as their 'compassionate pragmatism', to accede to inappropriate or ill-informed requests from a resident or family. Such requests may be founded on families' lack of relevant information, giving rise to confusion surrounding decision-making, reflected in these not uncommon responses.

> 'We didn't want to send him to hospital but the family insisted.'
> 'The GP wanted to make a referral for palliative care but the resident's daughter refused to consent.'
> 'We wanted to increase the pain medication but the family disagreed.'

These and other similar comments reflect a lack of clarity about the family's role and GPs' and professional staff members' responsibility to provide best practice care. 'Medical teams need to take into account, but are not under a duty to comply with patients' or family's requests for treatments that are not clinically indicated' (Ashby & Mendelson, 2011, p.7).

Anecdotal accounts suggest an increasing propensity for nursing homes (and, at times, other health institutions) to request 'permission' from families before offering treatments clearly within their professional duty of care, as in the following response from a resident's daughter:

> 'The nurse rang me last night at 11.30 pm to ask if they could give dad some medication, as he was up wandering into other peoples' rooms. I'm not a nurse, and I don't know one drug from another, so what was I supposed to say? I couldn't get to sleep after that, worrying that I'd said 'yes' when I should have told them to ring the GP. And, this is not the first time this has happened.'

With a well-informed understanding of duty of care, the nurse would not have phoned either the family member or the doctor. With appropriate authority the nurse would have been guided by a comprehensive documented care plan with clear directions regarding the resident's management. Such a plan would include the goals of care, with options to try before resorting to medication.

3.3 Clinical Care

The key, as for all resident care, is to maintain accurate documentation, commensurate with duty of care and avoidance of unwanted, unnecessary interventions. Nurse education should cover the circumstances within which 'permission' is required before a nursing response, reinforcing the legitimacy of the care plan in most circumstances. Communicating with families is important while differentiating discussion and information from seeking 'permission'.

> Nursing documentation is evidence that the patient received proper care. In a court of law, the patient's health record serves as the legal record of the care provided to that patient to defend against allegations of malpractice, negligence, or failure to meet standards of care. (Paulo et al. 2017).

It is evident from the anecdote (above) and the legal reference (Jacoby and Scruth) that increased confidence is needed for nursing home health professionals to practice with due (legally prescribed) care supported by clear documentation.

3.3.3 General Practitioners (GPs)

Nursing home residents' medical care is the responsibility of their general practitioner (GP). In one innovative variation, the benefits of a group medical practice being associated with a local nursing home are highlighted. The team approach greatly increased the ready availability of medical staff, including geriatricians, maximising the expertise applied to residents' care. This model is in stark contrast to an advertisement for recruiting GPs which stated in bold letters: '**in this practice you will NOT be required to visit aged care facilities.**' GPs' reluctance to visit nursing homes is prompted by many factors, not the least of which is the poor remuneration. Others are reluctant to supervise the care of residents whose medical needs are more suited to a geriatrician or palliative care physician.

What preparation is made for doctors' visits to nursing homes? One GP commented: 'I spend the first ten minutes trying to find my patients, then another twenty minutes trying to find a nurse who can tell me something about them.' While busy GPs cannot always commit to a regular time, and to being on time, it is not impossible to make mutually agreeable arrangements whereby the nurse anticipates the GP visit and prepares accordingly. This may include undressing the resident ready for full examination, in the privacy of their room, relevant charts accessible, family member/s present if feasible and up-to-date details regarding the resident's condition well documented. Without appropriate preparation, GPs may have no option but to conduct their visit in the public lounge area where it is impossible to speak privately with the resident or complete a comprehensive physical examination. Contrasting practice is evident in the following account:

> The newly appointed nurse manager was shocked to see the inadequate medical oversight by most of the GPs visiting the nursing home. He listened carefully to the doctors' comments ranging from 'I can never find a nurse when I visit' to 'How am I supposed to examine the resident who is fully dressed, sitting in the lounge?' The doctors also emphasised the lack of appropriate remuneration, especially for a comprehensive visit requiring private

consultation and physical examination. The nurse's response was to call a meeting of a representative number of visiting GPs to discuss some reforms to the system he, as nurse manager, had 'inherited'. As a result, most GP visits were by prior appointment, specifying date and time. This allowed family to be present if they wished. When needed (and not necessarily for every visit) the nurse would arrange for the resident to be in/on their bed and with easily removable clothing to facilitate a physical examination. The visit would include time for discussion with the nurse and/or family regarding the resident's condition and treatment, as well as comprehensive documentation.

Such preparation is perceived by some as too time-consuming, if not unworkable. However, when introduced as a trial, the results, including time-saving, may come as a pleasant surprise. The example is not intended to imply such a procedure for every doctor's visit to every resident: it is an option to consider when a nurse considers a resident's condition or symptoms warrants further attention. Another benefit for the residents is saving them unnecessary visits to hospital. These, and other benefits of partnership with GPs, deserve greater attention (refer to 'CMA's below).

Whether it concerns a discussion of nursing home policies and procedures or the care of a particular resident the cry goes up: 'We can't get the doctors to come to meetings.' The following scenario shows a positive response:

The nurse manager had scheduled a routine meeting for a resident and family; well-planned, with a comprehensive agenda, and expected to last no more than one hour. Unsurprisingly, the GP refused the invitation to attend, citing lack of time. When the nurse suggested she attend for the first ten minutes, when the resident's medications and one or two of the most urgent clinical matters would be raised, she reluctantly agreed. 'Okay, for ten minutes max'. The GP was pleasantly surprised to see the key issues addressed with efficiency, so she stayed longer. 'I'd always considered these meetings would be a waste of my time, but I learned so much about Jessica's general needs, and it was great to meet the family.'

Lack of time is often cited as a reason for either not having meetings or key personnel failing to attend, which can be overcome by thorough planning, advance notification and competent leadership. A new primary care model is advocated for aged care from 2024, whereby GP practices may apply for accreditation as approved providers (Commonwealth of Australia, 2021, p. 287). This proposed model would, presumably, ensure that the attending doctors were there from choice, backed by clinical expertise in gerontology.

Gerontology is the study of ageing and older adults… Gerontology is multidisciplinary and is concerned with physical, mental, and social aspects and implications of ageing. Geriatrics is a medical speciality focused on care and treatment of older persons. Gerontologists include researchers, educators, policy makers and practitioners in health, allied health and aged care, as well as others engaged in ageing issues. (Australian Association of Gerontology).

Nursing home residents have every right to be referred to a gerontologist or geriatrician when the GP considers such expertise is warranted.

3.3.4 Telehealth

Access to the latest technology is available, particularly for rural and remote areas, through the use of telehealth or telemedicine. Videoconferencing enables liaison between doctors and nurses, maximising nurses' treatment for residents and saving unnecessary hospitalisation. An increasing number of tools are also available, for example, to reduce falls risks, thus minimising residents' need for emergency surgery.

Other advantages of telemedicine and telehealth in nursing homes include early detection of clinical deterioration, treatment offered 'in place', reducing emergency admissions, access to specialists and financial savings. Ready access to specialist psychiatric care, palliative care and geriatric care is also listed in an 'integrative review', together with family and staff satisfaction (Groom et al., 2021, p. 1784). Further research is expected to confirm the benefits of these technological advances.

3.3.5 Comprehensive Medical Assessment (CMA)

Research is sparse regarding the number of comprehensive medical assessments (CMAs) completed in nursing homes, and there is little incentive for GPs to become involved. Ideally, a nurse should be present for the assessment to assist the resident with transferring and undressing and to provide the GP with additional information, particularly if the resident has poor cognition or verbal skills. Although in one report less than 50 per cent of residents had received a CMA, significant medical problems were uncovered in documented assessments. One review reported 'a female resident with undiagnosed breast cancer, several instances of polypharmacy, and residents with hearing, sight or allergy complaints impacting on their health and well-being' (Westbrook et al., 2011, p. 8). This research also found that regular monitoring would potentially reduce the risk of unnecessary hospitalisation. However, without appropriate remuneration, it may be impractical to expect busy GPs to incorporate regular CMAs into their workload.

The following account serves as a reminder of the importance of CMAs for the purpose of diagnosing and treating a resident's severe pain.

'Miss A' was often heard to cry out (including obscenities), resulting in unprofessional judgements from staff. 'Here she goes again! Just ignore her!' 'It's just her dementia!' Fortunately for Miss A, an observant RN suggested the resident may be in pain due to osteoporosis secondary to breast cancer. Other carers, including the GP, had failed to note the significance of the absence of one of her breasts. Following more thorough examination, Miss A was prescribed slow-release narcotic pain relief after which she ceased calling out, living the remaining weeks of her life with all the benefits of palliative care.

Protestations about Miss A's 'non-compliance' ('She won't swallow pills: she just spits them out!') were countered by comprehensive medication assessment, resulting in transdermal analgesia and additional options for her care.

3.3.6 Medication Management: Pharmacist's Role

Medication compliance is a challenge for nursing home staff, particularly where the majority of residents have cognitive impairment, preventing them from understanding the process of taking tablets or mixtures. Time constraints do not necessarily allow for staff administering medications to ensure the resident has swallowed them; hence the common finding of 'tablets in the bed'. One way of addressing this challenge is for regular medication reviews with GPs, pharmacists, families and residents, checking each resident's ability to take all prescribed medications, together with regular review of their purpose and effect. Personalised pharmacy services can lead to efficiencies in costs and time by reducing the number of drugs, often resulting in residents' improved alertness, mobility and general well-being. Allocating time for these meetings may well result in saving nurses' time, for example, by reducing the drug round.

> Such was the experience of a newly appointed RN, shocked and alarmed to find how much time was needed for the medications. After painstaking review, including liaison with prescribers, and checking negative interactions between some drugs, she was delighted to report to the manager: 'I've reduced the drug round time by thirty minutes!'

A timely word regarding medications comes from the sixteenth-century Swiss alchemist, Paracelsus: 'All drugs are poisons. It is only the right dose that differentiates a poison from a remedy.' The commentator notes: 'In this role, we need to learn from the stalwart Paracelsus the insistence on relying on facts rather than authority alone to protect against chemical hazards' (Grandjean, 2016, p. 126). According to one pharmacist: 'The problem is an ageing body may not respond to drugs as the side effect profiles indicate, especially when someone has multiple medical conditions, is taking multiple medications, and also has multiple prescribers.' This pharmacist recommends the wider use of case conferences, including family members where possible, as in the following example:

> The GP was reluctant when invited to a case conference concerning 'Miss M', which the RN had requested because of Miss M's reluctance to swallow the sixteen medications given to her on each drug round. Miss M would often 'store' the pills under her pillow or elsewhere in her bed, after 'promising' to swallow them. Miss M had multiple medical conditions and the GP wanted to be proactive in preventing any complications and minimising her discomfort. The RN persuaded him to attend the meeting: 'The pharmacist will be here, and the family, who are worried about all the pills she takes. This will be a good opportunity for us all to discuss the options.' Within twenty minutes, and following input from all key players, Miss M's list of drugs was reduced to six. Miss M's compliance improved, the RN's time spent on the drug round was reduced and the GP was satisfied: 'I was reluctant to reduce them off my own bat. It was a good meeting.' The family were also well informed, relieved to see the pharmacy bill significantly reduced as well as fewer medication side effects.

The wide-ranging assistance and advice from community pharmacists are not always appreciated or understood by nursing home staff, residents or families. Personal contact and regular meetings are recommended.

Polypharmacy is best addressed through medication advisory committees (MACs) discussing definitive strategies for each resident (Jokanovic et al., 2017).

Others recommend that MACs should be multidisciplinary, including residents and families, placing a high priority on education to identify adverse drug events (Picton et al., 2020).

Although many nursing home residents are prescribed multiple medications they are reluctant to raise concerns with their doctor (Thompson et al., 2020). These authors suggest that discussions with residents would be beneficial, particularly when focused on reducing their medications.

Deprescribing includes measures to address other issues, such as behaviour modification in dementia. Unfortunately, some doctors seem more ready to *add* another drug rather than to *remove* any medication whose benefit is doubtful. Some researchers have found that a change of GP after a resident's admission to the nursing home often leads to an escalation in psychotropic medications. Such prescribing may be due to lack of attention to the resident's pre-admission history, lack of time for a thorough assessment, lack of psychogeriatric review and lack of detailed nursing documentation regarding the resident's 'behaviour'.

At the forefront of all discussions/actions about medications is the need to focus on the *goal,* with appropriate review as the resident becomes less able to swallow or as they approach death. Nurses play a key role in such reviews.

3.3.7 Staffing the Nursing Home

Staffing issues are given prominence in various places throughout this book. Here, particular attention is drawn to staffing patterns, with descriptions of various roles. It is also important to note that approximately one-third of nursing home staff were born overseas, with implications for future demand in Australia's increasingly diverse population. Employing bilingual staff, together with access to interpreter services, provides benefits for those not proficient in the English language. Staffing the nursing home is a subject deserving wide attention: discussion is focused here on the descriptions of full-time and/or part-time nurses and carers.

3.3.8 Registered Nurse (RN)

The Australian College of Nursing (ACN) describes the role of the nursing home RN as follows:

> RNs play a vital role in residential aged care service management, planning and delivery of services. They can hold key management roles in RACFs having direct influence on the operational planning and are, typically, the clinical leaders involved in coordination, delivery and monitoring of evidence-based practice and continuous quality improvement within the RACFs. (Australian College of Nursing, 2016).

RNs' key range of direct care activities includes nursing care procedures, restorative care, safe behavioural management in dementia, health emergency responses, identification of acute deterioration in residents after a fall, infection control and complex care compounded by a resident's comorbidities. The role also includes palliative care and comprehensive pain management, medication administration consistent with quality use of medicines guidelines and oversight of unqualified staff. The RN's role includes collaborating with other health professionals and initiating new models of care. All of these requirements indicate that RNs 'cannot be substituted by any other health care category' (Australian College of Nursing, 2016). The implications of this statement cannot be ignored, given the current propensity for employing relatively unskilled workers, and with no current (2021) government mandate for employing even one RN 24 hours a day.

Noting the wide range of responsibilities, it is also clear that RNs in nursing homes require appropriate, ongoing educational support provided by senior management, within working hours and without compromising resident care.

Afterhours RNs have a particular responsibility for leadership when the manager is not readily available. Skills and continuous oversight are needed to ensure and enhance their supervisory role (Nhongo et al., 2018, p. 3879). Without a mandated RN presence every day and every night, including weekends, residents are likely to receive suboptimal care. Evidence shows staffing levels to be dangerously low, placing residents at serious risk, only receiving two-thirds of the care they need every day. Results which can only be described as 'shocking' show that residents are 'being left wet or dirty or hungry and thirsty 90% of the time' (Butler, 2018, p. 9). Butler notes that this is not due to lack of care; it is due to lack of *staff*. Put succinctly, 'Residents receive barely half the care they need' (Butler, 2018, p. 9).

While noting other recommendations made over recent years, the absence of Commonwealth standards and regulations regarding the employment of RNs in nursing homes remains a major policy deficit. Given the fact that many (if not most) nursing home residents have several serious, life-threatening comorbidities, it is astonishing that their care is not always and, in every place, monitored by appropriately qualified nurses. A fresh investigation into the complex list of residents' diagnoses should alert those in authority that most of the residents in their care are seriously ill, will not recover and deserve care commensurate with their significant needs.

During the COVID-19 pandemic hundreds of Australian nursing home residents were transferred to acute hospitals. The lack of time for professional nursing input to planning and decision-making seriously compromised continuity of care, not to mention the absence in many cases, of comprehensive details of their diagnoses, medications and nursing needs. Conversely, when additional RNs were employed as a temporary measure to fill staffing gaps in the nursing homes, residents benefited from excellent clinical care. Nurses found immense satisfaction from giving individualised care to a small number of residents. Unfortunately, the welcome effect of this significant staffing increase was short-lived.

Careful scrutiny of the RN's role compounds the incredulity of a management system that refuses to employ even one full-time nurse. This attitude is reminiscent

3.3 Clinical Care

of former decades when 'wellness and lifestyle' were the main focus of residential care. Today, residents require increased emphasis on *clinical* care, including *palliative* care and *terminal* care. That is not to underplay the importance of 'lifestyle' activities; however, most nursing home residents are anticipating declining health rather than maintaining 'wellness'. Quite properly the Royal Commission into Aged Care Quality and Safety Final Report (2021) has addressed the urgent need for an RN presence on every shift in every nursing home.

3.3.9 Enrolled Nurse (EN)

'Enrolled Nurses play a key role in Australia's health system, providing care and treatment in a range of settings and under the supervision of a Registered Nurse' (Australian Industry and Skills Committee, 2020). Top priority skills for ENs include 'emotional intelligence, teamwork and communication, critical thinking, resilience, stress tolerance and flexibility'. ENs represent a critical component of resident-centred care; their increasing knowledge and skills, including medication management, are a critical asset to the aged care team, especially when working in partnership with an RN. The presence of ENs on night shift also ensures much-needed professional oversight including supervision of unqualified carers.

Increasing the number of ENs (appropriately qualified and supervised) would enhance resident care and allow these nurses to achieve satisfaction through optimising their skills. Unfortunately, some management systems are governed by the mistaken belief that lower-paid, under-qualified staff are essential ingredients for a balanced budget, if not a sizeable profit margin. The number of nurses in nursing homes is *decreasing*, while the pressure of all aged care workers is *increasing*, leaving vulnerable residents to endure unsafe practices and life-threatening deficits in their care. Increasing numbers of ENs would also provide much-needed oversight and support for PCAs.

3.3.10 Personal Care Assistant (PCA)

Highlighting the inadequate numbers of nurses in nursing homes does not mean that PCAs (or otherwise named) should not be employed in this sector. Many PCAs, especially those with significant experience and formal education, are competent at recognising and reporting observations from their daily 'hands-on' resident care. Many develop deep, meaningful relationships with residents, arising from their close, intimate contact and conversations, demonstrating the breadth and depth of their care. When their role is focused only on tasks, the PCA's humanity is denied, together with that of the resident (Aged Services Industry Reference Committee (ASIRC), 2020, p. 5).

On the other hand, providing emotional, social, spiritual and psychological care in many instances, PCA's are 'the heart and soul of the aged care system' (Aged Services Industry Reference Committee (ASIRC), 2020, p. 11). This view is shared by international researchers highlighting the role of 'healthcare assistants' who often perform work beyond their role description, providing end-of-life care and 'acting as extended family and ensuring residents do not die alone' (Just et al., 2021). It remains a serious anomaly, however, that many care assistants are unqualified, lacking sufficient knowledge to inform their work, compared, for example, to child care staff who require appropriate certification for their role.

Unregistered carers have expressed fear that the introduction of nurse/resident ratios would mean losing their jobs. On the contrary, health forecasters predict the need for a steady increase in the numbers of all direct care workers in the coming years. Following this prediction, it is incumbent on all nursing home managers to ensure staff are appropriately accredited and suited to the task of caring for increasingly frail residents.

3.3.11 Nurse Practitioner (NP)

> A Nurse Practitioner (NP) is a Registered Nurse with the experience and expertise to diagnose and treat people of all ages with a variety of acute or chronic health conditions. NPs have completed additional university study at Master's degree level and are the most senior clinical nurses in our health care system. (Australian College of Nurse Practitioners, 2020).

NPs are a relatively new addition to residential aged care 'comprising only 0.2 per cent of the workforce' (Belardi, 2014). NPs' oversight and interventions are of significant value, resulting in improved resident health outcomes, quality of life and reduced hospitalisation. GPs in Belardi's research were also enthusiastic in their praise for the high level of skills among NPs, particularly in collaboration and communication.

One GP said of the NP's role that he would 'never work in an aged care residence again without the support of a nurse practitioner'. He appreciated the positive impact on his previously overwhelming demands and workload. NPs report significant professional satisfaction from their role, and some statistics show that where NPs are employed the admission of older people to acute care is minimised almost to zero.

> 'A recent national evaluation of Aged Care NP practice models showed a NP in an aged care facility reduced hospital leave days for residents aged over 80 by 12%. If each facility had an aged care NP it could mean a potential savings of $97 million.' (Dragon, 2016).

NPs clearly have a significant role to play in optimising outcomes for residents' clinical care. Board members and managers would be well advised to explore the benefits of the NP's role, rather than dismissing the option as 'too expensive'. NPs have the competence to address most, if not all, of the issues discussed below.

3.3 Clinical Care

The following sections are, of necessity, brief and not intended to provide a full, detailed coverage on each of the very important clinical issues described.

3.3.12 *Skincare*

Studies have shown that nursing home residents' skincare is haphazard and lacks an appropriate evidence base (Kottner et al., 2019). Others suggest that pressure ulcers could be significantly reduced by twice daily moisturising of the residents' limbs. Aside from the physical therapeutic benefits, the psycho-emotional stimulus from human touch cannot be overemphasised. Many family members are willing to assist: their contribution to skincare may prove an emotionally satisfying and clinically effective preventive measure.

At least one much-publicised example of horrific pressure wounds was brought to the attention of the Royal Commission into Aged Care Quality and Safety (2019), signalling incredulity as to how such serious neglect can occur in twenty-first-century aged care. This example, and others, also points to the need for carers' continuous, contemporary well-researched education on all matters related to skincare. Evidence of poor skincare is, regrettably, often covered up (literally!) when a nursing home intentionally showcases fleeting positive examples to impress the inspectors.

The decades-old practice of 'repositioning' the patient in the bed has been questioned. Does it prevent pressure ulcers or does it constitute elder abuse? Researchers (Sharp et al., 2019) found that two-hourly repositioning failed to prevent pressure ulcers in a third of at-risk residents. They queried whether this practice breaches residents' rights or whether it may even be unlawful. By contrast, the use of well-designed mattresses is recommended to relieve pressure over the whole body continuously and gently. Benefits include (a) the resident does not need to be disturbed or wakened and (b) apart from the initial purchasing outlay, use of such mattresses would be more cost effective than (labour intensive) repositioning. This research confirms the many benefits of evidence-based care, rather than pursuing 'old practices' without question.

Managers who highlight and publish examples of impeccable skincare, with statistics to show the whole community, can rightly be praised for placing residents' comfort at the centre.

3.3.13 *Nutrition and Hydration*

The manner in which food and drinks are delivered in nursing homes could be vastly improved with careful review and at no extra cost. In the nineteenth century, Nightingale wrote of nurses' pivotal role in the nutritional care of patients as well as screening for malnutrition, stating that 'every careful observer of the sick will agree in this that thousands of patients are annually starved in the midst of plenty, from want of attention to the ways which alone make it possible for them to take food' (Nightingale, 1969, p. 63). She also advised that meal trays should not be left beside the patient's bed, waiting for the patient to take the initiative. Her wise counsel continues to have a place in contemporary aged care, where adequate supervision is lacking.

Given inadequate staffing levels, there is insufficient time to properly assist all of the very dependent residents with their meals and drinks. Some residents may decline to eat at mealtimes and become hungry later. With some imagination, a selection of finger foods could be left within easy reach of mobile residents or those in wheelchairs (restricting access for residents with strict dietary needs). Families, volunteers and other visitors may be prompted to offer a snack to the resident they are visiting (ensuring safety precautions where needed). While offering a wide choice at mealtimes is considered unrealistic, creative options have produced beneficial results, increasing residents' interest in food and actually achieving welcome weight gain in some (Wang & Villarosa, 2017). More education is needed for chefs and managers on the strong relationship between good food and the emotional and physical health of residents, together with an emphasis on the *purpose* of food beyond mere nutritional factors.

Careful attention is also needed for residents who require full assistance with their meals: ensuring their posture is upright if possible, offering small bites and ensuring they can actually see the food on the spoon before it enters their mouth. Carers new to this experience may need supervision initially, especially if they have had no education on how to feed a frail resident.

'She's stopped eating' or 'he refuses all his meals' are comments calling for a multidisciplinary discussion with experts in nutrition. The use of supplements has not been fully researched, to indicate their benefits or otherwise, or their influence on a resident's weight. Given the alarming statistics – 22–50 per cent of residents are malnourished (Commonwealth of Australia, 2019, p. 7) – the need for evidence-based responses remains urgent. The subject of malnutrition deserves increased continuous attention until the matter is properly addressed. How is it that in twenty-first-century Australia almost half of nursing home residents (who are paying for their accommodation) are hungry?

Awareness of the risk factors for malnutrition may assist in its prevention. These include medications, cognitive and physical decline, depressive symptoms, dental problems, among others. Careful scrutiny of these factors indicates that they are represented in the majority of residents.

On the subject of greater scrutiny at mealtimes attention is drawn again to 'the lady with the lamp' for her detailed research.

> I have known a nurse in charge of a set of wards, who not only carried in her head all the little varieties in the diets which each patient was allowed to fix for himself, but also exactly what each patient had taken during the day. I have known another nurse in charge of one single patient, who took away his meals day after day all but untouched, and never knew it. (Nightingale, 1969, p.112).

3.3 Clinical Care

If asked today, it may be assumed that Nightingale would insist upon policies and procedures for the detailed monitoring of residents' intake at mealtimes, with associated documentation of all relevant data, and remedial action where needed.

Unintended consequences from lack of mealtime supervision have arisen during the COVID-19 pandemic, resulting in at least one unfortunate scenario.

The nurse attendant, with twenty residents to 'supervise' at meal times, observed that the meal trays were arriving with every item, including cutlery, sealed. Noting a resident's meal which had remained untouched since delivered half an hour previously, she knew the resident was incapable of removing the packaging from the utensils and food. 'The meal was cold, and I didn't have time to do anything about it so I had to throw it in the bin.'

It is no surprise that malnutrition statistics are increasing.

Careful attention to a resident's poor nutritional status would include screening for depression and cognition, meticulous oral assessment, auditing their medications (especially those known to affect appetite and/or weight), cultural considerations, religious observances and specific food preferences. Failure to address this significant issue may lead to serious consequences such as cachexia (causing extreme weight loss and muscle wasting), sarcopenia (age-related muscle wasting) or increased anorexia (appetite loss).

Investigators gathering data from several hundred nursing homes across Australia in 2019 found that spending on food, including fresh fruit, had *decreased*, with concomitant increase in supplements and food replacements. The outcome was an escalation in malnutrition, leading to 'a cascade of adverse outcomes, including increased risk of falls, pressure injuries and hospital admissions' (Hugo, 2019). This research prompts a call for every nursing home to regularly review their food budget and its effect on residents' health and well-being. It also prompts the question about mealtime procedures and ways in which the residents' food *experience* may be enhanced, including the timing of meals, which may have more to do with the staffing budget than with residents' preferences and needs.

3.3.14 Mealtime Experience

Evidence has revealed that funding structures, including cost-cutting, are not conducive to positive mealtime experiences. Residents are not being given a voice, and many other factors influence the alarming number of malnourished residents remaining constant over several decades. On a more positive note, the benefits of replacing supplements with staff training and offering high-quality food in the right mealtime environment have been shown. 'This approach significantly reduced malnutrition (44% over three months), saved money and improved the overall quality of life of residents' (Hugo, 2019). Validated tools are an essential component of malnutrition monitoring.

In consultation with the professional chef and/or dietician a 'food committee' could be formed in each nursing home, charged with the task of improving residents' mealtime experiences. While appropriate oversight may be needed for residents with diabetes or other health conditions impacting their food choice, creative, colourful food and drinks may entice others to increase their calorie intake. A variety of snacks may include fresh fruit, popcorn, granola bars, iced cupcakes, marshmallows, chocolate chip cookies, peanut butter cookies, jellies, ice creams, to name a few. However, for many residents, the choice of food is dependent on their swallowing capacity, necessitating other interventions. Many families also make a valuable contribution to residents' mealtime experience, welcoming the opportunity to be involved.

Research into the effects of the COVID-19 pandemic has yet to determine the outcome of mealtime experiences on residents' weight, appetite, dining alone, as well as the loss of taste and smell for those infected with the coronavirus.

(Other matters relating to nutrition and hydration are discussed in Chap. 4 on 'dementia'.)

3.3.15 Enteral Feeding

Enteral feeding for nursing home residents (via a tube from mouth or nose to the stomach) is a controversial subject, with claims that, for some, the risks and burdens outweigh the benefits (Carter, 2020). Suffice it to say that careful assessment and multidisciplinary discussion should accompany any decision regarding a resident's artificial nutrition and hydration. Accredited practising dieticians play an important role in nursing homes, assessing and managing the oral intake of residents with chronic diseases, providing advice on food service, tube feeding, training and education of staff. Research on this subject appears scant, indicating the need for careful, individualised decision-making, including appropriate professional advice, in every instance.

3.3.16 Alcoholic Beverages

The president of the Royal Australian College of General Practitioners recommends a common-sense approach to the use of alcohol in nursing homes, acknowledging that serving a glass of wine with meals can enhance residents' enjoyment (Egan, 2021). Government guidelines recommend no more than two glasses per day, and only after appropriate individual resident assessment. This option, among others, encourages residents to continue as many routines as possible from their preadmission lifestyle. The experience can be enhanced by the use of wine glasses and an attractive meal tray, providing a welcome alternative to the otherwise rather ordinary dining ware.

3.3 Clinical Care

3.3.17 Dehydration

One of the most common ailments nursing home residents experience is dehydration: occurring when the amount of fluid taken in is less than the amount released through the cells, resulting in loss of normal bodily functions. Many residents are at risk of choking if they are fed meals or drinks too quickly, and it can be time consuming to provide the appropriate assistance safely. When residents are served drinks by non-nursing staff, who is responsible for ensuring they are delivered in a safe, timely manner, and their consumption monitored? Water jugs may remain unused due to residents' incapacity to reach them or to safely pour themselves a drink. On the other hand, residents' fluid intake may be enhanced by a greater variety of tasty, attractively presented drinks such as milkshakes or other variations in coloured glasses and jugs, and by encouraging families and/or volunteers to assist residents with drinks of their choice.

Digital technology can aid dehydration monitoring, together with hydration charts, particularly for residents with cognitive impairment: evidence shows that increased dehydration is linked to decreased cognitive function (Sherwood, 2021). Sherwood's analysis supports other evidence relating to the serious implications of poor hydration, leading to serious illness. Using recently designed software, nursing home managers can ensure their residents are 'getting enough to drink'.

3.3.18 Elimination: Bowel and Bladder Management

Bowel and bladder management require careful scrutiny, especially for residents who cannot readily indicate their need for the toilet. Unmanaged constipation can lead to very serious complications, requiring oversight and amelioration by appropriately educated staff. Continence management is, in many nursing homes, well managed by a continence adviser. However, when PCAs are responsible for this aspect of residents' care, their limited anatomical and clinical knowledge may lead to failure in reporting serious signs and symptoms (such as faecal impaction or changes in voiding patterns). Poor continence planning can also involve inappropriate or overuse of continence aids, especially when residents are 'forced' to wear pads 24 hours a day. 'Don't worry, dear, you've got a pad on' may be an incontrovertible statement while taking no account of the resident's dignity or preference for maximising their independence.

Unfortunately, there is a dearth of detailed research regarding the contributing factors leading to incontinence, its link to depression and mental health issues, consequential financial factors and the worsening of symptoms after nursing home admission. The key to comprehensive bowel and bladder management is a carefully crafted individualised plan with goals to meet the resident's needs, coupled with prompting and advice for all staff. However, the subject matter pertaining to this important issue of incontinence requires further analysis by qualified advisers.

3.3.19 Quality Continence Care

A broad, multifocal approach to continence care begins with identifying causes and treating underlying factors. An appropriately qualified continence nurse/adviser can revolutionise this vitally important aspect of residents' care by relevant investigations, comprehensive assessment and planning. Expert oversight does not focus on 'cleaning, containing, concealing' incontinence; rather, attention is drawn to the resident's values, goals and feelings (Ostaszkiewicz et al., 2018, p. 2425). With more attention given to prevention and individualised assistance rather than mere 'management' of incontinence, residents may find their comfort and dignity restored.

3.3.20 Breathing Difficulties

The cycle of 'dyspnoea, panic, dyspnoea' is a distressing, if not frightening, experience, more often witnessed by family or staff rather than described by the resident. Dyspnoea is a subjective experience of breathing discomfort and can be an important predictor of quality of life and mortality (Laviolette & Laveneziana, 2014). Hypoxia is a state in which insufficient oxygen is delivered to the tissues. Dyspnoea does not necessarily mean hypoxia, and correcting hypoxia will not necessarily lessen dyspnoea. The use of oxygen therapy needs to be carefully assessed as residents may obtain the same benefit from adjusted posture, cool air movement from an electric or handheld fan, open window or cool face washer. Oxygen therapy and other medications are not without risk; contraindications include 'restricted mobility, social isolation, pressure areas from oxygen tubing and impaired communication' (Palliative Expert Group, 2016, p. 245). Staff may need to be reminded that oxygen is a drug requiring prescription, with associated limitations and recommended dosage.

Breathing difficulties are often frightening for the resident and alarming for those witnessing their distress. However, hospitalisation for a frail resident, particularly with a diagnosis of dementia, may exacerbate the problem, not to mention increasing their anxiety (amounting to terror in some circumstances). A careful plan for dyspnoea attacks, including advice from a palliative care consultant, may prevent an emergency transfer to hospital. As for other breathing difficulties, the *goal of care* guides the plan.

3.4 Infections

Infections are best managed by a health professional with relevant contemporary knowledge and with the capacity for additional timely response to specific outbreaks. Such was the case with the COVID-19 pandemic, where nursing home residents were shown to be particularly vulnerable and raising many questions about infection control policies; from failure to enact them to total absence of the same. Where such policies were followed, together with a referral for expert advice, residents' transfers to hospital were unnecessary, and lives were saved.

Much debate has arisen within the COVID-19 pandemic concerning hospitalising infected nursing home residents. Lacking clear, consistent guidelines, many different scenarios have been reported. The very real tragedies included the number of older people dying in hospital without another human person beside them. Their untold stories go with them to the grave, leaving relatives to lament the manner of their departure and no one to accept responsibility. In the absence of a clear plan of care, the outcome has been devastating for many residents, particularly when transferred to hospital without consent.

It is clear that, in circumstances such as a serious pandemic, managers need to seek expert advice in order to consider options for each resident, particularly focused on risk analysis. Depending on the severity of their illness, some will require hospitalisation. For others, nursing home accommodation may be adapted to allow the infected resident/s to be cared for in their familiar environment, albeit with due diligence according to expert infection control advice, and with expanded medical services where available. In such instances, the resident's personal contacts and relationships are given attention equal to their clinical status. In other words, the deeply human element is rated above a manager's uninformed decision akin to a mere transaction or relocation.

Infection control in a serious pandemic is, of course, particularly important for residents with dementia, who are incapable of understanding the need for isolation; being startled, confused, if not terrified, by the presence of unfamiliar staff clothed head to toe in personal protective equipment (PPE). Added to this bewildering, if not frightening, sight is the absence of visitors, even their closest relative. As one family member described it: 'The lockdown is worse than the virus.'

Decisions arising from such an unprecedented crisis need to be addressed on an individual resident basis wherever possible, involving families and others in the process. The serious implications of a resident being infected with the coronavirus arise from their weakened immune response and multimorbidity, resulting in a higher risk of long-term complications such as increased frailty, poorer communication and cognitive decline. Family members providing commentary on their firsthand experience of the effects of COVID-19 have expressed the view that the element of risk needs to be balanced with the resident's mental well-being and quality of life, including the impact of hospitalisation. Discussion with families is one of the keys to better outcomes.

The pandemic has given rise to the need for clear protocols; for example, allowing a close family member to visit a resident who is dying. With relevant infection control advice, ensuring adequate precautionary measures are observed, some managers have encouraged such visits, to the immense satisfaction of residents and families. Other advice between agencies has, unfortunately, been lacking: for example, inadequate infection control measures for staff, increasing the potential for serious harm. For others, false and ill-conceived boundaries have been erected, literally preventing family access to residents close to death.

Rather than a reactionary approach once an infection is evident, a comprehensive infection control policy, guided by a relevant expert, effected through regular staff education sessions, ensures optimal response when and if such a situation arises. The pandemic has shone a light on the serious, in some cases *deadly*, outcomes when nursing homes lack focus, standard procedures and evidence-based protocols for all aspects of infection control.

Response to other types of infections requires expert clinical assessment. Minimisation of risks includes implementing standard policies, procedures and precautions to prevent and control infection. Guidelines for antibiotic prescribing also need to be followed to minimise antibiotic resistance. These and other measures can be recommended, adopted and monitored by an appropriately qualified infection control health agent: in most cases a nurse.

Another major factor in infection control is the nursing home's building design which, in light of the COVID-19 pandemic, has informed the need for structural changes. It may mean fewer and/or smaller communal areas, to lessen the risk of cross-infection, together with a thorough review of all ventilation systems. Creative options would allow residents to mix with others, from a safe distance and in a less-crowded environment. Similarly, visitors may be safely accommodated with appropriate, well-informed protocols. Rather than applying an impromptu 'no visitors' rule, unsupported by evidence, managers should seek advice from relevant experts and develop a concise, user-friendly checklist for visitors, available in various languages. Understanding the devastating consequences in some cases, the unexpected prevalence of the coronavirus has resulted in vast improvements in infection control in many nursing homes.

3.4.1 Oral and Dental Care

Older people are more vulnerable to oral diseases due to their multiple medications, limited mobility and dexterity, as well as lack of cognitive capacity to perform their own oral hygiene. 'Poor oral health can impact quality of life and contribute to life-threatening conditions such as malnutrition or pneumonia' (Matthews, 2019). As there is no mandated requirement for dental services in nursing homes, Matthews offers some solutions including the introduction of a qualified dental practitioner's oral health assessment and care plan for every resident on admission, and for nursing homes to employ a dental practitioner to provide staff education, preventive dental services and culturally appropriate care. Some dental experts have recommended that a thorough oral and dental examination should be mandatory prior to every person's nursing home admission, including a comprehensive report, with a plan for regular, ongoing care. Such care could be provided by an oral health practitioner employed by the nursing home or across several nursing homes to rationalise costs. Private dental care cover for those without government assistance is prohibitive, another factor leading to poor oral health. A preventive approach, together with routine, regular dental assessments, may prove to be cost effective in the longer term, preventing complications such as the following:

> The daughters of one nursing home resident recounted their experience of holding their distraught mother's hand while she had teeth extracted 'due to neglect by an aged care facility'. They described the 'gum erosion and rotting teeth their mother had suffered as a result of poor oral care, which left the dentist "disgusted".' Only six months previously, a dentist had identified no problems with their mother's teeth. Apparently, the nursing home staff had not been removing and/or cleaning the resident's dentures, leading to the need for urgent, very costly dental care.

On the same theme, a PCA recounted her experience.

> As a new staff member, I was given no instructions about residents' dental care. While assisting one resident into bed, I instinctively offered to clean his dentures. His wife was present and said, with a mixture of pleasure and regret: 'You're the first person I've ever seen cleaning his teeth, and I'm here every night. I've been so worried about it but I didn't like to complain.'

It is impossible to know the extent to which nursing home residents suffer painful gums, ulcerated mouths, ill-fitting dentures, tongue lesions and painful dental caries; no doubt contributing to the appalling rate of malnutrition. A comprehensive dental care plan would address this important need, based on thorough, regular assessment, review and timely referral.

3.4.2 Pain Management

The significant subject of pain management is covered in Chap. 5. This is not to suggest that pain is confined to the end of the resident's life. On the contrary, it is considered such an important issue for most residents from the day of admission as to require a separate chapter section.

3.4.3 Mental Health and Depression

Medical oversight of nursing home residents is provided largely by GPs who may not all have skills and experience in mental health care, particularly those requiring specialist psychological or psychiatric intervention. A significant number of residents are prescribed antipsychotic drugs in the absence of a diagnosed psychiatric condition. For example, as dementia is not a psychiatric illness, it is prudent to question such prescriptions, while it is important that any resident with a diagnosed psychiatric condition is cared for by a qualified clinician.

In spite of the prevalence of depression or other psychiatric illnesses including suicide ideation, little psychological care is available in nursing homes. Most aged care workers are unqualified to distinguish between depression and psychosis, so follow up is essential for a resident who is observed to be sad, lonely, isolated and/

or withdrawn. Rather than falsely attributing these signs to 'part of normal ageing', specialist attention and diagnosis are needed.

> This issue has been hidden for too long, and met with minimal efforts targeted at prevention.... More than half of nursing home residents suffer symptoms of depression. This is compared to 10–15% of adults of the same age living in the community... Research has shown one in every seven residents exhibits self-harming behaviours on a weekly basis, such as cutting, hitting, or eating foreign objects. (Murphy, 2020).

Specialist attention would, potentially, prevent the following 'handover' scenario where staff were discussing a resident who seemed less responsive than usual.

> *'I think she's depressed'*, said one nurse.
> *'I don't think she's depressed. I think she's just being difficult'*, said another.
> *'Did you know she recently had a major tragedy in her family?'* asked the PCA.
> *'Oh, we can't delve into residents' personal lives'* was the RN's response.

Rather than idle speculation based on uninformed opinion, comprehensive care would involve referral to a psychologist or psychiatrist. The accurate diagnosis of depression may lead to treatment which ameliorates, if not cures, the problem. It is every resident's right to have access to such expert diagnostic advice. It is every staff member's responsibility to avoid using the term 'depression' in the absence of a qualified clinician's assessment.

Many other references to depression throughout the book underscore its prevalence and its relationship to other factors in the lives of residents.

3.4.4 Cardiopulmonary Resuscitation (CPR) and Not for Resuscitation (NFR)

Many people regard CPR in the light of popular television depicting a person who has stopped breathing being brought back to life, happy and smiling, restored to their former disease-free status. In the context of aged care, the positive and negative aspects of CPR need to be carefully, sensitively and thoughtfully discussed with each resident and their families, leading to informed choices.

Residents have the right to request CPR; however, the facts should be carefully explained, such as for any other invasive procedure where the benefits would be weighed against the burdens. Where CPR is offered, what education, explanation, advice (included written) is available? Are residents and families made aware of its comparative futility in this population? Is the health professional obligated to explain to the resident and family the risks associated with CPR, including infection and the infliction of severe physical trauma? One writer describes the offering of CPR to frail, elderly people as 'cruel'. 'Administering CPR when there is no medically reasonable chance that a distressed patient will recover from the underlying illness amounts to physical torture' (Smith, 2011, p. 493). In direct flouting of such facts, a common practice is for all residents or families to sign consent forms without appropriate information and discussion. Concomitantly, managers need to be well informed, and their right to dictate 'blanket' orders questioned. Readily accessible 'NFR' orders should be available for every resident (where applicable), every day and every night, so that no one is in doubt about their preferences or their rights.

Rather than offering CPR as an option in end-of-life care, Smith argues for increased emphasis on the right to refuse treatment – a more compassionate and enlightened way of caring. 'This right of refusal is not a right to hasten death, but merely a right to resist unwanted physical invasions' (Smith, 2011, p. 509). This subject would benefit from regular, evidence-based staff education leading to an informed discussion with every resident and/or family on admission, resulting in carefully documented wishes.

Researchers (Ibrahim et al., 2016) note that 'not for resuscitation' orders (NFRs) are legal documents completed by physicians but patients 'cannot demand treatment that the medical team deems to be futile' (p.119). Questions of dignity, consent and compassion require complex discussions generally disregarded when decisions are made 'on the spot'. Regarding NFR, the alternative is not to 'do nothing'; it means to pursue the strong goal of *comfort*. If these alternatives are discussed prior to the end of life, with both resident and family, the need for 'urgent' decision-making should be obviated, not to mention avoiding untold harm. In the context of dementia, providing dignified, compassionate care 'is a daunting task profoundly impacted by diminished patient capacity at the end of life' (Yeaman et al., 2013, p. 499). The Hippocratic Oath is an ethical reminder for doctors to 'avoid doing harm by useless effort'.

3.5 Care for the Whole Person

3.5.1 Who Makes the Decisions?

An astute aged care worker in the United States captures the essence of a nursing home resident denied decisions about his own care.

> Harold is still in his right mind, but owing to his weakness we make all his choices for him. Harold wants to stay in bed and sleep, but we make him get up and sit in his La-Z-Boy twice a day. We know what's good for him. That is, we know what's good for his biology. We focus on keeping his body intact, but we don't let his protests hold much sway. We count his heart-beats while we ignore what great rivers of sentiment may course through his veins. We monitor and measure all sorts of body parts – the kinds of things that the state inspectors can check easily – but we leave Harold the person off the charts.... He's at our mercy. We see his protests as just another obstacle to his care. We overpower his spirit to treat his colon, his skin, and his blood count. (Gass, 2004, p.34).

Gass's astute description of 'Harold' exemplifies the issue of whole person care. The main focus was on the resident's bodily functions, while failing to address his spirit. 'Harold' was capable of making his own decisions, but his carers decided what was best for him. He was certainly discouraged from questioning, not to mention protesting, staff priorities for his care. What difference would it have made for the focus to be primarily on Harold the *person,* and thereafter his biology?

One way of addressing residents' holistic, person-centred needs, especially the non-clinical aspects of care, is through creative activities, particularly when a choice is offered.

3.5.2 Allied Health

Music therapy, occupational therapy, physiotherapy and speech therapy are some of the essential components of residents' personal and communal care, each deserving their own descriptions and associated personal stories. These therapies, together with a variety of others, contribute to residents' holistic care, rather than perceived as 'luxury add-ons' for those who can afford to pay for them.

Well-planned activities achieve more than relief from boredom, especially if the 'one-size-fits-all' approach is superseded by social engagements that exercise residents' minds and bodies, according to their preference. Many residents are capable of learning new skills or being involved in activities beyond their normal experience, including using computers, social media, music devices, smartphones, a range of new technologies and communication platforms. With suitably qualified therapists, augmented by volunteers and supported by families, creative programmes can be designed which are offered every day, including evenings and weekends. Well-designed, imaginative activities also include engagement with the local community: churches, schools, kindergartens, libraries, elderly citizens' clubs and others.

Aromatherapy is another activity which may not be routinely, regularly offered with an invitation for all residents to be involved. Aromatherapy embraces both physical and mental health, using ingredients to enhance a person's overall well-being, taking its place among other complementary therapies whose benefits include treatment of circulatory conditions, muscular pain, anxiety and depression. Massage, relaxation and therapeutic touch have also retained their place in healing and well-being, adding to the array of therapies using aromas.

Whatever therapies are involved, the emphasis should be on *restoration and rehabilitation* wherever possible: concepts sadly lacking in many residents' care plans, denying them the chance to regain lost functions (Commonwealth of Australia, 2021, p. 21). *Reablement* is the term preferred by some exercise therapists, where the concentration is on regaining lost capacity, for example, after surgery. Rather than accepting diminished function as irreversible, reablement offers hope for renewed activity, described by one resident: 'I feel like a new woman. I didn't know my legs still had life in them!'

Another therapy bringing happiness and joy to residents is art; whether or not the residents can produce such work themselves. Some nursing homes offer a skilled artist space to work, inspiring the interest of staff, residents and the wider community. Exhibitions and cultural events are fostered; opening the nursing home to the 'outside world', dispelling the stereotypical views of nursing homes as 'islands of

3.5 Care for the Whole Person

the old'. Sad to say, 'happiness and joy' are terms seldom associated with residential aged care. Engagement with artists can be transformative: the basic human right to pleasure and meaning in each day are realistic goals even for those whose physical health and cognition is poor.

The need for major reform of allied health services receives significant attention in the Royal Commission into Aged Care Quality and Safety report which recommends providers employ, by 1 July 2024, the following:

> an oral health practitioner, a mental health practitioner, a podiatrist, a physiotherapist, an occupational therapist, a pharmacist, a speech pathologist, a dietitian, an exercise physiologist, and a music or art therapist and have arrangements with optometrists and audiologists as required. (Commonwealth of Australia, 2021, p. 179).

The report acknowledges the substantial amount of funding required, suggesting some options for blended models: a highly recommended aim to strive for. Some nursing homes, well ahead of the projected 2024 requirements, already serve their residents well by providing these services.

3.5.3 *Music, Singing and Dance*

Knowing the neuroscientific benefits of singing, a suburban community-based choir has visited one nursing home regularly for 5 years. As one member commented: 'Singing is a game changer. It is a miracle drug and it is free' (Egan, 2018). At another home regular playgroup sessions are held where children and residents interact with musical activities enjoyed by both age groups. The programme helps to reduce isolation, according to an occupational therapist. The delights of music and dancing are described in the following scenario:

> It was the evening of the masquerade ball. The nursing home's lounge was a blaze of colour, the musicians were playing, and the wheelchair dancing commenced. Ethel and Doris were sitting together, their respective husbands (residents) too ill and frail to join the festivities. Each in her early eighties, they visited their husbands daily, returning to their empty homes, experiencing the emotional and physical impact of 'pre-widowhood'. 'Would you like to dance?' one of the aged care team asked Ethel. 'Who, me?' Ethel replied, overcome by surprise. Linking arms, staff member and relative danced to the old tunes, Ethel flushed with excitement, stimulated by her memories. 'Do you know, Fred and I used to go dancing every Saturday night? Since his stroke, we've never even been able to embrace. I think that's the first time anyone has put their arms around me for months. Thanks so much.' Doris plucked up courage to join hands with a male resident in a wheelchair. Joe, usually passive and uncommunicative, tapped his feet as Doris twirled him around, daring to give him a hug at the end. 'Thanks for the dance', she said. Joe seemed to grow taller and straighter in his chair, beaming a rare smile. (Commonwealth of Australia, 2006, p.139).

Carefully tailored to meet individual residents' needs and preferences, musical activity can lead to surprising outcomes, deserving a place in their documented care record. Families may need prompting to provide specific music and equipment such as a musical instrument or headphones to enhance the resident's enjoyment. Visiting music/dance groups also inspire some residents to rise above their low mood, transforming their day beyond the mundane. Evidence for such transformation comes from a research project involving people with Parkinson's disease, whose motivation and quality of life improved with their involvement in dancing classes (Shanahan et al., 2017). This project exemplifies the focus on 'healthy ageing', even within the context of serious disease.

3.5.4 Pastoral, Spiritual and Religious Care

Understanding and appreciation of pastoral, spiritual and religious care are enhanced by the following definitions, showing their intrinsic relationships as well as their differences:

> 'Spirituality is a dynamic and intrinsic aspect of humanity through which persons seek ultimate meaning, purpose, and transcendence, and experience relationship to self, family, others, community, society, nature and the significant or sacred.' (Puchalski et al., 2014, p. 643).
>
> 'Spiritual care … is the umbrella term of which religious care is a part… Religious care is given in the context of shared beliefs, values, liturgies and lifestyle of a faith community. Spiritual care is not necessarily religious. Religious care should always be spiritual.' (Meaningful Ageing Australia, 2016, p.8).
>
> '(Spirituality is) the aspect of humanity that refers to the way individuals seek and express meaning and purpose and the way they experience their connectedness to the moment, to self, to others, to nature, and to the significant of sacred.' (Curtin, 2021, p. 8).

Curtin notes that the expanded meaning of spirituality now includes the role of the community, the importance of experiences and its multidimensional character, emphasising also that 'spiritual care isn't the exclusive domain of the chaplaincy' (Curtin, 2021, p. 8). Another writer notes that, while spirituality may defy universal definition, it is inherently relational (Lepherd et al., 2019, p7), enhanced by connections with others (Lepherd et al., 2019, p.26). Such connections are described in this account:

> 'Alan' watched with interest from his nursing home bed, as his roommate 'Fred' received regular visits from the chaplain. Homeless, deaf, with no family, and now close to death, Fred showed no meaningful response and Alan wondered why the chaplain bothered. 'Anne' would always try to chat with Fred, finishing her visit by praying loudly and unapologetically into Fred's ear. After witnessing one of these visits Bert summonsed his trusted nurse: 'Would that chaplain come and see me?' The nurse, a little perplexed, responded: 'I can ask her, but you told me when you first came in here to put "nil" for "religion" and you had no need for a chaplain and didn't want a funeral.' 'Well, a bloke can change his mind, can't he?' Alan responded. After several visits from the chaplain Alan asked for his details to be changed, signifying his desire for a funeral service in the chapel, conducted by the chaplain, followed by a formal burial.

With no formal process for assessing and reviewing his spiritual, religious or pastoral needs, Alan's increasing inner yearnings may have easily been overlooked. He would not have described himself as 'spiritual', and had he not observed the chaplain's regular, comforting pastoral care and heard her words to Fred, he would have remained ignorant of the role.

Such ignorance in others may be overcome by relevant assessment. 'It is recommended that a spiritual history be taken at the time of admission', and residents with specific spiritual needs should be referred for expert assessment and care (Best et al., 2020, p.3). This recommendation goes beyond mere box-ticking, which may be the only reference in a resident's file to matters of religion and/or spirituality, and prone to misinterpretation, evidenced by this scenario:

> *The night RN was concerned about a resident's sudden health deterioration. 'Quick, we'd better call the chaplain', he said. Unable to find any other relevant documentation he noted 'RC' in the 'religion' box on the resident's file. So, he called the nearest Roman Catholic diocese, requesting an emergency visit. When the priest arrived, the resident's wife exclaimed: 'O, Father, you know my husband would not welcome your visit, since his recent "falling out" with the church. He made me promise not to call a priest. I'm sorry for the inconvenience.'*

This episode reinforces the need for thorough assessment: for religious/spiritual care as well as physical/clinical care.

Management has the responsibility for setting the tone for spiritual care, including relevant publicity. What provision is made for this subject to be included in the nursing home's mission statement and promotional material? What priority is given in the budget for employing appropriately qualified staff to provide spiritual care? Spirituality is a basic human right. Managers have both a sober and creative opportunity to ensure effective spiritual care is provided in ways that are meaningful to those in their care. Seen in this broader light, residents may be given opportunities to discuss their spiritual needs both on admission and throughout their life in the nursing home. Meticulous assessment processes would also include documentation regarding the refusal of spiritual care, reinforced by one resident's strong assertion: 'I haven't needed a chaplain in eighty-nine years and I'm not about to call for one now!'

Assessment should focus on the deeply human, individual nature of spirituality. Older people may be aided to express their spirituality in their own words by responding to a well-designed spiritual assessment tool, aptly described as a 'rigorous process' (Lepherd et al., 2020). Assessment may also identify the residents' culture and traditions, including religious dietary requirements, arrangements after death and other matters indicative of their religious diversity (Best et al., 2020, p.7).

Spiritual factors often inform and influence physical care. As one very active family carer stated: 'I was willing to do what needed to be done ... but it would have been good to have felt that my role was valued ... that it was seen' (Hudson, 2009, p.44). For this carer, the tasks involved her whole person: body, mind and spirit. Merely taking a minute or two to ask her how she was feeling could have been a way of meeting her spiritual needs. A simple question may have sufficed: 'what would lift your spirits?' or 'where do you find solace or comfort?' or 'how are you feeling?' or 'are you getting tired?' Spirituality is *embodied:* it does not belong to some realm above, outside or beyond the human person and their lived experience.

Spiritual care has many different emphases within specific cultures. For example, the spirituality of people from Aboriginal and Torres Strait Islander communities is focused on the land and the associated life force within their culture. However, there are multiple 'cultures within their culture'. It is important, therefore, that residents belonging to these communities have access to care which is consonant with their specific cultural group.

Similarly, for residents from CALD backgrounds, it is essential that staff do not expect a generic approach to spiritual care. For many whose lives are steeped in cultures outside the perceived stereotypical Australian way of life, their unique beliefs and practices need to be acknowledged and respected rather than deemed an optional extra 'for those who like that sort of thing'.

When a chaplain or pastoral carer is recognised as an essential member of the healthcare team, their contribution is more readily acknowledged. A comprehensive care plan would include the resident's specific needs for a chaplain or pastoral care worker, rather than staff making assumptions. The following scenarios demonstrate why some chaplains may at first be reluctant to enter the strange environment of the nursing home. Others, with encouragement, are drawn by the unexpected responses.

> The new chaplain was a little nervous, having been asked by a resident's son to visit and pray with the resident, who found it difficult to hear. 'You'll have to speak up' he warned the chaplain. Looking around at the other resident in the room, the chaplain wondered what the gentleman in the other bed would think about her rather loud praying. When it came to the 'Our Father ...' the chaplain invited the resident to say the prayer with her. Much to the chaplain's amazement the resident from the other bed joined in. Describing this incident to one of the nurses, the chaplain's story was met with incredulity. 'But he never says a word'. From that day on, the resident in the other bed always recited the Lord's Prayer when the chaplain came to visit.

Another scenario illustrates the need for sensitivity towards specific religious celebrations, rather than assuming a broadly ecumenical approach will meet all needs:

> 'Ana's' religion was important to her, but her Russian Orthodox priest could only visit on rare occasions and there seemed to be no easy way of getting Ana to the Russian Orthodox church. The Protestant nursing home chaplain discussed this with Ana, took her to the nursing home chapel and explained the difference between Ana's own ornate, sensual experience of worship involving sights, sounds and smells, and the comparatively plain 'multi-purpose' chapel. The thoughtful chaplain introduced an unfamiliar practice. She lit some candles and an incense stick – a symbolic gesture towards the sharing of meaning. Ana was eager to attend chapel, saying 'we all believe in the same God'. She was delighted to report that she 'almost felt at home'.

It was evident from Ana's buoyant reply that faith was a matter of warm, human, reciprocal relationships. She enjoyed the experience of being lifted above the ordinary contours of her rather isolated nursing home life: the paradox of the transcendent coming to expression in human communication and action.

The COVID-19 pandemic has resulted in unprecedented challenges for spiritual care, given the increased numbers of residents experiencing isolation, loneliness and vulnerability, not to mention the fear of imminent death. In this new environment, where chaplaincy services and spiritual carers were needed more than ever,

3.5 Care for the Whole Person

restrictive isolation measures have often resulted in the absence, rather than the presence, of such personnel. In these circumstances 'spiritual care is not a luxury, it is a necessity for any system that claims to care for people – whether the people are in the bed or draped in protective gear' (Ferrell et al., 2020, p. e8).

In spite of the restrictions, some proactive chaplains have indicated that they can 'with innovation and skill, continue to fulfill appropriate spiritual care interventions even under the difficult conditions of COVID-19' (Drummond & Carey, 2020). Rather than assume chaplaincy services would not be available during the pandemic, some creative solutions have come to the fore, while others were uncertain of their role.

> .... many chaplains from countries around the world felt valued and understood by their employing organisations, adapted to using technology for communicating where necessary, got the right support from their professional associations, knew what to do to look after themselves and were very clear about their place in the healthcare team both before and during the pandemic. At the same time, a substantial proportion experienced the opposite... most respondents weren't clear about their role during the pandemic. Perhaps more surprisingly, the survey showed that chaplains weren't clear about their role before the pandemic either. (Snowden, 2021, p. 12).

These chaplains' descriptions are a salutary reminder that their role within the nursing home needs to be formally proscribed, not only during a pandemic.

3.5.5 Sexual Expression

Whether or not the subject of sexual expression is addressed on the resident's admission to the nursing home, it is a vital assessment component.

> When a person moves into an aged care facility, the need to express sexuality and experience intimacy with another person can remain important and can offer comfort and pleasure... Even if sexual intercourse or masturbation is no longer satisfying or possible, the need for affection, closeness and intimacy can still remain. (Bauer & Fetherstonhaugh, 2016, p. 2).

Nursing home residents are often deprived of touch, related to the often-untouched subject of sexual expression, together with poor understanding of their need for intimacy, physical contact and the corresponding contribution to their general health and well-being when these needs are met. Sexual expression should not be viewed as a problem; neither should it be left to 'whoever's on' to approve or disapprove of a specific activity. A holistic care plan would include this important element, and a written policy, developed after discussion with staff and families, should be the norm.

> When 'Norman' raised the issue of the intimacy he craved, and the privacy he longed for when his wife visited, he was encouraged to confide in a sensitive senior staff member. The matter was raised about the serious nature of Norman's unmet needs and the home's obligation to try and address them. Following discussions, a discrete notice on the resident's

bedroom door ensured Norman and his wife enjoyed uninterrupted privacy when they chose, and this part of his care was documented accordingly.

The response described above involved careful preparation. Staff training was provided, emphasising residents' need for privacy when appropriate, and a policy developed which took account of residents' rights and the need for a formal meeting of relevant parties when individual circumstances warranted.

While some staff remained uncomfortable about this aspect of care, others appreciated acknowledgement of its relevance, albeit for only a small number of residents. Without such a policy, differences of opinion are given full vent, particularly when it involves residents who are not a formal 'couple'. 'When I'm on duty I'm going to make sure his door is kept locked so he doesn't go into Mavis's room', said one staff member. 'I respect his right to get into bed with Mavis, providing no harm results', said another.

'Sexual activity encompasses intercourse, emotional intimacy, close companionship, flirting, affection, hugging, kissing, arousal, desire and self-pleasure' (Palliaged Palliative Aged Care Evidence, 2019, p. 1). Sexuality describes the way in which a person lives as a sexual being, including biological sex, gender identity/diversity and intimacy. The literature on sexual expression in nursing homes is sparse, signalling the need for increased education and awareness, including assessment and documentation. Within this cohort, sexuality may more commonly be expressed as intimacy rather than explicit sexual activity. 'Relationships based on intimacy create a safe space for both people and actively try to maintain that sense of security for one another' (Palliaged Palliative Aged Care Evidence, 2019, p. 1).

Older people are commonly viewed as asexual. They are generally reluctant to raise the subject for discussion, and aged care staff mostly lack the skills and experience to initiate such conversations. Some residents may welcome the invitation to discuss their sexual needs, including honest acknowledgement of their sexual orientation and preferences. Others state definitively that they have no interest in the subject. Both responses deserve to be acknowledged and documented with professional competence.

3.5.6 Psychosocial Care

One writer (Gawande, 2014) tells of an 89-year-old woman who, while mentally alert, recognised her increasing physical frailty and booked herself into a nursing home. She found that all the attention was related to her 'safety'. 'Basic matters, like when she went to bed, woke up, dressed, and ate, were subject to the rigid schedule of institutional life. She couldn't have her own furniture or a cocktail before dinner, because it wasn't safe' (Gawande, 2014, p.75). This feisty, previously independent, creative woman who made her own jewellery now found her activities confined to bingo, videos and 'boring' passive group entertainment. Gawande's insightful book prompts the question of comprehensive admission assessment,

3.5 Care for the Whole Person

including residents' psychosocial preferences. Regrettably, this aspect of holistic care is not routinely acknowledged.

A prominent newspaper quotes a former senator's comments during his final days in a nursing home. Citing one example of many, the reporter quoted the 84-year-old resident:

> 'Here is your coffee.' There was no greeting or even mention of my name as the staff member placed the cup on a stand well beyond my reach. She did not inquire as to how I would reach it or assist me to do so or indicate that she would arrange for another staff member to help me, although I was clearly incapable of moving the bed-like chair to which I am confined. (Silvester, 2019).

'Barney' (the retired senator, Bernard Cooney, consenting to his name being used) also reported that there was no system for those like himself, psychologically astute and cognitively aware, while physically unable to use a call bell. 'Staff assistance is irregular and lengthy periods will elapse before anyone comes to check on me. My voice is very weak and I cannot call out. I experience such times as a form of torture.' He also described the careless, insensitive communication.

> Mentally we may be quite alert but are still infantilized when spoken to by some staff… The obvious and continuing failure of our system to address seriously the mental and emotional wellbeing of our aged population shames our community. Neglect at this level can be as destructive as physical abuse and contribute to great distress. (Silvester, 2019).

Cooney's account (reported by Silvester, 2019) of life in the nursing home continues: 'There are programs and residents welcome them to help pass the time, but few provide any stimulation. They are generally tired and unimaginative.' Cooney also laments the absence of sufficient social workers to assist those experiencing separation and loss of family and homes, as well as social connections and familiar activities. As he told Silvester:

> There must be recognition and support to assist us at this level as we struggle to maintain our human dignity … when, in consequence of our physical circumstances and diminishing personal autonomy, it is constantly being stripped from us. (Silvester, 2019).

Silvester's narration of 'depersonalisation' and lack of autonomy highlights the outcome of what he and Cooney perceived as management's emphasis on saving costs rather than providing personalised care. His account is a sharp reminder of the inadequacies, particularly the ready availability of psycho-social care, even in the best of nursing homes.

One way of providing friendly, personalised care, particularly for those without their own networks, is the Community Visitors' Scheme (Australian Government, 2019): available to older persons receiving government-subsidised aged care, adding variety to what, for some, is a monotonous and boring life. Nursing home staff, many of whom are legitimately concerned about the lack of stimulation experienced by many residents, may need to be reminded of the external agencies willing to offer such assistance, particularly for those who are isolated and lonely.

3.5.7 Loneliness

Loneliness is endemic, not only in the broader community, but in many nursing homes. Left unchecked, loneliness becomes a serious issue with implications for diseases such as cardiovascular disorders and dementia. Loneliness is not merely a state of mind or a passing phase; it is a health condition requiring urgent attention.

Communication is the key to unlocking the 'loneliness door'. Researchers provide some valuable, albeit disturbing, insights, describing nursing homes as lonely places (Casey et al., 2016). Interviews with residents showed that, although they had clear concepts of what personal interaction meant to them and how they would like to engage with others, few friendships were developed, leading to 'high levels of isolation' (Casey et al., 2016). This research also highlighted the relationship between social engagement, chronic illness and reduced cognition. Loneliness is, therefore, not a 'side issue' but the key to a resident's health and well-being, deserving a place in their care plan, including comprehensive assessment, identifying goals and strategies with regular evaluation and review.

Loneliness is associated with real or perceived isolation. Isolation has been referred to as 'dangerous solitude', emphasising the serious outcomes when it is not appropriately and sensitively addressed. 'I think she's a bit lonely' deserves a professional response rather than a casual remark which fails to address the issue.

While some residents may find difficulty in forming new relationships the valuable support of volunteers (discussed earlier in this chapter) is one of the keys to ameliorating loneliness. Discussion with families paves the way for other interventions drawn from their personal circumstances and experience.

Death from loneliness has been attributed to the COVID-19 pandemic, highlighting the effects of decreased communication when nursing home residents were deprived of regular visits from family. Chronic loneliness is not a new phenomenon, however, particularly when linked to social detachment and isolation, identified by the British government as a public health crisis, leading to the appointment of a minister for loneliness in 2017 (Editorial, 2020). In twenty-first-century Australia, it is astonishing to learn that there is more loneliness in nursing homes than in the general community. Managers need to examine the evidence, with assistance from researchers, to gauge the extent of the problem in their particular context and then seek expert advice regarding ways to address it. Imaginative communication, coupled with regular well-targeted activities, can halt an escalation of this crisis. Social workers' contribution is invaluable; offering resources to assist care planning and working with colleagues in a multidisciplinary team.

3.5.8 Rituals, Celebrations and Laughter

Another avenue for shared meanings beyond the physical and in which the boundaries of caregiver and resident are broken down is the way in which the nursing home celebrates both ordinary and extraordinary events.

3.5 Care for the Whole Person

> Celebration is the form of interaction in which the division between caregiver and cared-for comes nearest to vanishing completely; all are taken up into a similar mood. The ordinary boundaries of ego have become diffuse and selfhood has expanded. (Kitwood, 1997, p. 90).

Within this meaning, 'ordinary boundaries of ego' are transformed into shared experience: parties, dances, concerts, other special events or mere spontaneous frivolity, as well as funeral wakes. As one resident commented, her otherwise uneventful life is transformed when nurses let down artificial barriers. In the following instance no offence was taken, for this resident's sense of humour and capacity for fun was well known, shared by her 'favourite' personal care worker.

> *Something about my electric chair, the way I load it up with all my personal effects when going for a shower, attracts a sense of the ridiculous. When emerging from the shower one morning I saw my chair parked with a lion's head 'eyeing' me from the headrest. Another morning I came out to see a strange figure sitting in my chair! Dressed in a sheet, with rolled-up towels for arms, cap over the lion's head and someone's shoes for the feet. I nearly fell over laughing. Another night I found a clothesline made of toilet paper stretching between the bed sticks with my washing draped on it. (Source unknown).*

For many residents and carers, the aged care environment seems humourless; a place where physical disability and cognitive impairment preclude playfulness. Others appreciate the everyday reality of humour, fostering a sense of the absurd and ridiculous. Older persons are often characterised by their wisdom and patience, while humour is not always acknowledged. However, it can be a gift older people offer those who take life too seriously.

> Humour is knowledge with a soft smile. It takes distance but not with cynicism, it relativises but does not ridicule, it creates space but does not leave you alone. Old people often fill the house with good humour, and make the serious businessman, all caught up in his [sic] great projects, sit down and laugh. (Nouwen & Gaffney, 1976, p. 74).

When spaces are occupied by sombre attitudes and unimaginative routines little room is left to 'fill the house' with laughter. When weightier moments are shared with mutuality and solidarity, when the lightness of every day is recognised and the community rings with joy, glimpses of shared meanings abound. Conversation that is not task related helps to establish bonds, both light-hearted and serious, between carer and resident. As Nay's (1993) study found, nurses believe that conversation with residents is an optional extra. 'Talking to residents was not defined as *doing something* in its own right. It was appropriate only when it occurred concomitantly with some physical action' (Nay, 1993, p.177). Gibb (1990) also found a serious lack of evidence regarding the art of conversation. Devoid of interest in personal interaction, nurses' speech is often more directive than responsive. 'Speech was not used to develop relationships but as a tool towards achieving a task, and more importantly, to maintain control' (Gibb, 1990). Celebrations can overcome this sense of hierarchy.

Recognition of birthdays and other significant events is not a matter for generalisation: not all residents wish to celebrate in the same way and the unwanted imposition of other people's expectations is to be avoided, as in the following account:

'Molly' was about to turn 101. This was a significant event, although she was not the oldest resident in the nursing home. Photographers were organised, the local paper alerted, special displays arranged for the lounge room and a great sense of anticipation was evident. Then, someone thought to ask Molly how she would like to celebrate the day. *'You mean I'm 101? Oh, how ghastly! No, I don't want my photo taken and I certainly don't want my name and face plastered all over the paper.'* After further discussion with Molly and her family it was agreed that a small number of family members would come for afternoon tea and Molly would agree to a photo for the nursing home and for the family but with a further warning: *'Mind you don't put my ugly face in the paper!'* When subsequently shown the nursing home photo Molly was unimpressed. *'I look terrible. It's awful to live this long.'*

Other everyday interactions with Molly did not reveal any further dissatisfaction in living so long. Perhaps it was merely the focus on the birthday, the photo and the fuss? Molly's story demonstrates the fact that symbols and rituals do not have the same significance for every resident: options should replace assumptions. Some staff who had assumed the occasion called for maximum publicity were disappointed in Molly's response; others acknowledged and praised her independence.

3.5.9 Laughter

Nursing homes are not universally regarded as places of merriment, mirth, happiness or light-heartedness; where laughter is regularly heard, even amidst the seriousness and sadness. Some places employ a humourist, clown or diversional therapist with the express aim of invoking laughter and joy, providing welcome relief from the pain and suffering experienced by many residents. As the 'best medicine', laughter may provide residents with relief over and above prescribed medications, as well as a welcome interlude for busy carers. Some staff have a natural ability to include appropriate merrymaking into their daily routine care, and some residents have an innate, keen sense of humour; all of which adds to a positive, pleasing nursing home culture. Laughter may not be entirely divorced from clinical care, given that a healthy mouth is essential not only for enjoying food, but for the ability to smile and socialise. Giving residents a reason to laugh, even if they forget why, provides a welcome contrast to an atmosphere of gloom, solemnity and task-centred routine.

3.5.10 Podiatry

According to the Australian Podiatry Council, the provision of podiatry to nursing home residents 'is an important contributor to these people's ongoing health, well-being, and quality of life'. It is essential, therefore, that formally qualified practitioners are employed to ensure basic foot and nail care, and proper diagnosis and treatment of lower-limb problems, with the aim of preventing falls and maintaining residents' ambulation. The section (below) on 'neglect' includes the horrific

3.5 Care for the Whole Person

description of residents' toenails left unattended. On the other hand (or foot!) the pleasure experienced by residents from a gentle therapeutic foot massage and regular pedicure is incalculable: thanks to the service of volunteers in some nursing homes. Podiatrists play a pivotal role in the comprehensive assessment of residents' feet and legs, identifying pathology that can reduce mobility. Their role needs greater attention by managers; considering the benefits are well worth the costs.

3.5.11 *Personal Clothing and Laundering*

In a bygone era, residents' clothing was 'pooled', with no capacity for personal choice, let alone care of their unique personal appearance. Today, welcome emphasis is placed on residents' clothing and the impact on their appearance, confidence and well-being.

The type of laundry offered, and the quality of the same, varies among nursing homes. Lack of care and due respect for residents' personal belongings is an important issue for quality control. Employing suitably experienced staff who are appropriately inducted is necessary for the care of residents' clothing, adding significantly to residents' pride in their appearance. Regular review of laundry practices is intended to guard against the incidence of missing and/or damaged clothes; another important area for residents' and relatives' satisfaction and confidence in overall care.

Careful and respectful handling of residents' laundry after death is also a significant sign of quality care, particularly when black plastic garbage bags are replaced by more aesthetically appealing, respectful alternatives.

3.5.12 *Hairdressing*

Hairdressing is often provided by a private contractor who comes to the nursing home offering a welcome, unrushed, pampering service including the freedom to engage with the residents in non-clinical conversation. As systems and costs vary, residents and families are advised to seek information prior to admission.

Who decides the frequency of shampooing residents' hair? It would seem that leaving this part of their hygiene and grooming to a qualified hairdresser would be a sensible option. In a comprehensive plan which covers 'whole person' care, attention to residents' hair deserves a legitimate place, ensuring consistency of practice. Families play an important role in this decision-making as for all other areas affecting the residents' pride in their personal appearance.

3.5.13 Bus Trips and Excursions

Anecdotal evidence is uncovering ways to enhance residents' bus trips. Such excursions are not necessarily confined to sightseeing; residents may enjoy travelling to a specific destination of their choice, alighting from the bus to experience a new venue as well as enjoying morning or afternoon tea, before the return journey. Creative bus drivers provide a variety of music, encouraging 'sing-alongs' for residents' enjoyment, as well as driving to venues of interest such as a historical museum, a resident's place of birth, 'old school' or expanding their horizons by visiting unfamiliar places.

Other transport modes have also been used by residents capable of moving 'outside the square', such as the following:

> *The thoughtful activities coordinator learned from reading the backgrounds of two residents that they were experienced motor cyclists. Contact with the local Harley Davidson outlet resulted in two enthusiastic salesmen bringing their bright, shiny new vehicles to the nursing home. With appropriate passenger safety measures in place, the residents were taken for a ride, giving them 'the time of their lives'.*

Residents' various responses to such outings are often documented, formally registering their satisfaction as a vital part of holistic care and prompting broad conversation from carers: 'Tell me about your bike ride.'

3.5.14 Dolls and Pets

Some residents derive great pleasure from having a doll to cuddle. One nursing home manager, upon reading of this innovation, purchased 30 dolls, one for every resident in her unit. Little imagination is needed to picture the various responses: negative, derisive and incredulous: 'They treat us like infants!' In another dementia unit one male resident, after careful assessment and discussion with family, was delighted to have his own 'baby', spending several hours each day tending his baby-like doll, changing her clothes, cuddling and settling her in the bassinet beside his own bed. Individual assessment comes to the fore again.

Ibrahim raises the subject of pets, especially dogs, asking how they are welcomed and integrated into the nursing home (Ibrahim, 2018). Examples abound, of the transformation in residents' lives when pets are introduced; bringing joy, happiness and health benefits to those in aged care (Young, 2019). Citing the 2018 Animal Welfare League report which found '… 64 per cent of Australian households have a pet, … only 18 per cent of residential aged care facilities allowed residents to live with a pet', Young enunciates the benefits of pets, including social support, relief from loneliness and ameliorating 'behaviour problems'.

Comprehensive, careful planning is the key to successful pet therapy programmes. Veterinarian and researcher Thomas (1996) introduced 'The Eden Alternative' which moves away from the hierarchical structure of aged care, towards a more homelike environment, including the presence of pets. He specifies that one dog for every 20–40 residents is an appropriate number, and that a variety of at least two dogs is ideal. It is important to seek professional advice from a veterinarian to choose wisely a dog which has a playful yet gentle disposition, and dogs of over 1 year of age are preferred. All dogs need a comprehensive care plan stipulating principles such as feeding, exercise, health checks and appropriate supervision (Thomas, 1996, pp.111–120). This text includes advice on cats, birds and other animals; the essential message throughout is to seek expert advice before any decision is made about introducing a pet to the nursing home.

3.5.15 *Sensory Loss*

How much space is given in each resident's care plan for assessing their sensory responses: sight, sound, touch, smell and taste? One writer emphasises the marked impact on quality of life when a person's sensory needs are recognised. *Sight* includes all visual surroundings such as furnishings and décor, select televised programmes or videos of activities within the nursing home and wider community. *Smell* is a powerful means of evoking memories, lifting spirits and enhancing wellbeing. *Sound*, predominantly via music, helps to recall memories, to entertain and, like the other senses, to calm the spirit. Music can actually 'light up the brain'. 'The use of ear phones may offer a practical solution in many instances' (Larner, 2018). On the other hand, unwelcome, disturbing sounds should be minimised wherever possible. *Touch* 'can express more deeply than words "I am here" or "I am with you"'. *Taste* can be enhanced by food and drinks of the resident's choice. 'For the "sweet tooth" issues such as fear of raising blood sugar should be set against the pleasure obtained' (Larner, 2018). The impact of COVID-19 on taste and small is referred to in the section on nutrition and hydration.

'Sensory impairment, explicitly vision and hearing impairment, among nursing home (NH) residents decreases their ability to socially engage' (Petrovski et al., 2019). The paucity of research data to expand on these findings inhibits further analysis. For example, it is not known how many residents have macular degeneration, cataracts or glaucoma; nor how this impedes their quality of life, and whether treatment would enhance their capacity for enjoyment. Referral and liaison with an orthoptist, optometrist or specialist ophthalmic nurse would assist in recommending appropriate devices for individual residents, together with advising staff on simple steps to aid communication (Entwistle, 2021, p. 263). The uncomplicated act of ensuring residents' spectacles are cleaned and worn each day would also help with social engagement.

Similarly, addressing hearing loss and associated care of hearing aids would benefit those experiencing this impairment, especially residents with dementia. 'While there is no causal link identified between hearing loss and dementia, there is a suggestion that auditory deprivation, social isolation and increased cognitive loading may contribute to a dementia disease pathway' (Chaffey & Slatyer, 2021, p. 244).

This text includes a comprehensive description of hearing loss, including its association with other functional impairments, and the recommendation for referral to an audiologist for a thorough assessment. (Refer also to other evidence pertaining to hearing loss and dementia in Chap. 4.)

3.5.16 Sleep

Several references to sleep are included in various places throughout the book; some additional points are made here to highlight the importance of promoting residents' healthy sleeping patterns.

The common image of residents sleeping in their chairs or bed for various periods throughout the day may be attributed to medication, catching up on sleep lost throughout the night, lack of exercise and other stimulation, confusion, boredom and apathy. Measures for encouraging night-time sleep include minimising noise and other distractions, warm drink and avoiding interruptions for non-essential nursing care. Many nursing homes have invested in high-quality mattresses which optimise comfort, rest and sleep. A calm, reassuring presence by a nurse or family can also influence sleeping patterns. Optimal pain management, including non-pharmacological therapies, may also assist, together with daytime exercise as tolerated. A comprehensive plan discussed with the resident, outlining the goals, actions and assessment procedures, with regular review, is an important factor in achieving an agreed outcome.

3.5.17 Information Technology (IT)

Information technology (IT) has increasing potential to influence nursing home life in a variety of aspects, including sensory experience. Much IT is beneficial, time saving and cost effective, such as computerised drug systems for administering medications. Used cogently in communication systems, human error and adverse events can be reduced. Assisted technology, both digital and mechanical, can enhance human communication when used wisely, as well as providing enjoyment for residents. For example, a 'virtual forest' on an enlarged screen proves immensely entertaining and interactive games can revitalise family members' visits. Opportunities are also increasing for reaching rural and remote areas; for example, virtual reality educational videos to teach staff, including new carers. However, when used for residents, the goals of IT need to be clearly documented and evaluated to demonstrate their effectiveness (or otherwise).

Another benefit deriving from IT is the capacity for surveillance, particularly of very vulnerable residents prone to falling or neglecting to eat and drink. While resisted by some staff, others realise the many benefits from having their care monitored, particularly when the focus is on optimising residents' care.

Adopting digital technology which replaces paper-based systems is proving to be time saving, cost effective, reducing errors and, of utmost importance, allowing more time to be spent 'away from the desk' to focus on the residents. However, the use of these IT tools for the benefit of the residents is widely varied across nursing home communities. In view of this discussion on IT, it appears that there is, at present, a paucity of research into the impact of IT in Australian nursing homes.

This chapter describes some of the ways in which residents' care might be enhanced: clinically, spiritually, psychologically, technically and socially. Increased automation, using digital platforms and electronic records, has the potential for streamlining care, transforming fragmented and inconsistent documentation, including more efficient funding systems and other resources.

3.5.18 *Robotic Care*

Another example of increasing technology is the use of robots replacing human care; albeit, not universally embraced. Benefits cited include the solution to staff shortages, avoiding the spread of infection (e.g. in a pandemic), ensuring social distancing is maintained. Obverse views include ethical concerns over the use of machines replacing human touch, ignoring residents' choice, and the lack of rigorous evaluation.

Some researchers have stated that any plans for using robots in aged care should be put to the test by conducting a poll of those involved. They suggest that the idea of robots is more about enthusiasm for the technology rather than addressing older persons' needs (Sparrow & Sparrow, 2006).

Other research claims that there are no proven benefits or otherwise of using robots in aged care: with no definitive outcomes cited (Broadbent et al., 2016). It seems that further discussion is needed to ensure that machines do not replace 'hands-on' care; at least without very clear proven benefits. Ethical enquiry is also needed to test the cost/benefit analysis for replacing human to human care with mechanical instruments.

3.5.19 *Independence, Privacy, Respect*

Taking up residency in a nursing home is usually characterised by a 'loss of independence', assuming that such a transition entails loss of agency, loss of control over one's life, lack of respect and privacy. These issues are referred to in several places throughout the book, deserving further emphasis here. For example, asking on admission about any privacy preferences the resident and family would like noted and respected as well as examples of the resident's capacity to maintain independence in certain areas of their lives.

When a resident's increasing frailty means an increase in dependency on others for most daily tasks it should also be noted that even the frailest of persons can maintain some form of independence. This might entail discussing how their room is arranged: placing of pictures, personal items close to hand, attending to aspects of their own hygiene or making their own decisions about personal care assistance when it does not conflict with other important routines.

Seemingly small, but comparatively significant for some residents, is the independence attached to mealtimes, bedtimes, rising times, hygiene care, where to sit, whom to sit with and many other idiosyncratic preferences. Simple gestures like a notice on the door 'Please knock before entering' would assist some residents maintain a small degree of independence, privacy and respect even within communal, dependent living.

Care for the whole person embraces, to varying degrees, all the factors mentioned above. With thoughtful attention to the vitally important aspects of clinical care, together with the seemingly incidental issues described in this chapter (together with others which may have been unwittingly omitted), the community focus as well as the individual nature of resident care is enhanced.

References

Abbey, J., Liddle, J., Bridges, P., Thornton, R., Lemcke, P., Elder, R., & Abbey, B. (2006). Clinical placements in residential aged care facilities: The impact on nursing students' perception of aged care and the effect on career plans. *Australian Journal of Advanced Nursing, 23*(4).

Aged Services Industry Reference Committee (ASIRC). (2020). *The re-imagined personal care worker*. The Centre for Workforce Futures at Macquarie University and SkillsIQ Limited.

Ashby, M., & Mendelson, D. (2011). Family carers: Ethical and legal issues. *Family Carers in Palliative Care: A Guide for Health and Social Care Professionals*. https://doi.org/10.1093/acprof:oso/9780199216901.003.0006.

Austin Health. (2019). ACPA position statement: ACDs within community and residential aged care.

Australian College of Nursing. (2016). *The role of registered nurses in residential aged care facilities: Position statement*. https://www.acn.edu.au/policy/position-statements.

Australian Government. (2019). *Community visitors scheme national guidelines*.

Australian Industry and Skills Committee. (2020). *Enrolled nursing: Overview*.

Bauer, M., & Fetherstonhaugh, D. (2016). *Sexuality and people in residential aged care facilities: A guide for partners and families*. Australian Centre for Evidence Based Aged Care (ACEBAC).

Belardi, L. (2014). Nurse practitioners 'fill a gap' in aged care. *Australian Ageing Agenda*.

Best, M., Leget, C., Goodhead, A., & Paal, P. (2020). An EAPC white paper on multi-disciplinary education for spiritual care in palliative care. *BMC Palliative Care, 19*(9), 1–10.

Broadbent, E., Kerse, N., Peri, K., Robinson, H., Jayawardena, C., Kuo, T., Datta, C., Stafford, R., Butler, H., & Jawalkar, P. (2016). Benefits and problems of health-care robots in aged care settings: A comparison trial. *Australasian Journal on Ageing, 35*(1), 23–29.

Bryant, J., Sellars, M., Sinclair, C., Detering, K., Buck, K., Waller, A., White, B., & Nolte, L. (2021). Inadequate completion of advance care directives by individuals with dementia: National audit of health and aged care facilities. *BMJ Supportive & Palliative Care*.

Butler, A. (2018). Residents get barely half the care they need. *The Lamp, 75*(3), 9.

Carter, A. (2020). To what extent does clinically assisted nutrition and hydration have a role in the care of dying people? *Journal of Palliative Care, 35*(4), 209–216.

Casey, A., Low, L., Jeon, Y., & Brodaty, H. (2016). Residents perceptions of friendship and positive social networks within a nursing home. *The Gerontologist, 56*(5), 855.

Chaffey, E., & Slatyer, S. (2021). Hearing. In C. Vafeas & S. Slatyer (Eds.), *Gerontolotical nursing: A holistic approach to the care of older people* (pp. 241–252). Elsevier.

Commonwealth of Australia. (2006). *Guidelines for a palliative approach in residential aged care.*

Commonwealth of Australia. (2021). *Royal commission into aged care quality and safety final report: Care, dignity and respect, Volume 3A The new system.*

Commonwealth of Australia. (2019). *Royal Commission into aged care quality and safety Volume 3, interim report: Neglect* Volume 3.

Cook, E. (1913). *The life of Florence Nightingale* (Vol. 1). The Macmillan Co.

Curtin, L. (2021). Spirituality and nursing care: The bond between caregivers and the dying. *American Nurse Journal, 16*(5), 1–12.

Detering, K., Hancock, A., Reade, M., & Silvester, W. (2010). The impact of advance care planning on end of life care in elderly patients: randomised controlled trial. *BMJ, 340*(7751), 847.

Dragon, N. (2016). Nurse practitioners: The road less travelled. *Australian Nursing and Midwifery Journal, 24*(5), 16.

Drummond, D., & Carey, L. (2020). Chaplaincy and spiritual care response to COVID-19: An Australian case study – The McKellar Centre. *Health and Social Care Chaplaincy, 8*(2), 165–179.

Editorial. (2020, Dec 29). Nursing home patients are dying of loneliness. *New York Times.*

Egan, C. (2021, June 24). Alcohol in nursing homes: How much is too much? *Hellocare.*

Egan, N. (2018). Making connections. *Australian Ageing Agenda* (Jul/Aug 2018), 28.

End of Life Directions and Aged Care. (2018). Facts sheet. ELDAC.

Entwistle, L. (2021). Vision. In C. Vafeas & S. Slatyer (Eds.), *Gerontological nursing: A holistic approach to the care of older people* (pp. 253–265). Elsevier.

Ferrell, B., Handzo, G., Picchi, T., Puchalski, C., & Rosa, W. (2020). The urgency of spiritual care: COVID-19 and the critical need for whole-person palliation. *Journal of Pain and Symptom Management.* https://doi.org/10.1016/j.painsymman.2020.06.034

Gass, T. (2004). *Nobody's home: Candid reflections of a nursing home aide.* ILR Press.

Gawande, A. (2014). *Being mortal. Metropolitan Books.*

Gibb, H. (1990). *Representations of old age: Notes towards a critique and revision of ageism in nursing practice.* Deakin University.

Grandjean, P. (2016). The dose concept in a complex world. *Basic & Clinical Pharmacology & Toxicology, 119*(2), 126–132.

Groom, L., McCarthy, M., & WitkoskiStimpfel, & Brody, A. (2021). Telemedicine and telehealth in nursing homes: An integrative review. *Journal of the American Medical Directors Association, 22*(9), 1784–1801.e1787.

Hudson, R. (2009). Responding to family carers' spiritual needs. In P. Hudson & S. Payne (Eds.), *Family carers and palliative care* (pp. 37–53). Oxford University Press.

Hugo, C. (2019, July 19). Why is nursing home food so bad? Some spend just $6.04 per person a day – That's lower than prison. *The Conversation.*

Hudson, R., & Richmond, J. (2000). *Living, dying, caring: life and death in a nursing home.* Ausmed Publications.

Ibrahim, J. (2018). What is quality in aged care. *The Conversation.*

Ibrahim, J., MacPhail, A., Winbolt, M., & Grano, P. (2016). Limitation of care orders in patients with a diagnosis of dementia. *Resuscitation, 98*, 118–124.

Jokanovic, N., Wang, K. N., Dooley, M. J., Lalic, S., Tan, E. C., Kirkpatrick, C. M., & Bell, J. S. (2017). Prioritizing interventions to manage polypharmacy in Australian aged care facilities. *Research in Social and Administrative Pharmacy, 13*(3), 564–574.

Just, D. T., O'Rourke, H. M., Berta, W. B., Variath, C., & Cranley, L. A. (2021). Expanding the concept of end-of-life care in long-term care: A scoping review exploring the role of healthcare assistants. *International Journal of Older People Nursing, 16*(2), e12353.

Kottner, J., Hahnel, E., El Genedy, M., Neumann, K., & Balzer, K. (2019). Enhancing SKIN health and safety in aged CARE (SKINCARE Trial): A study protocol for an exploratory cluster-randomized pragmatic trial. *Trials, 20*(1), 302. https://doi.org/10.1186/s13063-019-3375-7

Larner, M. (2018). How to enhance sensory elements in aged care facilities. *Aged Care Insite* (November 19, 2018).

Laugaland, K., Billett, S., Akerjordet, K., Frøiland, C., Grealish, L., & Aase, I. (2021). Enhancing student nurses' clinical education in aged care homes: A qualitative study of challenges perceived by faculty staff. *BMC Nursing, 20*(1), 111. https://doi.org/10.1186/s12912-021-00632-0

Laviolette, L., & Laveneziana, P. (2014). Dyspnoea: A multidimensional and multidisciplinary approach. *European Respiratory Journal, 43*(6), 1750–1762.

Lepherd, L., Rogers, C., Egan, R., Towler, H., Graham, C., Nagle, A., & Hampton, I. (2019). Exploring spirituality with older people: (2) A rigorous process. *Journal of Religion, Spirituality & Aging*. https://doi.org/10.1080/15528030.2019.1672236

Lepherd, L., Rogers, C., Egan, R., Towler, H., Graham, C., Nagle, A., & Hampton, I. (2020). Exploring spirituality with older people: (2) A rigorous process. *Journal of Religion, Spirituality & Aging, 32*(3), 288–304.

Matthews, K. (2019). Stop the rot in oral care for elderly Australians. *Australian Ageing Agenda*.

Meaningful Ageing Australia. (2016). *National guidelines for spiritual care in aged care*. Meaningful Ageing Australia.

Mosby. (1983). Editor in chief. In *Mosby's medical & nursing dictionary*.

Murphy, B. J. (2020). Preventing suicide in nursing homes is possible. Here are 3 things we can do to make a start. *The Conversation* (21 January). https://theconversation.com/preventing-suicide-in-nursing-homes-is-possible-here-are-3-things-we-can-do-to-make-a-start-128212

Nay, R. (1993). *Benevolent oppression: Lived experiences of nursing home life* [doctoral dissertation, University of New South Wales]. , Australia.

Nhongo, D., Hendricks, J., Bradshaw, J., & Bail, K. (2018). Leadership and registered nurses (RN s) working after-hours in residential aged care facilities (RACF s): A structured literature review. *Journal of Clinical Nursing, 27*(21–22), 3872–3881.

Nightingale, F. (1969). *Notes on nursing: What it is, and what it is not*. Dover Publications, inc. (First published by Appleton and Company in 1860).

Nouwen, J. M., & Gaffney, W. J. (1976). Aging: the fulfilment of life. Image Books.

Ostaszkiewicz, J., Tomlinson, E., & Hutchinson, A. M. (2018). "Dignity": A central construct in nursing home staff understandings of quality continence care. *Journal of Clinical Nursing, 27*(11–12), 2425–2437.

Pallative Expert Group. (2016). *Palliative care* therapeutic guidelines.

Palliaged Palliative Aged Care Evidence. (2019). *Intimacy and sexuality*.

Parks, J. A., & Howard, M. (2021). Dying well in nursing homes during COVID-19 and beyond: The need for a relational and familial ethic. *Bioethics*.

Paulo, M., Scruth, E. A., & Jacoby, S. R. (2017). Dementia and delirium in the elderly hospitalized patient: delirium is a medical emergency. *Clinical Nurse Specialist, 31*(2), 66–69.

Petrovski, D., Sefcik, J., & Hanlon, A. (2019). Social engagement, cognition, depression, and comorbidity in nursing home residents with sensory impairment. *Research in Gerontological Nursing, 12*(5).

Picton, L., Lalic, S., Ryan-Atwood, T., Stewart, K., Kirkpatrick, C. M., Dooley, M. J., Turner, J., & Bell, S. (2020). The role of medication advisory committees in residential aged care services. *Research in Social and Administrative Pharmacy, 16*(10), 1401–1408.

Preston, A. (2020). The lady with the wrong lamp! Marking the 200th anniversary of Florence Nightingale's birth – Her life in fascinating objects. Daily Mail, 1.

Puchalski, C., Vitillo, R., McCullough, M., Larson, D., Hull, S., & Heller, R. (2014). Improving the spiritual dimension of whole person care: Reaching national and international consensus. *Journal of Palliative Medicine, 17*(6), 642–656.

Sean Morrison, R. (2020). Advance directives/care planning: Clear, simple, and wrong. *Journal of Palliative Medicine, 23*(7), 878–879.

References

Sellars, M., Chung, O., Nolte, L., Tong, A., Pond, D., Fetherstonhaugh, D., McInerney, F., Sinclair, C., & Detering, K. (2018). Perspectives of people with dementia and carers on advance care planning and end-of-life care: A systematic review and thematic synthesis of qualitative studies. *Palliative Medicine, 33*(3), 274–290.

Shanahan, J., Morris, M. E., Bhriain, O. N., Volpe, D., Lynch, T., & Clifford, A. M. (2017). Dancing for Parkinson disease: A randomized trial of Irish set dancing compared with usual care. *Archives of Physical Medicine and Rehabilitation, 98*(9), 1744–1751. https://doi.org/10.1016/j.apmr.2017.02.017

Sharp, C. A., Moore, J. S. S., & McLaws, M.-L. (2019). Two-hourly repositioning for prevention of pressure ulcers in the elderly: Patient safety or Elder abuse? *Journal of Bioethical Inquiry, 16*(1), 17–34.

Sherwood, T. (2021). How dehydration influences cognitive status in aged-care residents. *Hospital and Healthcare*.

Silvester, J. (2019, February 23). *A statesman uses his last words to address aged care royal commission*. The Age.

Sinclair, C., Nolte, L., White, B. P., Detering, M., & K. (2020). Advance care planning in Australia during the COVID-19 outbreak: Now more important than ever. *Internal Medicine Journal, 50*(8), 918–923.

Siu, H. Y., Elston, D., Arora, N., Vahrmeyer, A., Kaasalainen, S., Chidwick, P., Borhan, S., Howard, M., & Heyland, D. K. (2020). The impact of prior advance care planning documentation on end-of-life care provision in long-term care. *Canadian Geriatrics Journal, 23*(2), 172.

Smith, G. (2011). Refractory pain, existential suffering, and palliative care: Releasing an unbearable lightness of being. *Cornell Journal of Law and Public Policy, 20*(3), 469–532.

Snowden, A. (2021). What did chaplains do during the Covid pandemic? An international survey. *Journal of Pastoral Care & Counseling, 75*(1_suppl), 6–16.

Spacey, A., Scammell, J., Board, M., & Porter, S. (2021). A critical realist evaluation of advance care planning in care homes. *Journal of Advanced Nursing, 77*(6), 2774–2784.

Sparrow, R., & Sparrow, L. (2006). In the hands of machines? The future of aged care. *Minds and Machines, 16*(2), 141–161.

Thomas, W. (1996). *Life worth living: The Eden alternative in action*. VanderWyk & Burnham Acton.

Thompson, W., Jacobsen, I. T., Jarbøl, D. E., Haastrup, P., Nielsen, J. B., & Lundby, C. (2020). Nursing home residents' thoughts on discussing deprescribing of preventive medications. *Drugs & Aging, 37*(3), 187–192. https://doi.org/10.1007/s40266-020-00746-1

Waird, A., & Crisp, E. (2016). The role of advance care planning in end-of-life care for residents of aged care facilities. *Australian Journal of Advanced Nursing, 33*(4), 26.

Wang, D., & Villarosa, A. R. (2017). Food choices: Letting aged care residents have their cake, and eat it as well. *Australian Nursing & Midwifery Journal, 24*(7), 39–39.

Westbrook, J., Georgiou, A., Black, D., & Hordern, A. (2011). Comprehensive medical assessments for monitoring and improving the health of residents in aged care facilities: Existing comprehensive medical assessments coverage and trial of a new service model. *Australasian Journal on Ageing, 30*(1), 5–10.

Yeaman, P. A., Ford, J. L., & Kim, K. Y. (2013). Providing quality palliative care in end-stage Alzheimer disease. *American Journal of Hospice and Palliative Medicine®, 30*(5), 499–502.

Young, J. (2019). Pets in aged care make health and economic sense. *Hellocare*.

Chapter 4
Dementia Challenges

4.1 Dementia Misconceptions

Commonly regarded as a 'fate worse than death' and those experiencing this dreaded disease as 'no longer persons', many would aver that death in this context would be preferable to living with dementia. Unfortunately, when suboptimal care is evident, such sentiments would seem well-founded.

It is ironic that, although it is the second leading cause of death for Australians after heart disease (Australian Bureau of Statistics, 2018), it remains poorly understood, leaving some family members (and, regrettably, some staff) to opine: 'The person is no longer there' or 'she's already dead as far as I'm concerned' or 'kill me if I ever get like that' or 'it would be good for him if he died, rather than linger on like this' or 'she can't communicate at all'. Death is not necessarily the only 'good' outcome for a person with dementia. Rather, enlightened, well-researched education and support provided by carers during a lengthy living and dying process are the keys.

Considering that, compared to death from other diseases, the final stage of dementia may take several years, there is ample time for planning a 'good death'. If a 'good' death incorporates awareness that the end is coming and the opportunity to make choices, adjusting one's goals and plans accordingly, those elements may not be available to a person with end stage dementia. If a 'good death' includes the active participation of, and full communication with, families and friends, those aims would also, in most situations, be rendered unachievable. (Further discussion on 'good death' is found in Chap. 7.)

Misconceptions about dementia lead to the 'better off dead' sentiment, so commonly expressed. On the other hand, much can be achieved by a comprehensive

---

While (some of) the stories are based on factual situations, real names and other details have been altered to protect the identity of the persons concerned. Resemblance to any particular person is therefore purely coincidental.

understanding of the disease in its manifold, diverse representations, including optimal end-of-life care. What is the best response for each type of dementia, and the concomitant symptom relief for each person with the disease?

4.1.1 Dementia Definition, Diagnosis and Prognosis

'Dementia is a clinical diagnosis requiring new functional dependence on the basis of progressive cognitive decline and representing, as its Latin origins suggest, a departure from previous mental functioning' (Cunningham et al., 2015, p. 79). Another definition and description:

> Dementia describes a collection of symptoms that are caused by disorders affecting the brain. It is not one specific disease. Dementia affects thinking, behaviour and the ability to perform everyday tasks. Brain function is affected enough to interfere with the person's normal social or working life (Dementia Australia, 2020).

Dementia diagnosis is often delayed due to public ignorance of the disease process, GP's lack of specific dementia training, patients' and families' fear of the outcome, and the services available. 'Australian services for people with dementia are fragmented, challenging to navigate and hard to access' (Low et al., 2021, p. 66). Although memory clinics provide an important multidisciplinary service for those seeking a diagnosis, lack of funding often means lengthy waiting lists for the inadequate number of clinics across Australia, together with lack of community awareness of the service. (Accessing the 'Find a Memory' website may assist in locating a clinic within your area.)

In 2021, 472,000 Australians were diagnosed with dementia and 'more than two-thirds (68.1 per cent) of aged care residents have moderate to severe cognitive impairment' (Commonwealth of Australia, 2020, p. 161). However, it is unclear how many have been given a formal diagnosis derived from comprehensive investigations, and the type of dementia is not always evident and therefore not documented. Misunderstandings frequently arise from the popular assumption that dementia is a normal part of ageing. Nursing home staff have a role to play ensuring every resident with 'dementia' has a proven diagnosis (wherever possible), including the type of dementia, acknowledging their right to information affecting their future. Comprehensive discussion includes the fact that only a small number of dementia types have a genetic link, and the average life expectancy is said to be 5 years from diagnosis, although some live for up to 20 years (Dementia Australia, 2021). However, because facts and figures vary, it is important to discuss with residents and families how they may focus on the best care available within full time care.

Without wishing to alarm residents and their families, and in response to the general community's lack of knowledge about the terminal nature of the disease; current statistics ought to be disclosed when the resident is admitted to the nursing home. Sensitive, honest and open planning for death needs to be discussed, including the availability of palliative care. Dementia is not always recognised as a terminal illness and timely discussion about prognosis is not readily, or regularly, offered. In the absence of a specific prognostic tool, families should be informed about the advantages of avoiding aggressive treatments and an emphasis on comfort in the advanced stages.

4.1.2 Types of Dementia

Among the 100 (more or less) conditions or disorders associated with dementia, the most common are Alzheimer's dementia, vascular dementia, frontal lobe dementia and dementia with Lewy Bodies (Dementia Australia, 2020). Knowing the type of dementia can influence (in some cases quite remarkably) the response. For example, with *frontotemporal dementia* antisocial behaviour is likely to ensue. The frontal lobe is responsible for changes in personality and behaviour, as well as forgetfulness; primary drives may be affected, leading to disinhibition. When the temporal lobe is affected, the person may lose the meaning of words or the ability to recognise faces and objects. Cognition is compromised: empathy, forming relationships, judging emotional states of other people are also factors, as well as the ability to recognise the lounge pot plant is not the toilet. A person with *semantic* dementia may lose the name of familiar objects, for example, the keen woodworker who cannot name any of their familiar tools. Depression is associated more with semantic dementia than frontal lobe dementia. As with all types of dementia, identifying the *type* may also lead to increased understanding of the *cause* of certain behaviours. Explaining such differences to staff and families enhances communication, prevents *blaming* the person, and clarifies the expectations of 'behaviour modification'. Different types of dementia require different types of responses: careful documentation and care planning are of the essence.

4.1.3 Alzheimer's Disease (AD)

'"Oh God, I have lost myself" was the response often heard by Karl Deter from his 51-year-old wife in a German mental institution and whose doctor Alois Alzheimer named the disease' (Salvatore, 2017). More than 100 years later, the clinical symptoms and resulting death remain the same. 'Along with cancer, it is among the most dreaded diagnoses to hear pronounced' (Salvatore, 2017). 'In spite of this knowledge and the persistently increasing statistics, there remains a high degree of ignorance in the general public even though it is the leading cause of death among Australian women' (McCabe, 2018). For many, the myths remain that dementia is a normal part of ageing, or that AD is not dementia. Nursing home staff need to be well educated about the various forms of dementia. While they cannot all be included here, AD is highlighted as the most common form of dementia (60–70 per cent) while Lewy Bodies dementia deserves particular attention.

4.1.4 Dementia with Lewy Bodies (DLB)

> Dementia with Lewy bodies (DLB) is an age-associated neurodegenerative disorder producing progressive cognitive decline that interferes with normal life and daily activities... similar to Parkinson's disease (PD) (Outeiro et al., 2019, p.1).

DLB is often misdiagnosed because its signs, such as hallucinations and body rigidity, are not usually seen in other dementia types, and nursing home staff may not be familiar with the distinctions. However, DLB has very painful personal consequences, including for some, a degree of insight into their behaviour. Symptoms include difficulty with concentration and attention, extreme confusion, misjudging distances often resulting in falls, as well as tremors and stiffness, fluctuating mental state, delusions and/or depression. Although both men and women can develop DLB, it is more common in men, and usually affects men younger than those with AD. While DLB prevalence is only 5 per cent of all dementias (Commonwealth of Australia, 2019a, p.3), it deserves particular attention due to its comparative rapid progression and unique characteristics. It also deserves increased focus in the nursing home when 'behaviour' can be misunderstood. 'Oh, it's just his dementia getting worse' may be the misguided response, when in fact staff need education on how to recognise the unique DLB symptoms and respond with the best advice available.

It should also be noted that some antipsychotic drugs can be dangerous, if not fatal, for a person with DLB. Because of its association with Parkinson's disease, drug regimens need to be carefully chosen (Ballard et al., 2011). The GP responsible for the resident's care may not be fully aware of these factors, or the need to seek a second opinion from a dementia expert. Advice from a palliative care physician may also be warranted, given that death from DLB usually occurs earlier than death from AD.

4.1.5 Younger Onset Dementia

Perhaps one of the most well-known exponents of early onset dementia is Christine Bryden who wrote of being diagnosed with younger-onset dementia at 46, describing her fading memory: 'like a grey wash over a blank canvass' (Bryden, 2015, p.218). From her lived experience, she warns of many misguided assumptions, together with the dangers of over stimulation. 'I've never yet met a person with dementia who wants to be overstimulated, especially by noise and motion, and often I notice that the television is on all day, even when nobody is watching it or even near it' (Bryden, 2015, p.267). Bryden's insights provide much stimulus for thinking more deeply about dementia, including paying more attention to the *person*, who may understand far more than we are willing to concede.

4.1.6 Government Responsibility

The National Institute for Dementia Research Special Interest Group in Rehabilitation and Dementia claims that 'Australian services for people with dementia are fragmented, challenging to navigate and hard to access' and a 'philosophical and societal shift' is needed to address the limited support for those with early dementia. Such a shift would emphasise the need for meeting their human rights, offer post-diagnostic support, address the variability across agencies, expand the role of memory clinics and empower people with dementia to 'direct their own lives' (Low et al., 2021).

A revolutionised model for health and aged care is needed to address the National Health and Medical Research Committee's (NHMRC's) concerns. Increased attention to the knowledge and skills required by health professionals and others who care for people with dementia would begin by addressing the questions below.

4.2 Who Is This Person?

4.2.1 Who Am I and Where Am I?

A tentative answer to 'who is this person' emerges from the nursing home dance.

> *Familiar tunes prompted the dancing, at first hesitant and self-conscious. While only one or two residents were able to dance in a more or less conventional manner, others danced with their feet, their hands and most noticeable of all – their eyes. What memories were provoked by the sheer physicality of dancing? What thoughts rekindled by the intimacy of touch? Staff danced with residents in wheel chairs, joining hands with those unable to move freely.*
>
> *Amongst it all, 'Phoebe's' demented thoughts were evident through her loud, strong and unforgettable cry: 'Would someone please tell me who I am and where I am?' Some staff thought Phoebe shouldn't have been there at all: 'It will only distress her more'. 'She's upset by too much stimulation'. Others did not want her to miss out, some were embarrassed and some were amused. Who knows what Phoebe thought? A single woman of 94 years, this may have been her cry at many dances. What is my identity? Does anybody know I'm here? Who am I?*

There is no more poignant challenge for the whole field of dementia than the question posed by 'Phoebe' at the ball: 'Who am I? Will somebody please tell me *who* I am and *where* I am?' We will answer that question not by listing Phoebe's defining characteristics in a meticulously accurate problem-solving exercise, nor by rehearsing with her for the hundredth time her name and geographic location. Her quest is not for a missing article or a reminder of her address but for existential belonging. We may stumble and hesitate; we may not immediately provide her with a satisfying answer. We will, however, be on the right path when we recognise her question as ontological. What is my *being* in relation to all these people? Who am I in relation to you, how may I recover my lost self, what is my purpose in this last chapter of my life? Where do I *fit*? Do I *matter*? These questions go to the heart of insightful dementia care: a journey for staff, residents and families.

The answers are not necessarily found in a medical textbook or on a treatment list. We discover our personhood not as monads but in mutuality. We find community not as a collective of individuals but as persons in relation.

4.2.2 Loss of Self?

To describe a person with dementia according to their deficits is to jeopardise their humanity. Such was the case in one (fictitious) person's medical consultation about her mother. She says to the doctor: 'You keep telling me what has been lost, and I keep telling you something remains' (Ignatieff, 1993, p. 58). Ignatieff's account continues in the daughter's voice.

> I want to say that my mother's true self remains intact, there at the surface of her being, like a feather resting on the surface tension of a glass of water, in the way she listens, nods, rests her hand on her cheek, when we are together. But I stumble along and just stop. The doctor tries to help me out. 'This seems to matter to you'.
>
> 'Because', I say, 'a lot depends on whether people like you treat her as a human being or not' (Ignatieff, 1993, p. 58).

Similarly, 'a lot depends' on whether nursing home staff treat a person with dementia as 'not there' or 'nobody home' or a 'human being' or having 'no communication'. Other writers enlighten us with their descriptions of the self. Sabat and Harré (1992) maintain 'there is a self, a personal singularity, that remains intact despite the debilitating effects of the disorder'. In other words, the social and publicly presented person can be 'lost' but only *indirectly* as a result of the disease (p. 444). These writers suggest one of the main problems is with 'us' (i.e. those who do not have dementia) who may describe the person with dementia as 'helpless' or 'confused' or 'wandering aimlessly' or 'away with the birds'. We are reminded, rather, that the self of personal identity is not lost 'even in the face of quite severe deterioration in other cognitive and motor functions' (Sabat & Harré, 1992, p. 459).

If I can no longer remember what I had for breakfast, if my memory for names is diminished, if I do not recognise close family, am I less a person? Have I lost my 'self'?

> Indeed, if memory is so important, one has to ask where those memories came from in the first place. And the answer is unarguable: they came from experiences that the person has had in and through his or her body... Is it not curious that many people today... consider the 'person' to be lost when memories are lost, with no regard for the value of the organism that permitted those memories to be made in the first place? (Sapp, 1998, para. 27).

Family members and staff may be encouraged by Sapp's description, to avoid describing what has been lost, in favour of what remains. The simple question 'Tell me about yourself' or, to family/carers, 'Tell me a little about this person so I can get to know them better' invites dialogue, prompting increased understanding when the resident lacks coherent speech. Having regard to their *bodily presence* reinforces the fact that the *person* is still there.

Swinton (2012) says the 'hypercognitive cultural story' which focusses on loss, tragedy and devastation results in 'treachery, disempowerment, infantilization, intimidation, labelling, stigmatization, outpacing, invalidation, banishment, objectification, ignoring, imposition, withholding, accusation, disruption, mockery, and disparagement' (p. 82). To counter all of these negative connotations, he recommends concentrating on rich relationships within a positive narrative. 'The experience of the person with irreversible and progressive dementia is clearly tragic, but it need not be interpreted as half empty rather than half full' (Swinton, 2007, p. 37). From this strong belief, Swinton continues to emphasise the *person* rather than the *disease*, refusing to speak of people with dementia in a way which defines them by their diagnosis. In a unique interpretation of the human condition, he says:

> Thinking about dementia in terms of citizenship allows us to think about community, the nature of solidarity and the centrality of organised communal relationship. In this perspective *we are all people living with dementia* (italics in the original). The individual, her community, her borough, her town, her country are all enmeshed in a way that means that we all in different ways, are people who live with dementia (Swinton, 2020).

Rather than defining people with dementia by their alienating condition, Swinton encourages us to regard ourselves as fellow citizens. To this end, we may regard each resident's *social health* as important as their physical and mental health. (Citizenship is discussed more broadly in Chap. 3). Such a focus is epitomised by the following hypothetical desire to recognise the self within the community.

> *Spending my last days in this community, I am able to be myself; dancing a creative tango with the staff and management. I am now in a place where my voice counts, my idiosyncrasies are acknowledged and having a 'mind of my own' is not regarded as deviant behaviour. My variant moods also are no cause for blame or shame. Though dependent on others I am not left in isolation; my family and friends are welcomed as part of this lively community. Here I need not fear abandonment; I will be cared for no matter how distasteful I may appear. I no longer fear death or dying. Here I am known not only as I am, but for who I have been and who I may yet become* (Hudson, 2002, p. 9).

The sentiments expressed above pose an opportunity and a challenge: (a) an opportunity to embrace rather than 'abandon' each resident with dementia and (b) a challenge to avoid any 'blame' or 'shame' by any person in the nursing home community.

4.2.3 *Focus on the* Person, *Not the* Disease

'Person-centred care' is a familiar mantra, intended to connote the utmost respect for each unique individual requiring care. While it may seem obvious that *the person* is at the centre of dementia care, it is not uncommon to hear 'the person is no longer there'. If this is the case, the person will be *treated* as being no longer there, reducing the human being – the person – to an object. On the other hand, when the person with dementia is acknowledged, their response (albeit unspoken) means: 'I am a person. I am still here' (Fetherstonhaugh et al., 2013, p.143). Person-centred care is exemplified in the following account of a nurse's interaction with 'Henry'.

> *At first glance, it appeared highly unprofessional behaviour. The nurse sat herself on the seat of Henry's 'walker'. 'Let's go, Henry!' and, laughing together, they proceeded along the corridor to the bathroom. In other circumstances this would be regarded as 'inappropriate behaviour' on the part of the nurse. On this occasion, a description was included in Henry's person-centred care plan, clearly articulating the reason for this rather unusual method of transfer. Henry's resistance to being 'taken' to the bathroom was overcome, with his dignity intact.*

Dementia is being reframed from a biomedical model emphasising the neuropathological effects to a person-centred approach, depicted with such poignancy by the illustration above. A biomedical model focusses on the progressive nature of the pathology and its effects on disability and function: physical safety (including restricting independent movement) is prioritised and medication is administered to control 'behaviour'. The person-centred model honours the person's individual history, distinguishes their personality from others' and listens for their 'voice' even when speech is no longer intelligible. Against all these principles, many people with dementia are left in isolation when what they desperately crave is relationships, intimacy and a sense of community.

This sad reality, of loss of community, came to tragic expression during the COVID-19 pandemic, when, in some instances, residents with dementia remained for hours alone in their single rooms without any human contact and when someone came to their bedside or chair, they were confronted by strange, unfamiliar personal protection equipment (PPE). While there is no simple response to such a lethal pandemic, careful thought and planning would question an 'automatic decision' regarding hospitalisation for a person with advanced dementia, particularly in situations where families are prevented from visiting and advocating on the resident's behalf. The noise, bright lights and anonymous faces are said to compound the effect of hospitalisation on patients with dementia.

Examples abound that knowing the person's background helps in understanding their behaviour, a key to person-centred care. Three different scenarios constitute the proverbial 'tip of the iceberg'.

> *Staff at one nursing home had no idea a resident whose obsession with managing and organising, which they found very irritating, was linked to the fact she had a PhD (doctor of philosophy) in public policy and a distinguished career as a people manager.*
>
> *Another resident panicked when he heard the meal bell; staff had not realised he'd been a fire fighter and assumed this was his call to action.*
>
> *A noisy resident was labelled as cantankerous and loud-mouthed until it was explained she had been a prominent civil rights activist, often using her loud voice at protest rallies.*

These details, when well documented, and reinforced at handover, may transform a generic approach to one which honours each resident's uniqueness.

It is impossible to fully describe the challenge of confronting, understanding and interpreting carers' encounters with a person with dementia. How might we interpret the reaction to our well-intentioned care, seen through that person's eyes?

> Who are you coming so quickly towards me with that look of menace? I don't know you. You look as young as my granddaughter. *Don't touch me, and not down there!* Only my wife is allowed to do that. Please don't take my pants off; you will uncover my shame. If only I had the words to explain to you that I feel so *frightened, confused and embarrassed* (Hudson, 2008, p. 20).

The dependency on others which characterises advanced dementia need not be contrary to meaningful existence. However, some health professionals find helplessness threatening, leading to the remark 'Shoot me if I ever get like that! (Hudson, 2012, p. 61)'. What is it about a person dying of dementia in a nursing home that provokes such a plea from one nurse to another? Guns are not among nurses' tools of trade, yet this normally gentle carer asks for the most violent means of ending her life rather than having to endure such a state of helplessness. 'I couldn't bear to be like that', she says. Those who are weak and vulnerable remind us of who we are in our own frailty and mortality, a fact we would prefer to ignore, prompting us to regard resilience as a more laudable trait than vulnerability. 'She/he never complains' is interpreted as 'fitting in well' or 'not rocking the boat'.

To regard a person with dementia as someone who has feelings as well as a personal history, rather than someone who has or does not have certain abilities is to take seriously the whole person: body, mind and spirit. While it remains largely a 'mystery disease', attention to the *person* is a helpful corrective to a single-minded focus on the person's diseased brain.

4.3 Evidence-Based Care

4.3.1 *Leadership and Education in Dementia Care*

Poor leadership results in increased stigma, insufficient attention to detail, and lack of sensitivity to the needs of residents with dementia. Lack of regular education, inappropriate recruitment, shortcomings in care, all result in poor outcomes. On the other hand, continuous education, sensitive staff support and increased understanding of the disease result in transformed, personalised care. Most relatives have received little education in dementia care: proactive steps to address this anomaly have significant benefits. Innovative schemes for community education in primary and secondary schools, as well as senior citizens' associations would also help to reduce the community's ignorance and stigma, prompting open discussion rather than leaving dementia as a 'hidden subject'. Several films about dementia have played a significant role in raising awareness, notably 'Still Alice' and 'The Father'. In the latter, vivid depictions of patterned floor coverings, bedding and wallpaper are used to illustrate confusing design features. Creativity results in, for example, distinctive coloured door handles, cutlery, toilet seats and clothing; low-cost changes now shown to transform dementia care.

Dementia care is caught as well as taught. When residents are referred to by name and treated with respect, their distinctive humanity is reinforced, sending a message to others who may prefer to ignore them or treat them with disdain. When education is prioritised, when person-centred dementia care is expected from all staff, the result can be transformative. For example, when staff learn of the evidence-based link between hearing loss and dementia (Lin & Albert, 2014), they may be prompted to initiate a hearing referral and to optimise the use of hearing aids when prescribed. One nurse observed: 'Since our new manager introduced all this education, we have far fewer "behaviour challenges" in the dementia unit'. When staff are valued and given relevant information, residents' care improves. On the other hand, when care-workers are poorly paid, poorly educated, and largely unsupported, the message is clear: people with dementia don't need high quality care.

One area requiring educational focus (particularly in the early stages of the disease) is on the real or perceived discrimination experienced by people with dementia, particularly when their rights are ignored. This includes the right to make their own decisions, the right to express their opinions and the right to direct conversation with their carers, rather than the common assumption that 'they wouldn't know the difference'.

Fortunately, and to the satisfaction and reassurance of families, many nursing homes provide person-centred, well-informed resident care, supported by contemporary research-based education. Staff in such an environment appreciate where the priorities lie, and their care is acknowledged.

I love working here! The residents are treated with respect, the managers listen to our concerns and suggestions, the families are very appreciative, and policies and procedures are regularly updated, using the latest research.

Where skilled leaders advocate contemporary, best practice dementia care, the benefits are enjoyed by the whole caring community.

4.3.2 Qualified Carers

It is unfortunate that nursing home residents with dementia are cared for, largely, by people with no formal education about this complex disease, resulting in poor assessment and lack of clearly enunciated goals. One survey has shown that 90 per cent of GPs had received no dementia-specific training, and use of tools such as the mini mental state examination (MMSE) for diagnostic purposes was not necessarily routine. One family member described her frustration thus:

I wanted to know what was behind my husband's strange behaviour. The GP dismissed my concerns: 'Don't worry, dear, it's probably just old age forgetfulness. After all, he is 89!'

Responding to such an unprofessional comment, this gentleman's wife may well have asserted: 'Well, I'm 91 and my memory is intact!' When a dementia diagnosis, including the *type* of dementia, is not considered important, other significant gaps are revealed, such as discussion of legal matters, the health of family carers, and an end-of-life plan. Disclosure of such a diagnosis need not prompt a catastrophic emotional reaction, and for many people may provide some relief once an explanation for symptoms is known and a treatment plan developed. Readily available literature about all forms of dementia provides families with reassurance drawn from contemporary knowledge, well catered for in many dementia units.

'Dementia care must be core business for approved providers . . . supported with specialist advice and services where people have complex needs' (Commonwealth of Australia, 2021, p. 104). Staff caring for people with dementia require continuous education to support and develop their skills and competencies. Increased confidence and pride in their work, in turn, helps to reduce the stigma and discrimination so often attached to this disease, evident in many nursing homes practising exemplary dementia care. Education in best practice also has the potential to reduce the unfortunate overuse of restraint for 'behaviour problems'.

4.3.3 Restraint Usage in Dementia

One of the most controversial issues relating to dementia is the subject of chemical restraint. Unfortunately, in some circumstances, the practice is driven by opinion rather than scientific evidence. What the science shows is that while pharmacological responses to disturbed behaviour may be affective for a small percentage of people with dementia, for others there is a high degree of risk of serious adverse events such as pneumonia, heart attack, stroke, cerebrovascular events and increased risk of death (Commonwealth of Australia, 2019b, pp.6–7). These data (together with the following statistics and discussion) should be readily available to staff, residents and their families.

About 80 per cent of residents with dementia receive psychotropic medication (as a form of restraint) while only 10 per cent might benefit. Although government guidelines limit the use of these medications to 12 weeks and only for severe behaviour management when all other responses have failed, many residents are given these medications for 2 years or longer. 'Limited evidence to guide prescribing and deprescribing for people with dementia makes this process even more challenging' (Burke, 2020). For others: 'Evidence-based care is enhanced by a pharmacist-led medication review, resulting in fewer medications and more appropriate use of psychotropics' (McDerby et al., 2020). Families need information about the benefits of deprescribing and reducing burdensome treatments, particularly at the end of life, described by some as 'healthy dying' rather than being damaged by harmful side effects of unwarranted medications.

Evidence also shows that physical restraints are not effective in preventing serious injury as a consequence of aggressive behaviour. In many cases, the resident's agitation, discomfort and anxiety is increased. (Restraint is discussed more broadly in Chap. 8, including alternatives.)

4.3.4 Language and Dementia

Dementia knowledge and understanding is revealed by the language used. Bryden (2005) counsels: 'Please don't call us "dementing" – we are still people separate from our disease; we just have a disease of the brain. If I had cancer you would not refer to me as "cancerous" would you?' (p. 143). Other insulting sobriquets such as 'the dements' need to be 'called out' and staff educated about the implications of such terms.

Health and aged care professionals are in the unique position of leading the way, becoming role models for broader society in speaking about dementia. Guidelines are readily available, such as the 'Talk to me' good communication guide (Guideline Adaptation Committee, 2016, p.7). These guidelines show that residents exhibit fewer problem behaviours and are more likely to interact with others if familiar language is used. For example, those living in ethno-specific nursing homes, where choice of language is a high priority, require fewer psychiatric medications. It seems that when treated like children or people with low intelligence, residents respond with aggressive behaviour. When treated as adult human beings with equal rights and status, their 'behaviour' improves (Williams, 2017, p. 707).

Internationally renowned dementia authorities challenge the label 'aggressive' or 'behaviour problem' for a person who is resisting care. Such descriptions miss the key point of focus, trying to identify the *unmet needs* of the resident whose behavioural responses lead to such labels.

> In summary, we believe that there is strong evidence that *aggressive* should not be used to describe behaviour of residents with dementia during care . . . Resistiveness may escalate to combativeness, but it is our task to prevent this escalation instead of restraining residents by chemical or physical means (Volicer & Mahoney, 2002, p. 875).

Although Volicer and Mahoney's (2002) research is from two decades ago, shifting the emphasis to the *caregiver's behaviour* may provide a new perspective for staff and families on these aspects of language. Discussing what the resident *needs* can lead to innovative, non-drug related responses, rather than trying to curb the resident's *behaviour.*

4.3.5 Pain and Behaviour

Learning from research can prompt a well-founded response to certain 'behaviours' related to pain. 'A systematic approach to the management of pain significantly reduced agitation in residents of nursing homes with moderate to severe dementia' (Husebo et al., 2011). Appropriate pain management reduced unnecessary use of psychotropic drugs; concomitantly, when the pain management ceased, the agitated behaviour recurred. The consensus is that, when witnessing agitated behaviour, *carers should assume the resident is in pain, until proven otherwise.* Attitudes and ignorance about pain play a large role in triggering ill-informed responses. 'Oh, it's just her dementia', said one daughter, 'Mum wouldn't know whether she's in pain or not'. A professional response would be to focus on assessment, including the resident's often significant comorbidities, any one of which may be experienced as painful. When old age is blamed for pain, a similar lack of logic is called upon.

> *The doctor said to the 91-year-old resident who had complained of a sore knee: 'William, you must remember you're bound to get some pain at your age, and painful knees are quite common. 'But' said William, 'my left knee is the same age and it's all right!'*

Another story linking pain with behaviour focusses on 'Olive' who had a history of osteoarthritis as well as Alzheimer's disease.

> *Olive was renowned for being 'difficult in the mornings' particularly when staff were trying to put her shoes and socks on. A new carer was assigned to assist Olive, and she took careful note of Olive's reactions. 'I think it's her painful arthritis, she screams when I bend her knees'. A regular low dose of analgesia was prescribed, with the first dose commencing before breakfast. The timing of the analgesia in relation to assistance with Olive's hygiene and dressing resulted in a dramatic change. She no longer resisted when the analgesic effects coincided with her personal care.*

Olive's story shows the importance of relying on staff insight and knowledge, rather than depending on the resident's descriptions of her symptoms. This nurse *assessed,* based on her understanding of the disease, that Olive's arthritis would be painful in the mornings, rather than *assuming* her screaming was 'behavioural'. Verbal response such as 'ouch' constitutes pain reporting for Olive, whereas another resident with full cognition would press the call bell, describing her pain as 'nine out of ten'. For Olive, relief and cooperation ensued from analgesic rather than antipsychotic medications. Nothing is lost and much is gained from a trial period of mild analgesia, *administered regularly and documented carefully,* as a means of assessing the link between pain and behaviour.

Education to reinforce this link between pain and behaviour may change carers' attitudes, as in the following.

> *Staff were advised to attend an education session on dementia where the topic of 'pain and behaviour' was addressed. Attendees were asked to describe their own behaviour, giving personal examples, when enduring pain: either mild or severe.*
>
> *To the astonishment of some, they realised their own reactions were not dissimilar to those they criticised in the residents with dementia. They were also reminded that everyone has 'behaviours' and it need not be a derogatory term.*

However, circumstances where the behaviour is cause for grave concern need to be addressed accordingly.

4.3.6 Violent Behaviour

A minority of residents with dementia exhibit aggressive or even violent behaviour. This may or may not be attributed to the type of dementia, for example, DLB (described above), in every instance. Without expert diagnostic skills and insight into the person's comorbidities, medications and other factors, staff may be left to manage the consequences, unaided. One staff member spoke of 'an unsupported, exhausted workforce, unable to provide decent care with the scant resources they are given'. When this nurse complained about 'being headbutted, punched, kicked and threatened, her superior shrugged and said, 'That's dementia' (McKinnon, 2019, p.17). This unprofessional, stigmatising, maligning, demeaning response reinforces the need for urgent, compulsory, evidence-based education for all nursing home managers and staff, with capacity to seek a second professional opinion for the resident when needed. In many instances, thoughtful, creative responses are preferable to medications; for example, the power and pleasure of music has proven benefits for moderating aggressive behaviour in many instances. Other distractions (sometimes based on 'trial and error') may also be effective.

4.3.7 Music for Pleasure

Music (also discussed in the preceding chapter) is sometimes regarded merely as an 'optional extra' and not intrinsic to a resident's daily care. However, in some circumstances, it can be literally life giving, as it was for 'Frank'.

> As a former professor of music, Frank's favourite instrument was the harp. He now lay in his nursing home bed with end stage dementia: no speech, little independent movement, hands always tightly clenched, fingers separated by carers with great difficulty. His daughter had an idea! She brought him his beloved small, hand-held harp and placed it on his chest, strumming the strings gently. Slowly, but very purposefully, Frank uncurled his fingers and caressed the harp. Would this action have been an appropriate inclusion in his daily care plan? 'Spend a few minutes several times per day, encouraging Frank to play his harp'. Unfortunately, this expectation was beyond the realm of nursing home routine. Fortunately, Frank's daughter had access to his funds, employing a private music therapist whose one role was to encourage Frank to remain connected to his harp 'until he goes to heaven and hears the angels playing'.

This story (paraphrased from Cox and Roberts (2013) demonstrates the theory that '... much processing of music takes place in the prefrontal cortex of the brain, which is among one of the last regions of the brain to atrophy in people with Alzheimer's' (Garrido, 2016, p. 46). Familiar music can help evoke memories and associated emotions, long after speech has gone, an ideal tool for alleviating associated distressing symptoms and bringing the person 'back to themselves' (Baird & Thompson, 2018, p. 827). Caution is needed, however, against a generalist claim that music always has a positive effect. It may depend on the particular type of dementia and the area of the brain affected as well as the *type of music* preferred by the resident. Comprehensive assessment by an appropriately qualified music therapist mitigates misguided assumptions.

Families may need encouragement to identify a favourite song or piece of music to include in the resident's daily routine; complemented by family or friends or staff singing along, overcoming self-consciousness while appreciating the benefits.

Nursing homes generally cannot afford to employ a full-time music therapist; however, their role cannot be overemphasised. Rather than resorting to psychotropic drugs for 'behaviour problems' in many instances, music of the resident's choice (with use of headphones where appropriate) provides the calming ingredient. Music in this context may be regarded as therapy, to be included in the care plan, rather than mere entertainment.

A music therapist provides this account of her engagement with an immobile, seemingly voiceless resident.

> One day, I felt moved to sing to him Brahms Lullaby in German, not his native language, but he was likely to know the lullaby and something of the language. It is probably a sentimental favourite of many Europeans. As I sang, he began to cry – I just held him and gently rocked with him. When we returned to the ward his wife was there and the resident said something to her in his native language. It was only later that I learned he had told his wife 'I've been dancing with tears in my eyes' (Hill, 2001, p. 50).

Music therapy for this resident prompted renewed insights from staff prone to describe his cognitive level as 'zero' or 'there's nothing there' or the ubiquitous 'he can't communicate'. A sensitive RN suggested to the GP that music might be a healthier alternative for this resident (and others) than psychotropic medication. Coupled with the discussion on 'touch' in the previous chapter, the story told by Hill (above) reinforces the power of being 'touched by music': a moving example of sensory stimulation.

Careful observation and measurement of a resident's *response* to music is a guide to whether or not it is appropriate. Here again, allied health professionals play an important role in dementia care.

4.3.8 Activities, Laughter and Play

Activities are discussed in Chap. 3, noting that staff are often very creative and skilled in this aspect of care. The topic of dementia, however, does not necessarily lend itself to humour and playfulness. Killick describes the many and various ways of incorporating hilarity, fun and amusement into the nursing home (Killick, 2013). He describes his dressing up, clowning, miming and improvising ways to encourage laughter among people with dementia, noting the way humour lifts their spirits as well as his own. Lifting the spirits of residents with dementia and the staff who care for them is a worthy pursuit, adding a much-needed dimension to care. Moreover, these activities may counteract, or prove an effective substitute for, psychotropic drugs which *depress the spirit*.

Assistive technologies also play a part in creative dementia care. For example, an electronic game has been described, recommended for use with several residents sitting at a table, using only their hands to capture different shapes, colours and light forms. Such equipment is costly but with management's support, fundraising events can sometimes provide the necessary path to purchasing.

4.3.9 Segregation of People Living with Dementia

Research into living arrangements of people with dementia appears to be fairly recent, introducing reforms that prompt new thinking about the meaning of living together and relating with one another (Steele et al., 2019). Taking a human rights law approach, these authors (including one who is living with dementia himself) set out a comprehensive case against segregating people with dementia from others, particularly within nursing homes. Highlighting the negative impacts of locked doors and gates, and the isolation of people with dementia from others in the community, the research compares the transformation within the disability sector in moving away from institutionalised care, advocating the same rights for people with dementia. The research refers to 'stigmatization, a sense of imprisonment, social exclusion and depression' and other forms of restraint which not only prevent those with dementia from mixing with others, but 'depletes the diversity and inclusivity of our communities more broadly' (Steele et al., 2019, pp. 2–3). Removal of barriers is called for, creating communities that welcome people living with dementia, questioning the moral, ethical, human rights and political foundations for segregating them (Steele et al., 2020).

Segregation is an important issue for Aboriginal and Torres Strait Islander people living with dementia, who may suffer greatly when separated from their community. While dementia is 'fivefold more prevalent than in non-Aboriginal Australians', the quality of care is poor (Smith et al., 2011). Notwithstanding the findings of Smith and colleagues, Dementia Australia provides dedicated services in rural/remote areas of Australia, staffed by skilled workers.

For inexperienced staff, advice is available from relevant healthcare specialists and Aboriginal elders. Further evidence is bound to uncover other inequities, prompting a call for urgent reforms in dementia care for specific population groups.

4.3.10 The Dignity of Risk

When safety considerations restrict a person's freedom to enjoy activities, their quality of life can be diminished. Are residents deprived of certain pursuits because 'they may hurt themselves'? Does this mean confinement or denial of the freedom they would enjoy if living elsewhere? While residents' safety is important, it does not necessarily imply removal of all risks. A pleasurable activity may be considered by some staff as 'too risky'; however, where there is a clearly documented plan, following discussion with resident and family where relevant, a scenario such as the following may ensue.

> 'Georgina' was admitted to the nursing home while still physically agile and with good manual dexterity but with rapidly deteriorating cognition. She had formerly been the chef in a boarding house and showed signs of extreme boredom when confined to a chair in the nursing home lounge. After careful discussion with family and the home's catering staff, a

4.3 Evidence-Based Care

corner of the kitchen was designated 'Georgina's space'. A small collection of utensils and ingredients provided sufficient attraction to keep Georgina occupied for several hours each day. Discussion and formulation of a care plan focused attention on risk management. 'What if she cuts herself?' 'What if she puts the wrong things in her mouth?' After a 'trial period' of one week, it became evident that Georgina's pleasure outweighed any of these risks, and catering staff very quickly adjusted to her presence. One unintended consequence was evident: kitchen staff became more alert to the complexities and the possibilities of creative dementia care.

An important consideration is how to balance safety with freedom of movement and protection from harm, or to weigh the 'dignity of risk' against unbending rules. What is the aim of having residents in 'secure' units? Are there legitimate reasons for placing the kitchen 'out of bounds?' What are the comparative benefits of removing some of the barriers? What is appropriate for each individual and how is it described in their care plan? Is the nursing home a place of incarceration or a long-term pleasant place of residence? These questions need to be asked of managers who may prioritise risk management over creative, evidence-based care. A disproportionate emphasis on risk is challenged by dementia researchers who prefer a human rights approach, noting the dangers inherent in institutionalised segregation (a style long since ceased for orphans and people with disabilities) in favour of accommodation which enhances emotional safety and well-being (Steele et al., 2019, p.2).

> Much the way that changes in work health safety have done for employees facing risk, Boards and Clinical Governance subcommittees are grappling to understand if they have a sufficient handle on the risks that have the greatest impact and harm to those receiving care (Ibrahim, 2020 p. 6).

Ibrahim suggests that governing bodies give due consideration to 'mitigation and minimisation', measuring their strategies against harm while also considering the importance of dignity in making decisions relating to risk. 'They must satisfy themselves that there is clarity around decision making and consent, that consumers have the opportunity to take risks, and that consumers are engaged in risk decisions that apply to them' (Ibrahim, 2020 p. 6).

4.3.11 *The* Person *Who Is 'At Risk'*

When one resident was found 'wandering' beyond the nursing home gate she was marked as a 'flight risk' and 'potential escapee', her personhood seriously compromised by these disparaging terms. By contrast, her caring family knew of their mother's inherent dignity, employing creative measures to restore her perceived 'loss'.

> Every six weeks or so, the family went out to a restaurant. At first, she took care over her appearance and put on her good clothes. Then dressing became confusing, and she mixed her clothes and wore two skirts at once. When going out, her son became her dresser, restoring the dignity she could not put on for herself (Dunn, 2021).

As the disease progressed, including loss of many of her functions, her life story was not eroded, described in the following account.

> Her conversation splintered into fragments – some light and meaningful, others random or irrelevant. She soon stopped speaking altogether, a daunting silence that turned away her visitors.. (T)he son read aloud the poems his mother used to read him, and her husband read her stories which he always did so well. The reading broke the silence with a laugh or smile of recognition. She seemed lost to the world at large, but stayed present to her family and her carers. They gave her what she needed – the rhythms of her language, stories in which to dwell and a regularity of life (Dunn, 2021).

Rather than *labelling* a person with dementia who is assumed to be losing all faculties, or prioritising risk over creative options, nursing home staff would do well to ask 'how may we restore this person's dignity'? Such an attitude goes to the heart of the nursing home's culture.

4.3.12 Nursing Home Culture

A culture based on knowledge drawn from evidence has the potential to translate into more effective dementia care. Culture in this context is described as 'the key determinant of an organisation's performance and ability to meet its objectives' (Commonwealth of Australia, 2021, p. 133). Effective dementia care is dependent on education for those prescribing the care.

What is the GP's knowledge level? In some undergraduate medical courses, there is a minimum of 1 hour's dementia education; in others, none at all (Ibrahim et al., 2016). How do the RNs relate to the GPs? It should be assumed that the former (particularly if they have benefited from advanced education) may have superior knowledge that can be conveyed to the GP in a spirit of advocacy for the resident, rather than antagonism towards the doctor. For example, a nurse who has access to contemporary, well validated research may propose a certain drug for the resident who is in pain. A strong request for a *trial period* of the drug may reassure families, as well as GPs concerned about over-medicating. Such recommendations will carry more weight if supported by the *whole team,* rather than relying on idiosyncratic opinions, as in the following:

> *GP: I heard from the RN this morning that you needed an antipsychotic urgently to calm down the resident's behaviour.*
> *RN: My strong belief is that her behaviour is directly related to her pain, so can we have a trial period of analgesia administered on a regular basis?*
> *GP: Well, whose request do I satisfy? Yours, or this morning's nurse?*

This scenario is characteristic of a culture based on unilateral, often ill-founded or uninformed, opinions, rather than a considered approach based on professional assessment subsequent to a team discussion. (This need not require a lengthy meeting but could arise from a 5-minute discussion at handover). It also assumes that key staff have advanced clinical skills in dementia, reflecting a culture of continuous, evidence-based quality care.

4.3 Evidence-Based Care

It is becoming more widely recognised and practiced in some dementia care environments that staff require not only a physically safe workplace but also an atmosphere where their own spirit is nourished, and where they are recognised for who they are as persons. This, in turn, has a profound effect on those in their care.

> Dementia nursing flourishes in an atmosphere of collegiality and good humour in which relationships of trust and professional fulfilment are fostered by management and permeate the entire care environment. When dementia nurses care for one another, their care for people with dementia is enhanced (Hudson, 2003, p. xxxi).

Management's role in creating such a culture cannot be overemphasised, profoundly influencing the care of staff, residents and families.

4.3.13 Nursing Home Design

Creative, insightful, evidence-based planning is evident in many nursing homes caring for people with dementia, whether in 'dementia-specific' areas or in their general design. Given the increasing proportion of residents with dementia, it seems indisputable that every nursing home should be designed with a view to its benefit for those with cognitive impairment. Such design would not focus primarily on the *disease* of dementia and its manifestations, rather on the *persons* with dementia and their holistic needs. Priority would then be given to maximising the *abilities* of the residents, rather than being influenced by their *disabilities*.

Planning would optimise safety while accepting the inevitability of some risks, creating space for freedom of movement, optimising stimulation and engagement while minimising potentially harmful obstacles. With this focus, designers would acknowledge the importance of lighting without shadows (the latter forming frightening vision for some residents); plain rather than confusing, patterned floor covering (the latter preventing some residents from safe walking); bathrooms incorporating a brightly coloured toilet seat to distinguish it from other white areas. When considering kitchen and dining design, including utensils, a dementia consultant may advise on coloured handles for cups, for example, together with other relevant 'tips'. These are a few of the many recommendations made by those who have researched the important area of 'dementia friendly' nursing home design. Any perceived cost factors may be readily outweighed by the benefits of enhanced safety as well as residents' pleasure.

4.3.14 Psychiatric Drug Prescribing

Advocates for caution remind health professionals that antipsychotic drugs have never been approved for general use in dementia, due to their potentially harmful effects, including death (Walsh et al., 2018). Only in situations where the person has been diagnosed with schizophrenia, major depression with psychosis, or bipolar disorder with mania is the use of these drugs warranted. The answer for nursing home residents with dementia is not medical therapy but environmental considerations, where staff look around and examine what may have triggered the 'behaviour', including pain. Geriatricians suggest that the physical and mental decline of an older person is often due to an underlying medical condition which has not been diagnosed but could be reversed if managed well. It seems geriatricians' advice needs to be sought more often.

Misguided prescribing includes benzodiazepines, hypnotics and anticholinergics, ignoring the risk of adverse events in this population (Hukins et al., 2019). These authors discuss the use of tools to identify inappropriate prescribing. Other researchers have found that the goals and preferences of the person with dementia are too often overlooked when prescribing certain drugs. Correct diagnosis, together with informed consent to any treatment, are also important factors to consider. Before resorting to 'behaviour modifying' drugs, comprehensive pain assessment is recommended, limiting antipsychotics to the treatment of agitation or psychosis when symptoms are severe, dangerous and/or causing significant distress. When/if such drugs are used, a comprehensive review is required after 4 weeks and, if there is no clinically significant response, the drug should be withdrawn.

In some nursing homes, comprehensive medication review is seldom, if ever, pursued. Others have reduced the use of psychotropic medications by collaboration with GPs, pharmacists, NPs, RNs and families, obtaining external funding for research and review, producing results which actually saved money, time and resources to the immense satisfaction of all concerned. Current prescribing patterns of antipsychotics would benefit from comprehensive review by practitioners with relevant knowledge of their side effects and associated contraindications. The benefits of informed, targeted reforms introduced in many nursing homes serve as an example to others.

4.3.15 Keeping Up with the News

It ought not to be assumed that a resident with dementia has no interest in the outside world, or in continuing their familiar routines, especially on moving into the nursing home. 'Mr H's' story is a case in point.

> *'Mr. H' was new to the nursing home and his wife arranged for his preferred newspaper to be delivered each morning. Even with his advancing dementia, he seemed to recognise the paper. But there were two different papers delivered and staff took little notice of Mr. H's preferences. Either they forgot to give him his paper (for which he had paid) or they delivered the wrong one. Or, on more than one occasion, a staff member took the paper to read during her morning break. 'He can't even read it!' was the excuse. 'He wouldn't know which way was up, let alone which paper he had'. These comments ignored Mr. H's past: a cultured gentleman with a keen eye for politics and world affairs, he had also contributed letters to the editor. Now that he seemed no longer able to read, his interest still needed to be acknowledged. Some staff appreciated this, ensuring he received his preferred paper, others either forgot or ignored the details. This seemingly insignificant issue was a total affront to 'Mrs. H', who perceived it as staff's failure to understand her husband and his*

needs. It would have meant 'all the world' to her, to know that he was receiving his preferred paper each morning, just as he did every day of their 62 years together.

It would seem a simple but important matter to include this daily ritual in Mr. H's care plan and ensure it was carried out until such time it was reviewed and discontinued, after family discussion.

4.3.16 Aversion to Water?

Many a 'tug of war' may be witnessed when a resident with dementia is directed to their morning shower. One astute carer observed: 'Water is invisible and disconcerting to the typical AD patient. They don't like to drink it, and they don't like to get in it' (DeMarco, 2016). Another observer notes that bathing is one of the top three problems associated with caregiving in this population, offering some simple cues to enlist a resident's cooperation.

Understanding that a person with dementia may perceive water differently, becoming afraid of water on their head or being sprayed on them from behind, leads to other simple options. A hand-held shower hose directed from the front, where it is readily seen, may not invoke fear. However, for some, a full body sponge is preferable to having 'a fight with the shower'.

Careful attention to this component of 'basic hygiene care' would include observation of each resident's response to the shower, followed by a detailed, individualised plan, in order to avoid this scenario.

> The care attendant was new to the unit and did not know the residents. 'You will need to shower the residents from rooms 3, 4, 5 and 6' the nurse instructed her. Having received no education about how best to shower a resident with dementia, she proceeded to use her own routine. She was dismayed by the loud screaming and physical resistance of the first resident. Unsure how to respond, she quickly formed the view, reinforced later by some colleagues, that this was 'normal behaviour for residents with dementia'.

This unprofessional idea permeating some nursing homes ignores the evidence that screaming signifies suffering. 'Screaming is related to vulnerability, suffering, and loss of meaning experienced by older persons (Bourbonnais & Ducharme, 2010, p. 1172). In other words, screaming is a self-soothing reaction to the constant lived-experience of distress, and should therefore, never be ignored.

A brief word of advice from a senior colleague to the newly appointed care attendant, and a prompt to read each resident's detailed care plan first, would have prepared her, increased her confidence, and reinforced her understanding of individualised dementia care.

4.3.17 Communal Lounge

Pointon compares her experience of caring for her husband at home with her observations in the nursing home: 'the damaged world of control.. of rigid institutional routines, of risk avoidance, of unhelpful environments, of eagerness to prescribe or run for the sedatives' all of which she describes as 'well-intentioned but totally inappropriate, even malign care' (Stokes, 2008, p.5).

> On numerous occasions I have asked carers and nurses: if you lived in a care home where would you like to spend your days? In your room among what is yours – possessions that resonate with memories, photographs that tap into a well of emotion, your music gently playing in the background – or in the unfamiliarity of a lounge in the midst of strangers who act in ways you rarely if ever comprehend? (Stokes, 2008, p.230).

In light of such examples (among myriad others), it would seem more appropriate to discuss this issue with each resident, and/or their family where relevant, before 'placing' them in the communal lounge, regardless of their preferences.

> If a person wants to stay in their room it should not automatically be seen as an unhealthy desire indicative of isolation, withdrawal and depression – instead it can be an understandable and appropriate wish for familiarity, continuity and privacy. We are not talking about bare bedrooms but rooms enriched with entertainment and the nostalgic trivia of people's lives (Stokes, 2008, p.232).

Stokes coined the phrase 'room centred care' as a place enriched by and resonating with the resident's personal preferences (Stokes, 2008, p.235). Infection control implications of the COVID-19 pandemic provided an opportunity for creative options to the bedroom versus lounge dilemma, such as creating a 'mini lounge' area where a small group of compatible residents could, at a safe distance, meet together.

As for many other areas of care, discussion with relatives, volunteers and staff may elicit surprising results, reforming long-held, unquestioned practices into innovative practice renewal.

4.3.18   Self-Care?

Rather than assuming a resident with dementia is incapable of self-care, a thorough assessment may reveal unexpected responses.

> 'Gladys', a new resident, was escorted to the shower by a carer who proceeded to provide all assistance, presuming the resident incapable of self-care. Gladys became increasingly distressed, refusing to be washed. With no words available to her, she screamed and cried, becoming increasingly agitated. The perceptive carer gave Gladys the washer and soap, inviting her to wash herself. The screaming stopped and, albeit painstakingly slowly, Gladys proceeded to manage her own hygiene.

'Self-care' is often inadequately assessed; assumptions were made that because 'Gladys' had no speech she was incapable of washing herself. The implications of this story are that if carers regard the person with dementia as a 'lost self' or label them with other insulting descriptors, the person will respond accordingly. In other circumstances, prompting self-care, even when conversation is limited, can have a seemingly miraculous effect on the person's independence and confidence. 'Look! Maria is cleaning her own teeth!'

4.3.19 Depression and Delirium

The statistics are cause for concern: '50 per cent of older Australians in residential care suffer from depression and anxiety, compared with 10% of older people generally' (Commonwealth of Australia, 2019a). Differential diagnosis of depression, delirium and dementia is vitally important, noting also they may coexist.

> Depression and delirium, particularly hypoactive delirium, may present with apathy, withdrawal and tearfulness. Delirium occurs suddenly (over hours or days) and the symptoms tend to fluctuate throughout the day; depression describes a negative change in mood that has persisted for at least two weeks; the onset of dementia is generally slow and insidious (Victoria State Government, 2020).

This brief summary is a guide only, requiring expansion in education programmes, noting that depression is referred to in several other contexts throughout this book. Further to the definitions quoted above, the Delirium Clinical Care Standard, 2021 (Australian Commission on Quality and Safety in Health Care, 2021), defines delirium as 'an acute change in mental status that is often triggered by acute illness, surgery, injuries or adverse effects of medicines'. This standard includes eight quality statements and a set of indicators, noting also that 'rates in residential aged care services exceed those in the general community'.

It is also evident that staff knowledge of delirium is generally poor, in spite of its high incidence (Buettel et al., 2017). Lacking such knowledge, a resident's symptoms of delirium and/or depression may be wrongly attributed to 'their dementia' or mere 'confusion'. Understanding that the incidence of delirium increases with dementia, staff are prompted to seek a diagnosis so it can be treated. The serious, complex nature of delirium is shown by the fact that it is 'the most common reason for older patient specialling in acute hospital wards' (Buettel et al., 2017, p. 22; Cook et al., 2020). Where the distinction between these 'three dreaded d's' (depression, delirium, dementia) is unclear, advice should be sought from a geriatrician.

4.3.20 Care at Night

Practical guidance is offered to those who provide night-time care for people with dementia (Kerr & Cunningham, 2016), noting that some residents may not be able to distinguish day from night. The challenge of night-time care is of course related to staffing issues, with far fewer personnel available than during the day and qualified staff not mandated. However, it is at night when the person with dementia is alone in their room with no distractions, that they may experience increased pain without words to describe it, and/or no capacity to use the call bell. Anxiety and confusion may give rise to 'challenging behaviours', calling for a creative response, such as this simple yet poignant example.

> 'Nurse A' enjoys great satisfaction from her three nights' work each week in the nursing home. The day staff also enjoy hearing of residents with dementia enjoying a settled night.

> *What is Nurse A's secret? First, she ensures there is an appropriate variety and quantity of food available. 'Josephine never knows what time it is and will often respond positively when given her 'dinner', even in the middle of the night. 'Billy loves to come and sit with me at the desk' says 'Nurse B', 'he seems to get very lonely at night and doesn't understand why his wife can't be with him. And, he loves a snack'.*

Night staff will, of course, have many other examples of creative care. These simple acts of imaginative, kindly, compassionate care can transform the night lives of people with dementia. Some families are more than willing to participate in discussions and planning, offering tips and advice based on their own experiences of caring for the person at home, bringing in favourite snacks and, in some circumstances, visiting throughout the night. The following example is antithetical to enlightened night-time care:

> *'Nurse B' prefers all residents in her care to be in bed and asleep throughout the night. Reporting at handover one morning she complained: 'Charles was up half the night, wandering around, so at 2am I gave him a risperidone'.*

These examples illustrate the need for comprehensive 24-hour planning, including appropriate medication use, so that residents receive consistent responses, based on their idiosyncratic needs. Creative, sympathetic food management is also called for, so at least a small range of food and drinks is available throughout the night.

4.3.21 End-of-Life Care

Most nursing home residents with advanced dementia will die there, calling for thorough planning initiated early in the disease process, reinforced by discussion soon after admission. Families may be ignorant about the life-limiting nature of dementia, having been told 'you don't die from dementia'. All residents and families would benefit from information about death and dying, and palliative care, before the resident's capacity for such discussions is diminished. Common perceptions of dementia as 'life not worth living' are related to negative aspects associated with loss and the portrayal of suffering as undignified.

> I argue that the seeming contradiction, in which people with dementia are grieved for even though the end of life with dementia is considered a life not worth living, can be understood in light of the acceptance of death as a form of care (Dekker, 2018, p. 322).

Death, from Dekker's perspective, is not perceived as a form of abandonment; rather, it provides an opportunity for attentive, affectionate care and loving accompaniment to the end. As death from dementia seldom occurs suddenly, opportunities should be provided for frank discussion and timely, thoughtful planning with families. As noted in the chapter on palliative care, residents with dementia ought not to be denied specialised care when needed, introduced well before their last hours or days of life. Concomitantly, they ought not to be burdened by unnecessary treatments such as (routine) antibiotics or enteral feeding, or other interventions which would not contribute to their comfort or match their goals of care.

Death certificates do not routinely reflect a dementia diagnosis as a terminal illness nor do care plans always reflect the individualised end-of-life care required. Many residents are therefore at great risk of suffering unmitigated pain, without the offer of opioids or other relief (Read & MacBride-Stewart, 2018), especially when close to death. At the heart of this serious lack of care is the emphasis on *'living well'* which, although well-intended, often omits any discussion of *'dying well'*. Such short sightedness results in a lack of end-of-life planning, ignoring the benefits of palliation and failing to recognise complexities of symptoms other than cognitive decline (Browne et al., 2021).

Lack of advance care planning, and ignorance about palliative care, results in residents missing out on options regarding spiritual care, place of care, and other matters of concern as they approach death. It is ironic that in the context of dementia, while there is ample time for such end-of-life planning, there is often a serious *lack of planning*. Decisions are left to those who may have little or no awareness of the wishes of the person with dementia, resulting in inappropriate hospitalisation at the end of life, or invasive treatments being prescribed. Increased education and skills training may lead to greater staff competence and confidence in this important area of care.

We also learn from others, such as this poignant account by mystery writer Louise Penny whose husband died from dementia. When the diagnosis was made, she says:

> 'I looked at Michel. His dear face, his clear blue eyes. This man who, as a pediatric hematologist, had held the hands of dreadfully ill and dying children, nodded. I took his hand and held it as the doctor went through the options. It didn't take long. There were none. . . He was a scientist. A full professor of medicine, a doctor with an international reputation. And now, he couldn't divide 2 into 20. And then he couldn't copy a triangle. Or write a full sentence. Or tie his shoes' (Penny, 2016).

This superbly written reflection highlights some of the anguish experienced by close family or spouse, watching the inevitable decline as dementia takes its toll. It also serves as a reminder to staff, that understanding the dying person's unique background may lead to increased empathy for families.

Contemporary research highlights the paucity of palliative care for people with dementia impacted by the COVID-19 pandemic, resulting in unnecessary hospitalisation and an over-reliance on invasive medical procedures (Kaasalainen et al., 2021). Other sequelae included lack of planning for end-of-life care, visiting restrictions, symptoms of delirium and pain under-recognised, and disenfranchised grief due to lack of opportunities to say good-bye or to be present at the death.

While the many references to the COVID-19 pandemic throughout the book may soon, in one respect, become 'dated'; they are included for their educational potential. Many lessons can be learned about unexpected infections, importance of care planning, and, as emphasised in many places, avoiding unnecessary hospitalisation. These issues are particularly apt for residents with dementia.

4.3.22 Bereavement Planning

Evidence shows that few bereaved caregivers of people with dementia have been prepared for the person's death (Hovland-Scafe & Kramer, 2017). Those who were well informed found it assisted them in 'accepting the reality': acknowledging death as the final outcome, they were not surprised when it happened. Acceptance of impending death helped with future planning, referred to as 'getting your house in order'. Knowing of death's imminence or at least inevitability also encouraged carers to be prepared in other ways, such as saying the things that needed to be said, and clarifying the person's wishes so they were documented in advance. Where goals of care are discussed in an interdisciplinary meeting with family carers present, the outcome can be effective and satisfying for all concerned. This is particularly important for issues such as resuscitation, tube feeding, treatment of infections and hospital transfers (Hanson et al., 2017). In the absence of a plan, the bereavement outcome may be seriously compromised.

Bereavement may affect other residents with dementia, who should not be excluded from rituals, funerals and similar practices that mark each death. False assumptions about residents' understanding of death should be replaced by opportunities to acknowledge the event, such as forming a guard of honour when the body is removed, drawing their attention to a photo and/or notice about the deceased resident, and encouraging family involvement.

4.3.23 Limitation of Care Orders (LCOs)

Some researchers (Ibrahim et al., 2016) advocate the wider use of limitation of care orders (LCOs) for people with dementia, preferably linked to their formal ACDs. Poor uptake of LCOs is associated with higher rates of aggressive, life-prolonging treatment, generally not in the best interests of the person with dementia. Ibrahim and colleagues acknowledge the fact that dementia is under-recognised as a terminal condition by clinicians, residents and families. Due acknowledgement is also needed for the multiple comorbidities affecting the mortality risk of people with dementia. Lack of understanding about dementia diagnosis and prognosis often results in poor planning for end-of-life care, giving rise to considerable regret after inappropriate, if not harmful, interventions, such as hospitalisation and resuscitation.

'Most persons with dementia die of acute illness and many are hospitalised at the end of life' (Ibrahim et al., 2016). Families and carers are not necessarily aware that the mortality rate of interventions associated with acute care is much higher for people with dementia, resulting in even greater suffering. When this is carefully and sensitively explained, action may be taken to incorporate the resident's wishes into an ACP, allowing the resident to die in the nursing home surrounded by those who know them well.

4.3.24 Social Worker Role

Social work research identifies the need for greater input from their discipline, highlighting the importance of their place in the multidisciplinary team. Social workers have identified four significant barriers to end-of-life care in dementia: hindrances to information, barriers to hospice, ineffective attempts to comfort and the nature of death from this disease (Hovland, 2019). Barriers can be lifted by increased attention to religious/spiritual beliefs, allowing caregivers to take initiatives, discussion about 'bearing witness to decline', what to expect from impending death, recognition and acknowledgement of family caregiving, and collaboration with a multidisciplinary team. Social workers have a significant role in nursing homes, helping to incorporate palliative care and to 'address spiritual and social aspects of care, thus contributing towards improving the quality of life of people with dementia and their families' (Van der Steen et al., 2014, p.197).

Dementia and palliative care may be an unfamiliar association for families, not to mention some staff. 'I thought palliative care was only for the last few days or hours' is a commonly expressed belief. When the dementia trajectory and its definition as an incurable disease are not well understood, the care is likely to be deficient. Given the goals of a palliative approach include maintenance of quality of life, comfort and dignity, it is regrettable that this option is not widely advocated. Misunderstandings and ignorance can be ameliorated by careful, well-planned family meetings, with input and guidance from a palliative care social worker.

4.4 Families Flourishing or Floundering

4.4.1 Information Deficit

Some family carers are unaware their relative has dementia: consequently, they lack advice on communication strategies such as shared decision-making and exchange of information (Penders et al., 2015). Lacking such knowledge, families' preparation for the resident's death is also compromised.

Others are misinformed or ill-informed about the possibilities of appropriate, holistic care, as in this account.

> An occupational therapist tells the story of a woman who had moved her mother into a dementia care facility – and was told there was little hope for improvement or quality of life. Her mother hadn't spoken a word in several years, she couldn't walk, and was described by her daughter as being a 'shell of her former self'. During the preadmission interview the therapist identified glimmers of the mother's capabilities that were overshadowed by negative comments such as: 'We were told your mother can't do anything at all for herself'. A little more probing showed that her mother could hold a magazine and turn the pages. Following transfer to a more forward-thinking nursing home the older woman was able to resume some of her beloved activities. Soon she was talking and walking. What had appeared like a failure to thrive now became a new lease of life.

This story reinforces the extraordinary responsibility of families choosing the 'right' care environment for their relative with dementia. It also emphasises the need for a strengths-based approach to assessment, emphasising what the residents *can* do rather than what they *cannot.*

A key to greater understanding of dementia is knowledge, and lack of knowledge is directly linked to stigma. Studies have shown that '. . . over one-third of the general population have been reported to hold stereotypical or discriminatory views about dementia', indicating a level of ignorance among some family members. Knowledge deficits include poor understanding of risk factors, the clinical course of the disease, the terminal phase, the harm caused by overprescribing psychotropics, appropriate medications to address pain or infections, and the benefits of a palliative approach (Annear et al., 2015). Without this knowledge, translated into action, dementia remains for many, associated with social and societal shame and stigma.

On a positive note, families can be encouraged to broaden their knowledge, such as the benefits of tactile communication when speech is compromised or absent. The importance of touch is highlighted by one researcher describing residents with dementia as 'hug hungry'. Families may need to be reminded that lack of speech does not mean lack of feeling. Correcting such misunderstandings may outweigh the small cost of time involved, evident in this brief exchange:

> A resident's daughter was invited to meet with the care manager soon after her mother was settled in the nursing home. 'I would like to spend some time with you and your family to discuss the impact of dementia and the likely course of the disease and how it may affect your mother'. The daughter replied: 'Nobody told me that mum has dementia. The GP called it "just a bit of memory loss", although I always suspected it was dementia. I'd be so pleased to have more information'.

Who made the diagnosis of dementia in this instance? What type of dementia was it, and what tests were offered? What information was given to the family and in what form? What ongoing support was needed? These are not generic issues fostering a uniform response: rather, they focus on the *particular person,* in the context of their *particular circumstances.*

Many other information issues not covered here can be discussed in support groups, with expert advice and resources (including written, and in various languages) available from Dementia Australia.

4.4.2 Family Assessment: Caring for the Carers

When a person with dementia lacks the capacity to provide detailed assessment data, time is needed for discussion with family or another caregiver, including the grief, loss and inevitable death associated with the disease. Statistics imply an increase in the number of males caring for relatives with dementia, associated with deficits in support. 'Males often have less stable social support networks and are less likely to seek assistance in dealing with grief than females' (Davies et al., 2018, p.17). Advice from palliative care experts may be advantageous in such situations.

4.4 Families Flourishing or Floundering

However, recent findings show: 'There is no systematic approach in place to assess and respond to the needs of family caregivers of people with serious illness' (Hudson et al., 2020, p. 7).

Those responsible for managing the nursing home admission process would gain an understanding of each family's unique situation by asking some basic questions such as: 'We understand there may be mixed emotions surrounding this decision to accept residential care. What are some of the issues you and your family are facing?' It is also appropriate to refer to evidence, such as: 'Research tells us families can feel relief as well as guilt. What has been your response to having your family member admitted to the nursing home?' 'Have you been offered support and advice?'

Caregivers of people with dementia face several challenges to their own health as they balance caregiving with other demands, including child-rearing, career and relationships. They are at increased risk for anxiety and depression (Brodaty & Donkin, 2009). More understanding is needed of the sustained stress resulting from long-term caregiving, often exacerbated by the bereavement process, especially when the carer's specific needs are not recognised.

In previous times, before comprehensive diagnostic interventions, many older people with dementia were isolated, often being cared for at home with one spouse/family member and no other support. For many reasons, accustomed social contacts may have ceased. With increasing knowledge resulting in timely, targeted, evidence-based care, and when dementia care is regarded as community care, best practice will be able to flourish. In this climate, nursing homes can become society's exemplars, inspiring confidence that people with dementia and their families need not be forgotten, even while living with the consequences of a 'forgetfulness disease'.

Family carers' needs, however, are not always given due emphasis: the negative impacts are only evident when and if an enquiry is made (Broady et al., 2018). Carers in this study noted that health professionals focused attention on safety issues while taking little interest in the resident's unique pleasures and activities, indicating the lack of specialised training in dementia care. Those families who claimed positive experiences welcomed staff recognition of the long-term nature of their caring, enjoyed an emphasis on interpersonal relationships and appreciated staff knowing the resident's idiosyncrasies. Such person-centred care transformed family carers' experiences: rather than being regarded as a nuisance, or irrelevant, families became partners in care.

4.4.3 Communication

For the many families who have received little or no assistance in responding to the challenging communication issues of dementia, a few simple rules, with examples, may be helpful.

Don't argue: *'No mum! I'm Betty, not Anne'*.

Don't contradict: *'You're wrong, dad. It was yesterday we went out'.*
Don't dismiss: *'Oh, don't be silly mum. She wouldn't hurt you'.*
Don't infantilise or patronise: *'Come on, pet. Let's go to the toilet, that's a good girl'.*
Don't attempt to convince or use logic: *'Dad, you know you can't go home'.*
Don't say: *'Don't you remember. . .'*

Positive comments and change of topic are preferable, such as the following.

'Your hair looks nice today'.

'Let's go outside for some fresh air'.

'I brought some of your favourite chocolate'.

'Come with me to say hello to Freda'.

'Here's a drawing from your grandson'.

These tips, and many others, encourage staff who are unaccustomed to relating to people with dementia, to discover the art of conversing with someone who is losing their language skills. Carers and families may be prompted to learn and practice this art, so it becomes natural rather than forced. For example, due to their reduced concentration, residents are more likely to understand short sentences rather than long, matching their own truncated speech.

Another basic rule worth reminding carers is to approach the person with dementia from the front, not from behind or at the side when speaking to them or offering personal care. This is important because people with dementia often lose their peripheral vision, a reminder also for their professional carers and assistants of the importance of eye contact. The old adage is apt: *actions speak louder than words.*

Another aid to communication for some residents with dementia involves digital tools. These may include ready access to favourite poems or songs, peaceful scenes or familiar voices. The latter has been found to quieten an agitated resident, for example, during a procedure such as hygiene or toileting. Advice from an occupational therapist or other dementia expert may provide further options for the use of enhanced information technologies.

4.4.4 Wandering: A Management Problem?

Wandering, regarded as a 'behavioural problem', usually invites 'management', far too often by chemical or mechanical restraint. Rather than responding with confinement, this so-called 'problem' calls for expert clinical skills in *assessment,* with advice from families where relevant.

- Why is this person 'wandering'?
- Is it always at a particular time of day or night? If so, what did the person usually do at this time?
- Is it related to hunger, searching for a familiar face, boredom, loneliness, looking for the toilet, lack of understanding where they are and why?

- Has the person always been active, for example, a cyclist or intrepid walker?
- Has a thorough pain assessment been made?
- Have non-pharmacological responses, such as music, food, conversation, pleasurable activities and companionship been tried?

Why is 'wandering' the term used for a resident with dementia who is merely *walking* from one place to another? Rather than seeing wandering as a 'problem', attention to physical and social environments may prompt safe outside walking as a pleasurable activity, especially when there are items of interest to stop and look at on the way. Many residents would benefit from increased physical exercise: a neglected issue in some nursing homes. Evidence shows significant improvements in cognition, agitation, mood, mobility, functional ability and independence, as well as behavioural and psychological benefits (Brett et al., 2016). Exercise is also recommended in some pain management regimens. British doctor, Richard Asher, put it this way:

> Teach us to live that we may dread;
>   Unnecessary time in bed.
> Get people up and we may save;
> Patients from an early grave (Asher, 1947).

Comprehensive planning with outcome documentation validates heightened emphasis on this important component of care: getting residents up and walking. Language is also important: 'wonder walkers' rather than 'wanderers' conveys a more positive picture. Walking can be plotted on a 'wonder walkers' map', noting familiar places as well as other locations to stimulate interest.

4.4.5 *Visiting*

'You don't need to visit every day; she doesn't remember one day from the next'. This unfortunate, albeit well-meaning, comment ignores the realities of interpersonal dementia care. Families, including children, grandchildren and great-grandchildren may need encouragement, if not direct education and advice, about how to visit the person with dementia. For residents whose response to gestures of affection is a joyful smile, even a 5-minute visit is worthwhile. Families may be encouraged to consider what can be achieved in a visit: assistance with standing, or changing the resident's position, providing an object to grasp, or looking out a window. Taking a few steps together with their visitor, where that is possible, can provide a welcome relief from boredom, not to mention a positive impact on the resident's circulation, and mood. One creative family member would regularly bring ingredients for blowing bubbles, which always amused her father. Concentrating on enjoyment in the 'now' can prove of inestimable value to both the person with dementia and their visitor. The 'witness' or 'example' should not be underestimated; other visitors or families may take note of the innovations and be encouraged to do likewise. Making the most of valuable moments means, for some, fewer regrets later. While some visitors prefer frequent short visits, others gain satisfaction from longer periods, some spending many hours each day at the nursing home.

> *'I've spent the last seventy years of my life with him; I'm not going to abandon him now. Also, I'm lonely and bored at home by myself. There's so much stimulation here, and I can talk to the other residents when he's asleep'.*

Family visiting during the COVID-19 pandemic has posed a significant challenge for many nursing homes, with some imposing strict 'no visit' rules from the outset. Others have weighed the risk of infection with the risk of isolation, aware of the implications of the former while appreciating the benefits of the latter, particularly at the end of life. Rather than making unilateral decisions, managers are advised to seek evidence-based information from those skilled in infection control.

4.4.6 Spiritual Care

This subject is covered more generally in Chap. 3 (care for the whole person); however, additional comments are apt in a discussion of dementia. Some people assume that as the dementia progresses the person's spiritual needs diminish. Experts in this area understand the opposite is often the case. Although there is a paucity of research into spiritual care in nursing homes, some theologians and pastors have taken up this challenge. Goldsmith sees his role as a pastor involved in dementia care in the following terms:

> Who are you – walking to the toilet every few minutes? Who are you – wiping bottoms, answering the same question time and time again, sitting with those who weep and absorbing the anger and the frustration of those who do not know where they are? Who am I? I am a minister of the Gospel of love, and this angry lady is my sister and this weeping man my brother (Goldsmith, 2000, p. 21).

While gerontic nurses, managers and other staff may not necessarily espouse Goldsmith's 'family of God' approach which sees all humanity closely related, nor be compelled by the 'Gospel of love' imperative, it nevertheless is incumbent upon each carer to ask: Who are you and who am I in the separateness and in the totality of our being? Who are we, together?

'Explicit attending to needs for spiritual care may fill a gap in dementia care practice, as spiritual care is nearly absent in most dementia guidelines and national dementia strategies' (Davies et al., 2018, p.5). Families can contribute to residents' spiritual care by direct involvement or discussion with a chaplain, some of whom have developed skills through additional training and access to a variety of resources. Such resources are available from organisations such as Meaningful Ageing Australia.

> Meaningful Ageing Australia is the Australian national peak body for spiritual care and ageing. We are a membership based not-for-profit, supporting organisations and groups to

respond to the pastoral and spiritual needs of older people, their significant others, and their carers (Meaningful Ageing Australia, 2016).

Advice from the resident, where possible, and/or their family, may prompt referral to this agency for assistance. It would be encouraging to see more references to spiritual care in each resident's overall (holistic) dementia care plan, subject of course to sensitive individualised assessment.

4.4.7 Involving Families in Care Planning

On admission to the nursing home, some comments and queries may be elicited from the residents themselves. In the absence of coherent speech, questions should be addressed to the family, including information about the person's pleasurable activities, likes and dislikes, previous occupation, hobbies, entertainment. These details are of equal importance as bowel and bladder habits or medication details.

Timely discussion between management and families may help prevent distressing experiences such as the following:

> 'N' asked if he could speak with the manager about his visits to his mother. 'I feel it's absolutely useless – a waste of time. I feel her spirit has gone and there's just a shell, and I don't know how to relate to a shell. She's probably punishing me for not visiting more often, then by making me endure these painful visits when she never utters a word'. The chaplain's assistance was offered, together with some practical suggestions such as playing music, sending flowers, and replacing some visits with a phone call.

While some staff were judgmental about 'N's' reduced visits; others saw value in this change of routine, particularly as it alleviated his frustration and added satisfaction to his visits, albeit less frequent. He also knew that, by sending the flowers, the staff were aware he had not forgotten his mother, nor neglected her.

Reinforcing earlier comments related to the COVID-19 pandemic, the absence of families' involvement led, in many circumstances, to staff failure to follow a plan of care, such as avoiding hospitalisation where possible, and ignoring a current ACP document. Family involvement is crucial in such a crisis, involving very serious life-and-death decisions, particularly when residents cannot speak for themselves.

4.4.8 Touch

Reinforcing the many references to 'touch' in other chapters, its inclusion here highlights the issue in relation to dementia. When a person with dementia lacks clear speech, touch is a powerful means of communication. Touch is associated with the production of oxytocin (sometimes referred to as the 'care and connection hormone'). While we may despair at the lack of curative treatments for dementia, it is literally within our hands to improve the person's life experience, particularly as death approaches (McCleary et al., 2018). Families may need encouragement and reassurance about touch being a meaningful source of care, love and attention. More emphasis needs to be given to the *skin* which is the largest organ in the human body; much can be communicated through skin-to-skin contact. It seems ironic, therefore, that the issue of touch remains a relatively *untouched* subject in dementia and seldom included in the care plan.

History, together with further research, will undoubtedly 'touch' on the subject of protective equipment such as gloves, being incorporated as a necessary protection against infection and the associated loss of intimate contact between carers and residents.

4.4.9 Sexuality

Sexuality and sexual expression are discussed in Chap. 3; brief comments are included here to highlight the subject in the context of dementia.

Is the expression of sexuality a need or a problem, or a source of embarrassment? What happens when a person with dementia seeks sexual satisfaction in public or with another resident who is not their partner? Staff education is needed, to expand knowledge and encourage a thorough understanding of the sensitive issues involved, including consensual intimacy, spousal and family attitudes, and residents' rights. Older people do not necessarily lose their sexual drive with the onset of dementia, neither do they lose capacity for forming new friendships. Some may have no memory of current or past close relationships. 'The need for companionship, relationships, intimacy and human touch does not change because someone has dementia. People living with dementia have lived with their sexuality for much longer than they have lived with dementia' (Bauer & Fetherstonhaugh, 2016, p. 3). It is important for the nursing home to have a policy about sexuality and sexual health in this context, and for the subject to be discussed with families where relevant. Managers and staff who lack confidence in addressing these issues may refer to an appropriate agency, such as Dementia Australia, for information and advice.

4.4.10 Mealtimes

The presence of devoted family carers at mealtimes means, in many instances, the difference between residents being well-fed or suffering malnutrition. The COVID-19 pandemic impeded some mealtime assistance due to visiting restrictions, leaving some residents unfed for many hours, sometimes days. Lack of sufficient staff leads to inadequate time for assisting residents, particularly those who are totally dependent and/or may be very slow eaters. As a result, residents may not only become malnourished but also miss out on the enjoyment associated with food and shared mealtimes. With restrictions on communal dining imposed by the pandemic, residents also miss such socialising opportunities. As noted above, families may welcome opportunities for greater involvement at mealtime, some already devoting many hours per week to ensure their relative is not denied the pleasures associated with eating and drinking.

Training is often lacking for assessing a resident's nutritional requirements, as well as specific instruction for preparing and serving food to residents with dementia. This should be standard practice, utilising the knowledge and experience of the Maggie Beer Foundation, as one example (Commonwealth of Australia, 2021, p. 103), noting the practical advice for creative, visually appealing meals at minimal extra cost. Food preparation and delivery should be regarded as an essential quality indicator, with specific application to dementia care.

Eating and swallowing problems are not uncommon in residents with dementia. Advanced dementia is often accompanied by dysphagia or anorexia or both, indicating careful assessment based on individual need, and the employment of a suitably qualified speech pathologist where indicated.

The importance of thorough oral and dental care, described in the section on clinical care, has particular significance in the context of dementia, where residents may be incapable of maintaining their own oral hygiene and unable to describe pain or discomfort arising from poor mouth care.

4.4.11 Beyond Stereotypes

Given the increasing statistics and prevalence, dementia care should be central to every nursing home's philosophy, education and practice, with *the person* at the centre. Residents and their families need ready access to information, appropriate diagnostic services, with open invitation for involvement in planning and evaluating the care. Increased focus is needed on measuring the care, based on the latest research, as the means towards continuous improvement. Every staff member should be provided with relevant, ongoing dementia education aimed at improving their knowledge and skills. Rather than ad hoc responses drawn from idiosyncratic views of individual carers, decision-making should derive from best practice research, articulated in a care plan followed by all staff. Rather than being described as 'basic', dementia care requires expertise, including palliative care, demanding no less attention than any other major, life-threatening terminal illness.

This chapter on dementia challenges us to look beyond the stereotypical understanding of a community as a group of likeminded people. We are reminded that in the 'dementia community' there are many different manifestations of the disease. Every person with dementia, whatever their precise diagnosis, is different from every other person with dementia. Caring for them means entering a space where frail and strong, needy and self-sufficient enjoy equal status. In the words of one person with dementia:

> '*If you can get it right for dementia you can get it right for everyone else*'.

References

Annear, M. J., Toye, C., McInerney, F., Eccleston, C., Tranter, B., Elliott, K.-E., & Robinson, A. (2015). What should we know about dementia in the 21st century? A Delphi consensus study. *BMC Geriatrics, 15*(1), 5.

Asher, R. A. (1947). Dangers of going to bed. *British Medical Journal, 2*(4536), 967.

Australian Bureau of Statistics. (2018). Causes of death, Australia, 2017.

Australian Commission on Quality and Safety in Health Care. (2021). *Delirium clinical care standard*.

Baird, A., & Thompson, W. (2018). The impact of music on the self in dementia. *Journal of Alzheimer's Disease, 61*(3), 827–841.

Ballard, C., Kahn, Z., & Corbett, A. (2011). Treatment of dementia with Lewy bodies and Parkinson's disease dementia. *Drugs & Aging, 28*(10), 769–777. https://doi.org/10.2165/11594110-000000000-00000

Bauer, M., & Fetherstonhaugh, D. (2016). *Sexuality and people in residential aged care facilities: A guide for partners and families*. Australian Centre for Evidence Based Aged Care (ACEBAC).

Bourbonnais, A., & Ducharme, F. (2010). The meanings of screams in older people living with dementia in a nursing home. *International Psychogeriatrics, 22*(7), 1172–1184.

Brett, L., Traynor, V., & Stapley, P. (2016). Effects of physical exercise on health and well-being of individuals living with dementia in nursing homes: A systematic review. *Journal of the American Medical Directors Association, 17*(2), 104–116.

Broady, T., Saich, F., & Hinton, T. (2018). Caring for a family member or friend with dementia at the end of life: A scoping review and implications for palliative care practice. *Palliative Medicine, 32*, 3.

Brodaty, H., & Donkin, M. (2009). Family caregivers of people with dementia. *Dialogues in Clinical Neuroscience, 11*(2), 217–228.

Browne, B., Kupeli, N., Moore, K. J., Sampson, E. L., & Davies, N. (2021). Defining end of life in dementia: A systematic review. *Palliative Medicine, 35*(10), 1733–1746.

Bryden, C. (2005). *Dancing with dementia: My story of living positively with dementia*. Jessica Kingsley Publishers.

Bryden, C. (2015). *Before I forget: How I survived a diagnosis of younger-onset dementia at 46*. Penguin Books Australia.

Buettel, A., Cleary, M., & Bramble, M. (2017). Delirium in a residential care facility: An exploratory study of staff knowledge. *Australasian Journal on Ageing, 36*(3), 228–223.

Burke, C. (2020). *New research identifies barriers to good medication management in dementia care*. INsite.

Commonwealth of Australia. (2019a). *Dementia in Australia: nature, prevalence and care* (Background Paper 3).

Commonwealth of Australia. (2019b). *Restrictive practices in residential aged care in Australia*.

Commonwealth of Australia. (2020). *Royal Commission into Aged Care Quality and Safety International and National Quality and safety indicators for aged care Research paper 8*.

Commonwealth of Australia. (2021). *Royal commission into aged care quality and safety final report: Care, dignity and respect, Volume 3A The new system*.

Cook, J., Palesy, D., Chenoweth, L., & Lapkin, S. (2020). Older patient specialling in acute hospital wards: What's your policy? *Australian Nursing and Midwifery Journal, 26*(10), 22–25.

Cox, H., & Roberts, P. (2013). *Harp and the Ferryman*. Michelle Anderson Publisher.

Cunningham, E., McGuinness, B., Herron, B., & Passmore, A. (2015). Dementia. *The Ulster Medical Journal, 84*(2), 79–87.

Davies, D., Klapwijk, M., & Steen, J. (2018). Palliative care in dementia. In *Textbook of palliative care* (pp. 1–23). Springer International Publishing.

Dekker, N. (2018). Moral frames for lives worth living: Managing the end of life with dementia. *Death Studies, 42*(5), 322–328.

DeMarco, B. (2016). Water is invisible and disconcerting to dementia patients. *Alzheimer's Reading Room*. https://www.alzconnected.org/discussion.aspx?WT.mc_id=enews2016_11_26&utm_

References

source=enews-aff-&utm_medium=email&utm_campaign=enews-2016-11-26&g=posts&t=2147525004

Dementia Australia. (2020). *What is dementia?* https://www.dementia.org.au/about-dementia/what-is-dementia

Dementia Australia. (2021). Dementia statistics.

Dunn, M. (2021). Yes, there is dignity in dying, even for people with dementia. *Mercatornet*.

Fetherstonhaugh, D., Tarzia, L., & Nay, R. (2013). Being central to decision making means I am still here! The essence of decision making for people with dementia. *Journal of Aging Studies, 27*, 143–150.

Garrido, S. (2016). Music and dementia: Hitting the right notes. *Australian Ageing Agenda*, 46–47.

Goldsmith, M. (2000). Through a glass darkly: a dialogue between dementia and faith. Ageing, spirituality and pastoral care conference, Canberra, Australia.

Guideline Adaptation Committee, C. P. G. f. D. i. A. (2016). Clinical Practice Guidelines for Dementia in Australia: A step towards improving uptake of research findings in health-and aged-care settings. *Australasian Journal on Ageing, 35*(2), 86–89.

Hanson, L., Zimmerman, S., Song, M., Lin, F., Rosemond, C., Carey, T., & Mitchell, S. (2017). Effect of the goals of care intervention for advanced dementia: A randomized clinical trial. *JAMA Internal Med, 177*(1), 24–31.

Hill, H. (2001). *Invitation to the dance: Dance for people with dementia and their carers*. Dementia Services Development Centre University of Stirling.

Hovland-Scafe, C., & Kramer, B. (2017). Preparedness for death: How caregivers of elders with dementia define and perceive its value. *The Gerontologist, 57*(6), 1093–1102.

Hovland, C. (2019). When death with dementia is "a memory seared in my brain" Caregivers' recommendations to health care professionals. *Journal of Applied Gerontology*, 1–8.

Hudson, P., Morrison, R., Schulz, R., Brody, A., Dahlin, C., Kelly, K., & Meier, D. (2020). Improving support for family caregivers of people with a serious illness in the United States: Strategic agenda and call to action. *Palliative Medicine Reports, 1*(1), 6–17.

Hudson, R. (2002). Creating caring communities. *ACHSE Health Manager* (Autumn 2002), 5–9.

Hudson, R. (Ed.). (2003). *Dementia nursing: A guide to practice*. Ausmed Publications.

Hudson, R. (2008). Do unto others. *Nursing Standard, 23*(8).

Hudson, R. (2012). Transforming communities in residential aged care. *Journal of religion, spirituality & aging, 24*(1–2), 55–67.

Hukins, D., Macleod, U., & Boland, J. (2019). Identifying potentially inappropriate prescribing in older people with dementia: A systematic review. *European Journal of Clinical Pharmacology, 75*, 467–481.

Husebo, B., Ballard, C., Sandvick, R., Nilsen, O., & Aarsland, D. (2011). Efficacy of treating pain to reduce behavioural distrubances in residents of nursing homes with dementia: Cluster randomised clinical trial. *British Medical Journal, 343*(7816), 193.

Ibrahim, J. (2020). Residential aged care communique. *The Communiques, 15*(1.2).

Ibrahim, J., MacPhail, A., Winbolt, M., & Grano, P. (2016). Limitation of care orders in patients with a diagnosis of dementia. *Resuscitation, 98*, 118–124.

Ignatieff, M. (1993). Scar tissue. Viking.

Kaasalainen, S., McCleary, L., Vellani, S., & Pereira, J. (2021). Improving end-of-life care for people with dementia in LTC homes during the COVID-19 pandemic and beyond. *Canadian Geriatrics Journal, 24*(3), 164–169.

Kerr, D., & Cunningham, C. (2016). *Night time care: A practice guide*. HammondCare.

Killick, J. (2013). *Playfulness and dementia: A practical guide*. Jessica Kingsley Publishers.

Lin, F. R., & Albert, M. (2014). Hearing loss and dementia – Who is listening? *Aging & Mental Health, 18*(6), 671–673. https://doi.org/10.1080/13607863.2014.915924

Low, L.-F., Laver, K., Lawler, K., Swaffer, K., Bahar-Fuchs, A., Bennett, S., Blair, A., Burton, J., Callisaya, M., & Cations, M. (2021). We need a model of health and aged care services that adequately supports Australians with dementia. *Medical Journal of Australia, 214*(2), 1–4.

McCabe, M. (2018). Dementia out of mind. *Aged Care Insite, 104*, 25.

McCleary, L., Thompson, G., Venturato, L., Wickson-Griffiths, A., Hunter, P., Sussman, T., & Kaasalainen, S. (2018). Meaningful connections in dementia end of life care in long term care homes. *BMC Psychiatry, 18*(1), 1–10.

McDerby, N., Kosari, S., Bail, K., Shield, A., Peterson, G., & Naunton, M. (2020). Pharmacist-led medication reviews in aged care residents with dementia: A systematic review. *Australasian Journal on Ageing*. https://doi.org/10.1111/ajag.12827

McKinnon, A. (2019). Aged care in crisis. *The Monthly* (June, 2019).

Meaningful Ageing Australia. (2016). National Guidelines for spiritual Care in Aged Care. Meaningful Ageing Australia.

Outeiro, T. F., Koss, D. J., Erskine, D., Walker, L., Kurzawa-Akanbi, M., Burn, D., Donaghy, P., Morris, C., Taylor, J.-P., Thomas, A., Attems, J., & McKeith, I. (2019). Dementia with Lewy bodies: An update and outlook. *Molecular Neurodegeneration, 14*(1), 5. https://doi.org/10.1186/s13024-019-0306-8

Penders, Y. W., Albers, G., Deliens, L., Vander Stichele, R., Van den Block, L., & IMPACT, E. (2015). Awareness of dementia by family carers of nursing home residents dying with dementia: A post-death study. *Palliative Medicine, 29*(1), 38–47.

Penny, L. (2016). The last promise. *AARP The magazine*. https://www.aarp.org/caregiving/stories/info-2017/mystery-writer-essay.html

Read, S., & MacBride-Stewart, S. (2018). The 'good death' and reduced capacity: A literature review. *Mortality, 23*(4), 381–395.

Sabat, S., & Harré, R. (1992). The construction and deconstruction of self in Alzheimer's disease. *Ageing and Society, 12*, 443–461.

Salvatore, M. (2017). "Oh God, I have lost myself:" palliative care and Alzheimer's dementia. *RACmonitor, 2017*. https://www.racmonitor.com/oh-god-i-have-lost-myself-palliative-care-and-alzheimer

Sapp, S. (1998). Living with Alzheimer's: Body, soul and the remembering community. *The Christian Century, 115*(2), 54–60.

Smith, K., Flicker, L., Shadforth, G., Carroll, E., Ralph, N., Atkinson, D., Lindeman, M., Schaper, F., Lautenschlager, N., & LoGiudice, D. (2011). 'Gotta be sit down and worked out together': Views of aboriginal caregivers and service providers on ways to improve dementia care for Aboriginal Australians. *Rural and Remote Health, 11*(2), 1–14.

Steele, L., Swaffer, K., Carr, R., Phillipson, L., & Fleming, R. (2020). Ending confinement and segregation: Barriers to realising human rights in the everyday lives of people living with dementia in residential aged care. *Australian Journal of Human Rights*, 1–21. https://doi.org/10.1080/1323238X.2020.1773671

Steele, L., Swaffer, K., Phillipson, L., & Fleming, R. (2019). Questioning segregation of people living with dementia in Australia: An international human rights approach to care homes. *Laws, 8*(18), 1–26.

Stokes, G. (2008). *And still the music plays: Stories of people with dementia*. Hawker Publications.

Swinton, J. (2007). Forgetting whose we are: Theological reflections on successful aging, personhood and dementia. *Journal of Religion, Disability and Health, 11*(1), 37–63.

Swinton, J. (2012). *Dementia: Living in the memories of god*. William B Eerdmans Publishing Company.

Swinton, J. (2020). Citizenship, Personhood and Dementia. *Meaningful Ageing Australia Newsletter*.

Van der Steen, J. T., Radbruch, L., Hertogh, C. M., de Boer, M. E., Hughes, J. C., Larkin, P., Francke, A. L., Jünger, S., Gove, D., & Firth, P. (2014). White paper defining optimal palliative care in older people with dementia: A Delphi study and recommendations from the European Association for Palliative Care. *Palliative Medicine, 28*(3), 197–209.

Victoria State Government. (2020). *Differential diagnosis – depression, delirium and dementia*.

Volicer, L., & Mahoney, E. (2002). Are nursing home residents with dementia aggressive? *The Gerontologist, 42*(6), 875–876.

Walsh, K. A., Sinnott, C., Fleming, A., Mc Sharry, J., Byrne, S., Browne, J., & Timmons, S. (2018). Exploring antipsychotic prescribing behaviors for nursing home residents with dementia: A qualitative study. *Journal of the American Medical Directors Association, 19*(11), 948–958. e912.

Williams, K. (2017). A communication intervention to reduce resistiveness in dementia care: A cluster randomized controlled trial. *The Gerontologist, 57*(4), 707–718.

Chapter 5
A Palliative Approach

5.1 What Is Palliative Care?

5.1.1 Dispelling the Myths

'Palliative' is derived from the Latin for cloak or mantle: 'pallium'. In the sixteenth century, doctors used the term 'cura palliativa' or 'palliation' to describe alleviation or mitigation of suffering (Stolberg, 2007, p.7). A prominently displayed definition, on nursing home notice boards and/or in newsletters, would help to sharpen understanding.

> A palliative approach aims to improve the quality of life for individuals with a life-limiting illness and their families, by reducing their suffering through early identification, assessment and treatment of pain, physical, cultural, psychological, social, and spiritual needs (Australian Medical Association, 2015).

The palliative approach can be practised by all nursing home staff, with assistance from specialised palliative care services when required. Palliative care includes symptom control, intimate physical care such as massage, psychological care such as reassurance and empathy, and spiritual care in line with the person's beliefs. It is not confined to the last days or hours of life, with experts in the area now calling for a 'proactive approach to early integration for improving quality of life' (Hudson et al., 2021, p.1).

Increasing knowledge and awareness of palliative care's advantages have resulted in the recommendation: 'Palliative care should be integrated with all medical care for frail older people' (Harwood & Enguell, 2019, p.2). While many nursing homes have incorporated palliative care principles, others are slower to acknowledge its broad application. Although the claim has never been made that palliative care can

---

While (some of) the stories are based on factual situations, real names and other details have been altered to protect the identity of the persons concerned. Resemblance to any particular person is therefore purely coincidental.

remedy all physical, psychological and spiritual suffering at all times, its holistic emphasis would benefit all residents.

What is the difference between aged care and palliative care? When palliative care was introduced in the 1970s, it was restricted to patients with a diagnosis of cancer and a prognosis of six months or less. The criteria have now broadened to include people living with an active, progressive, advanced disease, regardless of the prognosis. A quick glance at residents' diagnoses would show this description effectively means most, if not all, nursing home residents.

Palliative care is strongly responsive to residents' needs, preferences and values, and those of their families and carers. Such a person- and family-centred approach is based on effective communication, shared decision-making and personal autonomy. In this context, autonomy is perceived not in absolute terms, rather as a partnership described by some scholars as 'relational autonomy' (Slife, 2004). This description affirms the belief that every person's health decision has an effect on others. In the nursing home context, absolute autonomy is seldom enacted, given that residents are there because they need the assistance of others. The principles of palliative care are founded on such partnerships.

Palliative care affirms life; it is not directed at either bringing forward or delaying death, nor is it about withdrawal of treatment. It is total active care based on impeccable assessment continually reviewed; involving comfort measures not drawn from the minimum but the maximum response to all needs; attaining, wherever possible, freedom from major discomfort until death.

In describing palliative care, it is important to note the more acceptable grammatical usage: 'palliative' describes the *care,* not the *patient or resident.* 'She's *palliative*' often implies 'she' is nearing the end of her life; whereas the person may benefit from *palliative care* months, if not years, before she dies.

5.1.2 Who May Benefit from Palliative Care?

'Would you be surprised if this person were still alive in twelve months' time?' This indicator gained some popularity when first recommended as 'the surprise question'. However, it should not be used in isolation from other assessments predictive of palliative care needs, giving rise to careful, timely planning. Evidence shows that those identified by the surprise question as at risk of dying, are more likely to receive palliative care interventions, even if they do not die within the twelve-month period (White et al., 2017).

Rather than waiting until death is imminent (within a few days or hours), early intervention prompts appropriately responsive care, reduces hospitalisation and improves quality of life.

A conceptual framework to guide health professionals includes (a) 'early palliative care' which may include the 'curative or life-prolonging' stage, (b) 'mid-stage palliative care' which anticipates death's probability within months, and (c) 'late-stage palliative care' when death is imminent (Hudson et al., 2021, p. 4). This framework encourages aged care health professionals to expand their understanding of palliative care, and to plan accordingly, rather than limiting it to the end of life.

End-of-life care includes risk of dying from an unexpected acute event, which may not necessarily fit the criteria for palliative care, adding a layer of complexity to assessment processes. Identification of needs also varies according to nursing home staff expertise, knowledge of palliative care, residents' common comorbidities and previous experience with a specialist service. Assessment is further complicated by the large number of residents with dementia who are missing out on palliative care through general lack of contemporary knowledge in both of these domains, and the relationship between the two.

Reports on palliative care services in nursing homes do not accurately reflect the total number of residents who may need such care: those with cancer (lung, bowel and prostate), circulatory and musculoskeletal disease, mental health conditions, dementia, and depression, to name a few. A report by the Australian Institute of Health and Welfare (2017) shows that 'only one in 25 appraisals of 250,000 residents' indicated the need for palliative care. These reports also acknowledge the difficulty of accurate diagnosis, reflected in the lack of comprehensive data.

The fact that most people entering nursing homes will die there is not readily acknowledged or spoken about with appropriate candour, honesty and respect. Lack of understanding about palliative care leads to a false desire to focus on 'the positive', resulting in limited views and misguided opinions such as 'she's not dying yet' or 'we don't want to hasten the death' or 'we don't want to talk about such morbid things' or 'in my opinion he's a long way from dying'. Rather than 'opinions' guiding a resident's care, greater emphasis is needed on facts derived from research. For example, delaying or avoiding palliative care have not been shown to extend life; encouraging realistic discussions, including early access to palliative care, are shown to *extend* life, particularly when symptoms such as pain are expertly treated.

Discussion in Chap. 1 includes a focus on disability, commonly associated with younger people in nursing homes. Further reference is made here to emphasise the fact that people of any age who have a disability may also require palliative care. Best practice suggests a strong partnership is needed within a multidisciplinary team, to address their physical and psychological needs. Aged care staff may not all have the experience or the skills necessary for such broad-based end-of-life care.

Similarly, staff caring for residents from LGBTIQ communities who require palliative care may need additional advice and assistance from those familiar with their particular needs.

Counterintuitively, the number of nursing home residents receiving palliative care is *declining* rather than increasing. Accordingly, the Royal Commission's final report proposes a new Aged Care Act to include: 'that all aged care staff be required to undertake regular training in palliative care' (Commonwealth of Australia, 2021, p. 119). Well-targeted, ongoing education is needed to address the discrepancy between the numbers of nursing home deaths and decreasing numbers receiving palliative care.

5.1.3 Misunderstanding Palliative Care

Unfortunately, palliative care's purpose is not universally acknowledged. 'Society is still afraid to speak its name, and specialised units are identified as "places of death" as opposed to "places of life" meant to treat suffering' (Reigada et al., 2020).

> A survey of Australians with advanced cancer found that people associate the term palliative care with diminished care, diminished possibility, and diminished choice. It isn't or shouldn't be. A survey of Australian general practitioners (GPs) found that many narrowly define palliative care as "non-curative care and pain and comfort relief". 'It is more than this' (Erny-Albrecht, 2019).

Erny-Albrecht argues that such misconceptions lead to fear, hopelessness and abandonment. 'Reading the menu, many note the omission of cure and skip directly to asking for the bill because they fail to recognise the choices available' (Erny-Albrecht, 2019). Rather than diminished choice, palliative care offers a range of care to meet the individual resident's needs, and those of their family: a menu conveying significant options.

Public perceptions of palliative care have been found wanting, with researchers claiming few articles have concentrated on palliative care's ability to relieve suffering and improve understanding of its other potential benefits (Kis-Rigo et al., 2021). Ignorance of palliative care's influence is evident in the comment: 'No thanks, I'm not ready for that yet!' Other terms such as 'just for comfort' or 'only nursing care' diminish the extent of well-planned, holistic, responsive palliative care.

An integrated model of aged care and palliative care begins with impeccable assessment, asking of *every resident* whether they would benefit from the least burdensome form of care and maximum comfort until they die. Such a focus would not be delayed until the very end of the resident's life; it would be introduced *on admission,* or even *pre admission,* and regularly reviewed. Timely discussion of these matters would help to overcome the disempowerment and lack of information experienced by families. To increase their understanding, direct encouragement and participation in decision-making would help to reverse this trend. In a conversational milieu, residents and families would be encouraged to raise their fears and to accept the implications of permanent residential care, including the inevitability of death (Halifax et al., 2020). Such a focus need not be alarmist; rather, it would emphasise the expertise available regardless of the resident's life expectancy.

Misunderstanding of palliative care is compounded by the lack of appropriately qualified staff to care for residents with increasingly complex needs. Addressing this gap, positive outcomes have been reported from regular palliative care study days for all staff (Anstey et al., 2016, p. 353), further consolidating partnership between aged care and palliative care. Where such partnerships are enjoyed, resident and family care are optimised and staff report increased job satisfaction.

Information about palliative care may take the form of a prominently displayed definition and description, and its inclusion in every nursing home's mission statement. Given the limited resources of external palliative care services, it is incumbent upon nursing home managers to ensure staff are equipped with the skills to meet this increasing need.

5.1.4 Comprehensive, Professional Reporting

Timely referral and reporting to palliative care services is far preferable to some contemporary practices where decisions are made by the staff on a particular shift meting out 'care' according to their own opinions, as in this handover scenario.

> *'I think Minnie won't be with us much longer. I haven't discussed it with her or the family, but I get the feeling she's failing fast.' 'Rubbish!' says his colleague, 'She's as tough as nails. I've seen her in a worse state than this and she came good.' The care assistant ventured to relate her experience from the previous day. 'Minnie's family asked me if I thought Minnie was dying. I didn't know what to say'. Unfortunately, these various opinions did not lead to any resolution, so the resident's care continued in an 'ad hoc' manner rather than in response to an individualised, carefully developed plan including the pain relief and family involvement she so clearly needed.*

It would be interesting to note how much handover time is given to such undisciplined conversations (as in the above), replete with personal predictions, rather than a team commitment to expert palliation based on the resident's needs, continually assessed and accurately reported. There are, of course, many examples of the latter in nursing homes where palliative care is routinely offered, and practised with confidence and skill.

Communication is the key to decision-making, particularly at the end of life. Reporting needs to be unambiguous and truthful. One survey found that older people preferred to entrust decision-making about their future to their family; although complications arise where there is family conflict and a denial of dying (Gerber et al., 2020, p.1). These findings provide further evidence of the benefits derived from family meetings accompanied by professional reporting, documentation and contemporary factual advice about the benefits of palliative care.

5.1.5 Multidisciplinary Team

Traditionally, palliative care is provided by a multidisciplinary team, unfortunately not always evident in the nursing home environment; too often leading to unilateral decision making. This deficit is addressed by seeking advice from a regional palliative care service, exploring how they may assist nursing home staff, and the resident's GP, to provide optimum palliative care. Rather than waiting for the final stages of life, it is far better for all concerned, that residents, families and staff have advance knowledge of all options. Staff can then initiate or continue discussion with residents and families at appropriate times. Such discussions will vary, of course, from team to team and from time to time; particularly in rural areas. Nursing home managers are advised to meet with their regional palliative care service to explore what assistance is available and when.

Geriatricians provide a welcome addition to general medical care, responding to residents' needs not met by GPs, who are mostly not palliative care clinicians.

However, not all geriatricians are willing or able to provide palliative care. 'Notwithstanding their expertise in caring for persons in the final decades of life, some gerontologists remain unfamiliar with the effective palliation of symptoms for older adults with life-limiting or chronic, debilitating conditions' (Mallinson, 2015, p.503). It is therefore, in certain circumstances, more appropriate to seek advice from a palliative care clinician: an effective use of multidisciplinary care.

Other examples of multidisciplinary support include timely referral for therapies such as physiotherapy, which, ideally, would continue until the resident's death. At times, however, the opposite is the case: physiotherapy assessments are either not obtained or their recommendations are not followed, they are poorly documented or not endorsed by the GP. Case studies (Commonwealth of Australia, 2019, pp.248–259) highlight the benefits of physiotherapy, recommending this important aspect of care is included in every palliative care plan, highlighting the goals; for example, to prevent the development of painfully contracted limbs and pressure wounds, and to maximise circulation. A renewed focus on best practice physiotherapy would involve restorative and reablement care (discussed in Chap. 3), potentially improving residents' quality of life and preventing early death. Other therapies (discussed under 'Allied Health' in Chap. 3) also find their rightful place in a multidisciplinary approach to palliative care.

It is evident that a multidisciplinary team is best placed to provide optimum palliative care to residents at the end of life. Who, among the various disciplines and support staff, belongs in the team? Some are not welcomed, as in the following report:

> *An activities worker stayed back following the palliative care education session, with a question for the educator. 'I saw this woman dying and I said to the nurse in charge "How can I help?" And she replied, "It's none of your business."' The activities worker believed it was her business, her office being opposite the resident's room. Concerned at the lack of response to the resident's constant calls for help, she addressed her concerns to the manager: 'I wouldn't want to die like that, and especially to be left alone.' The manager tried to persuade the worker that the resident was not alone, although the worker knew otherwise. She also knew the resident's family appeared to have abandoned her. The palliative care education session had sharpened this worker's concerns, but she received the strong impression that it was not 'her business', and that if she persisted, she may find her number of shifts reduced.*

This scenario (above) highlights the rights of all staff to raise issues of concern regarding residents' care. Some staff wrote on their evaluation forms that the time given to a palliative care education session was inadequate, opining that 'management needs to come to these education sessions'. No space/room had been prepared for the session, no person available to introduce the visiting educator, and no drinks provided. This description may strike a familiar chord, indicating that palliative care education is not always prioritised by management, or educators acknowledged as members of the team.

In many other instances, palliative care education is taken seriously, widely advertised, all staff as well as family members encouraged to attend, visiting educators acknowledged appropriately and reports made available after the event. Such education helps to dispel the many myths surrounding palliative care, firmly embedding education within the multidisciplinary team. Specific input from a palliative care social worker is also essential to a holistic approach to education, particularly for addressing aspects of care outside the competency and experience of aged care nurses. Another means to a productive engagement with palliative care is to designate a 'palliative care champion' whose role within the nursing home is to promote engagement and institute referrals to specialist services where needed.

5.1.6 Specialist Palliative Care

'Specialist palliative care' is a subset of palliative care provided by clinicians who have advanced training, delivering direct care to those with special needs as well as consultation, support, advice and education for non-specialist clinicians. As a 'sub set' of palliative care, specialist palliative care is not necessarily needed for every nursing home resident approaching death. However, when a direct relationship is developed with the regional palliative care service, appropriate referral processes can be optimised and hospitalisation minimised (Chapman et al., 2018, p. 102). While the palliative approach can be practised by all residential care staff, assistance from specialised palliative care services may be required in some circumstances, particularly if engaged well before the resident's death. GPs not necessarily skilled in palliative care may also be unaware of the benefits of specialist care for those residents who have complex needs.

Palliative care, when regarded as a core skill for all staff, means that *specialist* palliative care is sought only for circumstances where complications are beyond staff's capacity. In one such situation, a family member commented:

> 'The palliative care nurses provided expert advice to the aged care staff . . . They helped to relieve her pain and distress and reassured the family they were making the right decision in abiding by their mother's wishes . . . The palliative care staff listened to my fears and doubts and reassured me we were not hastening Mum's death by medicating her to keep her calm and pain free and by not sending her to hospital. That reassurance helped tremendously when self-doubt set in. Having access to them any time, day or night, gave me the courage to stay true to Mum's wishes. . . Mum had the best care and a comfortable passing, which is all we could ask' (Shaw, 2019).

The author of this account believes there is a need for more awareness of the role of palliative care services in nursing homes, particularly in providing expert advice for residents' complex needs. 'This enables them to provide better care, avoid hospital admissions and support people to die well in a familiar environment' (Shaw, 2019).

5.1.7 Managers' Knowledge

A direct correlation has been found between a manager's knowledge of palliative care and the way residents die: one survey revealed more than half the managers interviewed lacked basic knowledge in this area. Although some nursing homes have successfully integrated palliative care into their daily practice, others are unaware of its benefits. Palliative care offers timely access to specialists such as physicians or NPs who can prescribe medications not readily available through the GP. These clinicians may also assist the whole nursing home community to normalise death, to avoid unnecessary hospitalisations and to promote conversations with residents and relatives (Johnson & Bott, 2016, p. 124). Led by a competent manager, this approach fosters dialogue, providing essential information regarding a resident's likely trajectory towards death, aided by a focussed plan of care.

Managers who are aware of the benefits of palliative care, advertise the details in their promotional material; not only to increase the general public's understanding, but also to raise the nursing home's profile as a place where residents enjoy the best available care until their life's end.

5.1.8 Palliative Care 'Needs Rounds'

'Needs rounds' include a checklist which prioritises 'the integration of specialist palliative care into residential care to drive up quality care, provide staff with focused case-based education, maximise planning and reduce symptom burden for people at end of life' (Forbat et al., 2018, p. 347). In nursing homes where 'needs rounds' have operated, a number of benefits have ensued: higher probability of the resident dying in their preferred place, unifying the care team, plans in place and families informed, identifying those most at risk of unplanned dying, and residents' symptoms better controlled. The rounds, led by a palliative care NP, includes a checklist to trigger who would benefit from a specialist palliative care referral, having their needs identified well before death. The system would be advantageous for those in rural and remote Australia, using telehealth and other communication systems, where one NP uses the checklist to cover a large area and identify those whose needs are most urgent.

Research conducted in 12 Australian nursing homes describes palliative care 'needs rounds' as monthly hour-long triage meetings where residents without an appropriate care plan are identified. Examples showed that when relevant documentation was addressed and protocols enacted, dying residents received much better care (Liu et al., 2019). Unfortunately, accurate data reflective of each resident's multiple diagnoses is not always available and the high burden of chronic disease does not necessarily result in the provision of palliative care, despite the concomitant symptoms and the incurable nature of their illness. Other issues, such as the lack of appropriate clinical indicators, are reflected in the inaccuracy of much of the data used to attract funding; pointing to the need for expert advice from specialists such as geriatricians with palliative care expertise.

Internationally, palliative care RNs in nursing homes are recognised for their capacity to prevent hospitalisation, optimise care, improve symptoms and work collaboratively with other clinicians (Hickman et al., 2020, p. 152). In Australia, the need for such collaboration is more sharply focused in the wake of COVID-19, where palliative care is 'stripped down to its basics', leaving bereavement support sadly lacking, and challenging *'what care can mean in the times of social distancing'* (Elsner, 2021, p.70) (italics in the original). These times of uncertainty highlight the need for renewed emphasis not only on residents' care but staff and families' care for one another.

5.1.9 New Technologies

Palliative care is optimised by the rapid advances in smart technology, where big and artificial intelligence monitors a person's changing health data. 'The emerging field of biotechnology and precision medicine will mean that life-limiting illness is easier to predict and biological markers will provide a clearer indication of life expectancy for many people with varying diagnoses' (Deerain, 2019). This report acknowledges the dominant illnesses accounting for growth in palliative care, including dementia and cancer, highlighting the significant factor of comorbidity. While on the one hand, increasing biotechnology options will be available, there will be other circumstances where further treatment would be inappropriate, considered too burdensome or lacking in benefit. Deerain (2019) notes that the distinct advantages of advancing technologies will need to be balanced with issues of human contact and acts of kindness. Optimal psychosocial, emotional and spiritual care, together with grief and bereavement support for families, will continue to be high priorities, indicating the benefits of including palliative care's holistic services more broadly into nursing homes.

Artificial intelligence (AI) and genomics are at the frontier of twenty-first century health care. It remains to be seen how this revolution will improve health systems, especially in regard to diagnosis, prognosis and optimum palliative care for nursing home residents.

5.1.10 Food, Feeding and Fluids

Nutrition and hydration are discussed more broadly in Chap. 3; however, the issue of food and fluids deserves a prominent place in end-stage palliative care. As death approaches, the resident's needs should be reviewed, with advice from a palliative care consultant if necessary. When a resident refuses to eat and drink, their decision needs to be respected; adjusting amounts and offering variety may be all that's required. While a speech pathology assessment may result in 'unsafe swallow' followed by the inevitable 'nil orally', in circumstances where the resident is dying, the aspiration risk needs to be seen in the context of humane care and maximum comfort. Managers who place 'risk' above holistic care have no need to fear from such a plan; particularly if agreed by a palliative care team, well documented and frequently reviewed. Many speech pathologists are thoroughly cognisant of these palliative care options, assisting other nursing home staff to achieve excellence in holistic care.

The role of clinically assisted nutrition and hydration (CANH) has been widely researched: the benefits include preventing thirst, delirium, hypercalcaemia, and opioid toxicity. However, artificial measures are also known to cause discomfort, as well as increased risk of aspiration, pressure ulcers, infections, and hospital admissions. 'In the final days and hours of life, the risks and burdens of CANH generally outweigh any potential clinical benefit' (Carter, 2020). Careful hand feeding, particularly by a trusted family member, is the preferred comfort measure when death is imminent. Strict dietary protocols may give way to offering small amounts of food and drink that bring delight; such as ice cream or a sweet beverage. As for all other care, the *goals* need to be continually re-assessed, with family involvement wherever possible.

5.1.11 Dementia Needs a Palliative Approach

Adding to the discussion in Chap. 4, other points are raised here to reinforce the inclusion of palliative care within dementia care. While increasing numbers of residents with dementia are dying in nursing homes, the offering of palliative care remains underutilised, leaving many to die without appropriate support. Advanced dementia has not always been recognised in health care as a terminal illness, depriving residents and families of palliative care's benefits. More than a decade ago, researchers found that 'dementia patients are significantly less likely to be referred to specialist palliative care, be prescribed palliative drugs, or have carers involved in decision making' (Barber & Murphy, 2011, p. 587).

Regular training is needed for those caring for residents with dementia, to raise awareness that it is a 'palliative condition' needing a palliative approach (Mataqi & Aslanpour, 2020, p.145). Others regard this issue as 'an urgent public health priority' with the strong recommendation that 'patients with dementia are given the chance for the best possible care at the end of life' (Hashimie et al., 2020, p. 329). Palliative care includes expert advice on whether symptoms often ascribed to 'worsening dementia' may actually indicate delirium, which can usually be successfully treated. This advice serves as a reminder of the need for thorough assessment by a competent, well-informed clinician who takes account of the residents' multiple comorbidities.

Research shows that patients with late-stage dementia are 'one of the most understudied populations in palliative care' (Ferrell, 2019), compounding the absence of skilled end-of-life care. Others note that services are inconsistent in the way they are delivered, lacking equitable care, unified standards and accepted definitions (Dementia Australia Policy Team, 2019, p.1).

Prognosis is also vitally important in this context. 'Given the high symptom burden and terminal nature of dementia, good prognostic awareness and integration of palliative care (PC) is needed' (Gabbard et al., 2020, p. 683). Regardless of this advice, discussions about dementia rarely include its incurable nature and the undeniable fact that every person with dementia will die of the disease. Consequently, families continue to be either misinformed or uninformed about the end-of-life benefits of a palliative approach, including alternatives to hospitalisation. Treatment options need to take account of the burdensome nature of invasive care, the unsettling effect of transfer to hospital, not to mention the cost of acute care, which may have questionable value. As noted in the previous chapter (on dementia): Although dementia is the second highest cause of death in Australia (after heart disease) (Australian Bureau of Statistics, 2018), and the leading cause of death for women, research-based care is not always accessed, and coordinating approaches with palliative care not always optimised.

5.1.12 *Quality of Life Questioned*

Readers will, no doubt, note the many references to 'quality of life' in preceding and subsequent chapters of this book. However, it is addressed here with particular reference to palliative care.

Quality of life is 'entirely a product of medical and nursing thinking in the second half of the twentieth century' (Randall & Downie, 2006, p.25) and no satisfactory definition has been produced. The concept remains, in large part, a value judgement. 'She's got no quality of life!' opines the nurse, who is keen to justify measures which may hasten the resident's death. Rather than promoting discussion about what *care* – including a focus on *comfort* – this particular resident needs, one nurse's opinion is an inadequate basis for developing a *plan*. A strong word of caution is issued about any attempt to *measure* quality of life (Randall & Downie, 2006, p.35); far better to discuss with the resident and family what their perception is, addressing concerns with appropriate information about options and referrals.

Sellick describes his unease with the use of 'quality' in relation to dying:

> Part of my unease is the easy use of this word that is strongly associated with consumerism and managerialism, in the context of the unimaginable extinction of the self. It seems that the devastation that death brings, the removal of our most loved and precious, is somehow reduced to the quality of a product that we might buy (Sellick, 2007, p.44).

Sellick's 'unease' is a salutary reminder for all healthcare workers to focus on *the person* and their *experience* of living and dying; rather than presuming we can make a measured judgement about *quality* which is more appropriate for industrial objects considered replaceable when damaged. This kind of thinking has no place in palliative care.

5.1.13 Fear of Becoming a Burden

When lack of autonomy, perception of burden or the fear of dependency are uppermost in the resident's mind, they may consider a request for hastening their death. Open discussion about personal expectations, changing relationships, and respectful caregiving may ameliorate their fears. Cultural and societal attitudes need to change so those in need 'recognize the good in receiving care and get used to the idea that they do not need to *do* anything to be valuable' (Fox, 2020). While the older person may not be easily persuaded of their value, staff can convey this attitude by a smile, a gentle hand squeeze, and care interventions which say in essence: 'You matter because you are you'.

Discussing the concept of burden, one writer (Meilaender, 1991) counsels that, rather than asking 'Is this life worth living?' or 'How big a burden will it be?' it is better to ask 'What can we do to benefit the life that remains?' Imagining his own impending death, he predicts he *will be a burden* to his wife.

> No doubt this will be a burden to her. No doubt she will bear the burden better than I would. No doubt it will be only the last in a long history of burdens she has borne for me. But then, mystery and continuous miracle that it is, she loves me. And because she does, I must of course be a burden to her (Meilaender, 1991, p. 16).

Meilaender's strong suggestion is that, while we cannot eliminate all burdensome care or associated suffering, we may press the question of how the burden may be lessened by appropriate, timely support. Bearing burdens has different connotations, depending on the nature of the relationship, and what each person is willing to 'bear' for the other. Assumptions should be avoided, replaced by careful assessment. What one person is willing to bear proves intolerably burdensome for another.

5.1.14 Acknowledging Signs of Dying

The subject of death and a 'good death' is taken up more fully in Chap. 6. It is included here to emphasise the place of palliative care in preparing for death. A significant rise in mortality from chronic, incurable diseases means more older people in nursing homes would benefit from a palliative approach to their care. Such conditions include heart failure, dementia, chronic obstructive pulmonary disease, diabetes and renal failure, to name a few.

> Dying itself has become, in many cases, a long-term condition. The 'good death' is a largely professional and ethnocentric construct, which takes no account of cultural diversity and the different values which may be espoused by different groups or individuals... (Pollock & Seymour, 2018, p.328).

Pollock and Seymour (2018) acknowledge the difficulty of 'prescribing a good death' for people dying from chronic degenerative diseases, incapacity, frailty and mental health issues which follow an ill-defined, uncertain illness trajectory. In their opinion, the increasingly slow death of frail older people requires a new model of palliative care based on need rather than diagnosis or prognosis.

Taking such an approach, preparation for death may include asking the person what matters most to them at the end of their life, as well as the nature of family and other support available. Such conversations may transform care through early interventions: physical, spiritual and psychological, according to individual need.

Death cannot always be 'sanitised' and the death rattle is a naturally occurring phenomenon at the end of life, which does not require 'treatment'. As Peskin describes it: 'We squirm and cry out coming into the world, and sometimes we do the same leaving it' (Peskin, 2017, p.4). While medication is sometimes needed, the most important therapy is emotional comfort; speaking soft, calming, reassuring words even if there seems to be no response. Simple expressions of love and reassurance by carers may need to be encouraged by staff who are competent and confident in their communication with families during the resident's dying process.

Discussions about what to expect may include a description of the changes that often occur when death is near: social withdrawal, confusion, declining of the senses (with hearing the last of the five senses to be lost), decrease in appetite and thirst. The dying person may spend more time sleeping and be less alert, body temperature may vary and skin become discoloured. Control of bladder and bowels may be lost, and less urine is produced. Saliva and mucus may cause unwelcome sounds, which are not usually distressing to the dying person; neither should noisy breathing cause alarm. Light massage or playing the dying person's favourite music may ameliorate restlessness.

Information for families, and for staff unfamiliar with the dying process, about what to expect includes:

- Breathing and heartbeat have stopped.
- The person cannot be woken up.
- Eyelids may be half open.
- Pupils are fixed and dilated.
- Mouth may fall or remain open as the jaw relaxes and
- Skin becomes pale and waxy looking (Palliative Care Victoria, 2018, p.5).

Not all of these changes are evident in every dying person, nor do they occur in the same order. The changes, which may perturb some families, should not be viewed in contradistinction to a 'good death'. Many accounts also testify to the fact that some residents who exhibit 'unmistakable signs of dying', surprisingly 'rally' and live much longer than expected: predictions are therefore to be avoided.

Similarly, rather than platitudes or presumptions such as 'it won't be long now' it is more appropriate to ask a family member about their preparation for the resident's death: 'Is there anything you would like to say or do'? 'Is there anything troubling you?' 'Do you have any questions or concerns?' Additional support and information may then be offered as needed.

5.1.15 Terminal Stage

International research has found that nursing home staff had difficulty in identifying early signs of dying; regarding it as 'a happening, not a process' (Persson et al., 2018). The later signs of dying when the body begins to shut down were more easily recognised than earlier signs. This meant that palliative care was not made available as early as it should be (Persson et al., 2018). Persson and colleagues also emphasise the important role of nursing assistants in observing and reporting early signs of dying, such as residents withdrawing or losing interest in the outside world. Reiterating other references (in other chapters) to nursing assistants or care attendants, their inclusion in discussion and care of residents who are dying is prompted by their previous experience, either positive or negative, and including any fears or anxieties.

Rather than a generic or routine approach, the complex nature of death, dying and palliative care needs to be discussed individually with all residents and families, offering relevant resources where needed, to help them prepare for the resident's final days or hours.

5.1.16 Final Wishes

Clear documentation of palliative care measures would normally be included in the resident's care plan, complementing routine care. However, factors such as social and spiritual needs are not always recorded; neither do issues such as 'final wishes' receive routine attention. A specific question such as 'What would you like to do before you die'? may elicit some unforeseen answers. While last wishes cannot always be granted, it is worth asking the question, providing opportunity for creative, responsive care, such as enjoyed by 'Brian'.

> *Brian's family were well aware that he was approaching 'the end' as they gathered at his bedside in the single room. It was a Friday night, prompting Brian's daughter to approach the manager. 'Friday night has always been fish and chips night in our family. Although Dad's no longer capable of enjoying his favourite meal, would it be okay for our rather large family to enjoy this routine at his bedside?' There followed a joyous feast, accompanied by music, creating an atmosphere they knew their dad would enjoy. The family knew their father would not want them to be sitting in sombre silence while he took his last breath. 'I'm sure he could smell the warm potato chip I wafted past his nose', said the daughter.*

Another family may prefer a quieter bedside vigil, or the resident may appreciate the close presence of a well-loved pet. Dying residents and their families should be given the opportunity to decide how they would like to spend such a memorable time and it would not be inappropriate to include relevant details in the resident's individualised care plan. Families may also need to be encouraged to include the dying resident in their conversations, even if there is no sign of response; rather than talking amongst themselves and ignoring the person in the bed.

5.2 Palliative Care in Partnership

5.2.1 *Liaison with Palliative Care Services*

Discrete palliative care services in Australia vary widely, including access to nursing homes. It may not be assumed that they all provide the same support; models of care differ across states and territories according to their funded services, local practices and differently structured healthcare systems. It is therefore wise for each nursing home to establish a liaison system beginning with a meeting of key staff from both agencies, confirming the partnership and avoiding the following:

'I rang the palliative care service last night but all I got was the answering service!'

The liaison system works well when trusting relationships are developed and staff (particularly at night and weekends) know how to make a referral or access additional care when needed, with current contact numbers and other details readily available. It should be noted that some palliative care services have suffered increasing strain during the COVID 19 pandemic, resulting in staff shortage and limited access. This situation serves as a reminder for nurses in aged care to take every opportunity for improving their own palliative care skills, referring to the specialist agency only when greater expertise and advice is required. One practical measure for improving palliative care is to liaise with GPs, palliative care physicians, NPs and/or pharmacists for retaining a ready supply of opioids, with associated formal medication orders, so residents are not left waiting for pain relief.

5.2.2 *Hospital in the Nursing Home*

Rather than frail nursing home residents and carers relying on admission to an acute healthcare facility, at least one programme developed by clinicians is keeping residents out of hospitals (Fan et al., 2018). Evaluation found a 47 per cent decrease in hospital admissions and significant cost benefits. The data also showed that many older people were sent to emergency departments for relatively routine procedures such as catheter change, wound care or blood transfusion, all of which could be provided in the nursing home. This education programme recommended appointing a contact person at the hospital emergency department to maximise liaison procedures and communication (Fan et al., 2018).

Other nursing homes may be prompted to establish similar education programmes involving partnerships with their local hospital or palliative care service. Managers who balk at the cost (financial and care hours) of this kind of liaison need to balance the benefits with the costs and staff time, not to mention residents' stress and anxiety, of unnecessary transfer to hospital.

5.2.3 *Core Business or Optional Extra?*

Palliative care's partnership with nursing homes is neither well understood nor universally practised even though, for many residents, dying is not far from their mind.

> 'This is probably my last move'.
> 'I suppose this is where I'll die'.
> 'I wonder what it will be like here, I'd rather die in my own home'.
> 'Will they move me out if I get really sick?'
> 'Will they know when to call the priest? (Hudson & Richmond, 1998, p.293).

These vitally important issues need to be confronted soon after the resident's admission to the nursing home. Prompt reassurance invites a positive response. 'Yes, this seems a caring place. I'll be all right here'. Other questions may arise from open discussion with families. 'Will she have to go to hospital at the end?' 'Will she be in pain?' 'What will happen to dad's body when he dies?' A comprehensive advertising brochure would include a statement about the nursing home's palliative approach, reassuring prospective residents and families of its benefits.

Palliative Care Australia (PCA) believes palliative care should be seen as core business for residential aged care. Given the 'reason for discharge' from nursing homes is mostly due to death, it is evident that residents in the final stages of life require the best available care. This may mean accessing specialist palliative care services in addition to 'core business' palliative care delivered by nursing home staff. As mentioned above, many healthcare workers believe that palliative care applies mainly to the last few days (or hours) of life. This common misunderstanding is to the detriment of residents, who deserve care congruent with their needs from the day of admission to the day of death.

Along with residents' needs is the important issue of their preferences; particularly for end-of-life care. 'Discussion and documentation of older people's care preferences needs to be further encouraged within the healthcare system. It is essential not to wait for a crisis at the end of life to begin these processes' (Williams et al., 2020, p.313). A case in point is the COVID-19 pandemic, where it was evident that many residents had decisions made *for* them, without any prior discussion *with* them, of their needs or preferences.

Discussions about end-of-life care have particular, idiosyncratic relevance for residents belonging to a distinctive cultural community. Palliative care within diverse groups may require nursing home staff to seek further advice from the relevant organisation, noting that residents from Aboriginal and Torres Strait Islander communities (for example) need care commensurate with their specific cultural milieu, especially at the time of death.

5.2.4 GPs Delivering Palliative Care

Doctors responsible for the care of nursing home residents are largely *generalists* with limited expertise in the specialist areas of palliative care, gerontology or complex pain and symptom management, resulting sometimes in unfortunate, ill-advised responses such as the following.

> *The nurse was concerned about a resident whose condition was deteriorating, albeit slowly. Weight loss, communication problems, inability to swallow the significant number of daily medications, signs of increasing pain, and family concerns prompted the nurse to contact the GP. As well as several other comorbidities the resident had Alzheimer's disease, compromising his ability to make his own wishes known. When the nurse requested a referral to palliative care, the GP responded: 'Oh, it's too early for that, he's not dying yet'.*

Noting the increasing demands of palliative care for conditions other than cancer, palliative care team members welcome early referral, establishing relationships with the resident, family and GP. When palliation is confined to a resident's last hours or days, protocols for referral are compromised. Closer liaison is of mutual benefit for GPs and residents, as in the following example:

> *When the GP expressed his reluctance to prescribe opioids for a resident in severe pain, the manager suggested a discussion with the local palliative care physician. Together, the GP and specialist sat at the desk while the latter wrote a brief outline of the steps to take, the recommended drugs and doses, and a plan for review. The GP responded: 'Now I've got it!' From that time the GP provided expert palliative care for residents requiring a strong pain regimen and other symptom management.*

While some GPs are reluctant to seek further advice, others welcome a partnership with palliative care physicians. Barriers to GPs providing palliative care include 'prognostication challenges and lack of confidence' (Damarell et al., 2020), capacity to provide terminal care, spiritual care, out-of-hours care, end-of-life care in dementia. GPs also cite lack of training in communication, 'impeccable assessment', deprescribing, and relational care (Damarell et al., 2020). As noted above, GPs are *generalists* rather than specialists with the competence or innate desire to provide a palliative approach to the care of nursing home residents. This raises the question of medical oversight should all nursing homes become 'palliative care units' as advocated in this chapter; a matter requiring further research.

5.2.5 Pharmacists' Role

Palliative care education for nursing homes focuses on the need for *anticipatory prescribing*, mainly for analgesics and especially for night time, weekends, and residents in rural/remote areas. Prescribing analgesics is dependent on comprehensive assessment skills, particularly for residents who are unable to describe their pain. Rather than waiting for a crisis, or a distressing escalation of symptoms, the resident's need for twenty-four-hour pain control should be based on their disease burden, often including comorbidities associated with *constant pain*. For readily available analgesia, particularly opioids, an effective storage system is needed to ensure the necessary drugs are to hand throughout the twenty-four-hour cycle, avoiding the following scenario.

> *'Nurse A' at Friday afternoon handover: 'Looks like Franco will die within 24 hours so I haven't ordered any more morphine'. Franco was still alive on Monday morning, having endured a painful weekend in a remote area nursing home with no ready access to opioids.*

Apart from the false assumptions about the resident's expected death, this account reinforces the need for a team approach. Contacting, via remote technology where necessary, a palliative care consultant or community pharmacist for provision of extra medication on the Friday may have prevented this unsatisfactory outcome. Pharmacists' role in palliative care is crucial for ensuring residents receive effective pain control. Continuous infusion of drugs widely used in dedicated palliative care units and some patients' homes would benefit some nursing home residents, emphasising again the value of collaboration. Accredited pharmacists can provide advice on complex medications regimens, involvement in family meetings and case conferences, as well as ongoing education for nursing home staff.

Many nursing homes pay particular attention to effective, efficient liaison with their local pharmacist, and GPs welcome such collaboration for the benefit of optimal symptom management for the residents.

5.2.6 Link Nurse

Staff allocation and appropriate funding arrangements would enable a 'link nurse' in every nursing home, whose role is to ensure residents' access to palliative care. The skills and expertise of a link nurse (or NP as described in Chap. 3) would pave the way for regular triage meetings to review the care of all residents, identifying those who require specialist palliative care. Such care is also *preventive* in that hospital transfers would be minimised and the GP would receive appropriate support and advice. Most important, it would optimise residents' care and comfort, and increase family satisfaction. In some locations, one link nurse or NP is employed to cover a region or a group of nursing homes; achieving very positive outcomes and reinforcing the professionalism of team work. Again, any additional cost is outweighed by the benefits to residents.

5.2.7 Involving Families

When families were asked in a research study what they would wish from nursing home care, they emphasised the following:

5.2 Palliative Care in Partnership

... recognizing and treating symptoms; assuring continuity in care; respecting resident's end-of-life wishes; offering environmental, emotional and psychosocial support; keeping family informed; promoting family understanding and establishing a partnership with family carers by involving and guiding them in a shared decision-making (Gonella et al., 2019, p. 589).

These components (cited above) reinforce the partnership approach to palliative care emphasised in this chapter. In contradistinction to evidence-based care, one of the mistakes/indiscretions common to nurses is to presume they know what families want or need, particularly at the end of the resident's life, leading to comments such as:

'You'll be wanting to come in for longer periods now'.

'We will call you the moment we notice any changes in his condition'.

'We will make sure we notify you in time, so you can be here when she breathes her last'.

These statements assume that the family of every dying resident will want to be present at the time of death; or that a nurse or carer can presume to make accurate predictions of death's timing. Such assumptions may cause distress, shame, anger, and confusion to a family who may have already planned their last visit. 'We've decided we want to remember mum as she is this afternoon. We've agreed we don't need to come in again'. Or, 'They said they'd contact me but they never did'. To avoid misunderstandings, staff are advised to check the documented family's wishes and preferences, bearing in mind they may change their mind, and various family members may make different decisions about visiting.

Another example highlights the need for greater staff sensitivity towards family visiting: in this case, promising the resident would not die alone.

An elderly woman had kept a bedside vigil for many hours, both day and night, as her husband was dying. She left the bedside briefly, during which time he died. She felt lasting, deep regret, as she'd promised to be with him until the end. The staff had also reassured her that he would not die alone.

This outcome may have been ameliorated by staff gently advising: 'We can't predict the time of death, and in our experience some patients "choose" to die when the family member has left the bedside. Others will await the arrival of a friend or family member and "choose" that moment to breathe their last'. Staff may reassure families that some people seem unable to 'let go' while family are present. Other staff may feel constrained to offer 'advice': 'He may be waiting for you to give him "permission" to go'. This statement is also presumptive: such 'advice' may have a negative impact. Advice such as: 'It's okay to let go' can put pressure on the dying person who may not be 'ready'. Staff are encouraged to lay aside any attempt to placate the situation by making assumptions about the dying process. Each death is unique and the best preparation is honesty, rather than predictions or presumptions, even by staff who have witnessed countless deaths.

Some families appreciate encouragement to participate in the resident's palliative care by attending planning meetings where their contributions may include issues otherwise overlooked. For example, one family member had taken considerable care to provide a well-illustrated 'social history/photograph album' for her mother, used to great advantage by family when visiting. They found this 'memory aid' assisted in sharpening their mother's interest, in spite of her increasing dementia. Although it only needed a few minutes of staff time, the daughter found it had been used only once or twice before it became 'lost' (Commonwealth of Australia, 2019, pp. 288–289). A more careful approach to planning would ensure the inclusion of this memory aid and its location, clearly documented as an essential component of care.

When well-meaning nurses and other carers focus solely on the resident, the needs of the family may be overlooked; assuming (wrongly) that they are well prepared for the death. Families' contribution to care is not always acknowledged; nor the physical and emotional exhaustion suffered by many, regardless of the satisfaction gained from their caring role.

> The markers of a transformed system will result in a society in which family caregivers have their own health and well-being considered, together with their rights and protections. They would also have access to evidence-based health information and support when they need it (Hudson et al., 2020, p. 16).

Support for families includes acknowledging the effects of death, particularly after a protracted period of care. While, in many circumstances, family communication is well documented within a carefully crafted care plan, in others, it is subject to the whim of isolated carers to highlight this important aspect of palliative care.

Family involvement is included in this chapter to complement other references in the chapter on dementia (Chap. 4) and the chapter on leadership (Chap. 7).

5.2.8 Palliative Care Education

Although Australian nursing homes provide care for those who are dying, they may not universally provide person-centre palliative care. Internationally, the first set of recommendations (twenty-eight in all) for palliative care in long term facilities focus on funds and capacity for ongoing education for all staff (Froggatt et al., 2020). Other studies reinforce the notion of clearly defined implementation strategies, embedding the education within day-to-day practice, thus 'sustaining ongoing change' (Collingridge Moore et al., 2020, p. 558).

It is interesting to note that as early as 1987, a government recommendation advised that palliative care education should be a significant and ongoing part of professional development programmes for all healthcare professionals, to change the single-minded focus on cure, aggressive and invasive treatments to a more acceptable education syllabus (Latta & MacLeod, 2018). Unfortunately, many years later, this focus is not always evident, particularly for 'support workers/care aides who spend the most time at the bedside and yet receive the least amount of education' (Kaasalainen, 2020, p. 555).

Creative managers, recognising the need, provide for regular palliative care education sessions, with the aim of optimising residents' care until their death.

5.2.9 *Palliative Care for Chronic Conditions*

Palliative care's original intent is now expanded to include patients with a variety of diagnoses and an unknown prognosis. It is well recognised that most of the chronic medical conditions experienced by nursing home residents would respond to a palliative approach; for example, Parkinson's disease (PD). In one study, five themes were identified: emotional impact of diagnosis, staying connected, enduring financial hardship, managing physical challenges and finding help for advanced stages (Hudson et al., 2006, p.87). It is evident that these themes fit the definition and application of palliative care. It would be interesting to note whether, after admission to a nursing home, a person with PD would be referred to specialist palliative care, in anticipation of their progressive, incurable disease trajectory. Early referral does not necessarily imply frequent palliative care input: an introductory assessment followed by monthly visits or regular phone contact is often sufficient, at least initially.

Another example is chronic obstructive pulmonary disease (COPD), which researchers have identified as fitting the criteria for palliative care. The advantage of such referral would be (a) to acknowledge COPD as a life-threatening illness not amenable to cure, (b) to recognise the need for psychological, social and community support and (c) to ensure chaotic, fragmented care with frequent emergency department visits is replaced by a proactive palliative approach (Philip et al., 2018, p.452). Staff may be reassured by receiving professional advice, translated into a specific plan of care for the frightening symptoms often associated with breathlessness. Families also may be reassured, as in the following:

> *'Ollie', a male resident, aged 89 and suffering from longstanding COPD, advanced dementia and other painful comorbidities was transferred from hospital to a nursing home catering for residents with palliative care needs. His family were distraught when recounting the number of emergency hospital admissions their father had to endure while in the previous nursing home, and with no speech to articulate his suffering. They were now reassured that a palliative care physician would assess his needs soon after admission, prescribing comprehensive crisis treatment for exacerbations of his COPD to be administered in his familiar surroundings rather than having to endure more unsettling, distressing hospital transfers.*

Optimum care for 'Ollie' followed recommendations that clinicians adopt as standard daily practice the integration of geriatrics and palliative care principles (Iyer et al., 2020), avoiding hospitalisation wherever possible. One example of this approach suggests the judicious use of regular, low-dose, oral sustained-release morphine for moderate to severe breathlessness in COPD, bringing welcome relief and with no adverse effects (Verberkt et al., 2020). Such clear recommendations confirm the benefits of seeking advice from those with relevant, advanced clinical knowledge and expertise in prescribing.

5.2.10 Compassion, Sympathy, Empathy

Palliative care emphasises aspects of a patient's experience other than physical or clinical, such as compassion, sympathy, empathy. Although used interchangeably, the differences have been described.

> Sympathy was described as an unwanted, pity-based response to a distressing situation, characterized by a lack of understanding and self-preservation of the observer. Empathy was experienced as an affective response that acknowledges and attempts to understand individual's suffering through emotional resonance. Compassion enhanced the key facets of empathy while adding distinct features of being motivated by love, the altruistic role of the responder, action, and small, supererogatory acts of kindness. Patients reported that unlike sympathy, empathy and compassion were beneficial, with compassion being the most preferred and impactful (Sinclair et al., 2017, p.437).

Compassion is a word needing careful application so it is not regarded as merely a passive sense of pity. To be compassionate, literally to 'suffer with', one's actions inevitably go beyond sympathy and empathy. Compassion in the aged care context is not, therefore, an optional extra, an irrelevant extravagance, or confined to those with specific expertise. Compassion lies at the heart of high-quality health care delivered by all staff: it is not merely one-to-one communication; it involves relationships throughout the entire nursing home community. The key is focussed leadership, including specific training and policy design, so that a nursing home may become a 'compassionate community' (Brito-Pons & Librada-Flores, 2018). Such was the case for 'Piero'.

> Piero had lived a solitary existence, leaving his squalid room in a boarding house only to replenish his alcohol supplies. When admitted to the nursing home, unaccustomed to women fussing over him, he became intolerably abusive. Knowing only the language of his fists to settle an argument, he often became physically aggressive after which he would apologise and appear humbly contrite until the next time. His alcohol affected brain seemed unable to discern right from wrong, nor adjust to any form of behaviour modification. We would not transform Piero into a gentleman.

> His final illness came suddenly and when called to his bedside the manager was moved by the terror in his eyes. 'Matron' he said (unaccustomed to modern first name terminology), 'am I in danger?' While searching for an appropriate answer the manager was reminded that the knowledge and wisdom of the dying person so often surpasses that of the professional health carer. 'I think I'm dying,' said Piero. As an Irish Catholic he knew about last rites, confident that this ritual would take care of his soul, but he had other concerns.

> 'What else is on your mind, Piero?' asked the intuitive manager. 'I'm scared, Matron.' It was evident from his contorted face and heaving chest that this was the fight of his life. Frightened of the pain and of being alone, he believed it was his lot to die this way, he did not deserve to die peacefully. While his soul was assured of a safe passage his body must bear the guilt of his past.

> Pleading not to be sent to hospital, Piero received all the benefits of palliative care in his last 48 hours. Lacking the energy to abuse staff, he accepted the close and intimate care his body required, even allowing the night nurse to massage his back and stroke his hand. With fragrant oils burning at his bedside, a tape of Irish lullabies playing softly and administration of regular small doses of opioids, the pain and terror gave way to a peace and calm

never before seen in Piero. He did not die alone or in fear, nor did he lose his sense of entitlement, commenting on the Irish Welfare Agency's intent on 'getting their hands on' his money.

'Piero', while not always easy to care for, was denied none of palliative care's benefits. In the absence of family, nursing home staff provided him with 'treats' and tender, compassionate, skilled, hands-on care in the final chapter of his life.

5.2.11   Compassion Fatigue

Although 'compassion fatigue' has been the focus of research since at least 1992, little has been written about its relationship to nursing (Turner et al., 2019) or how it may affect a partnership approach between aged care and palliative care. As one article describes it: compassion fatigue is the 'loss of the ability to nurture'. This can be evident in the most experienced nurses and other carers, arising in unexpected ways and with various physical and psychological indicators. Increasingly, the focus is on nurses and carers to be aware of their own health.

> Compassion fatigue progresses from a state of compassion discomfort to compassion stress and, finally, to compassion fatigue, which if not effaced in its early stages of compassion discomfort or compassion stress, can permanently alter the compassionate ability of the nurse (Coatzee & Klopper, 2018, p.235).

Another writer explains: 'Self-care is not selfish-care. It's what makes you a better nurse' (Williamson, 2019). Compassion fatigue may also occur among non-professional and informal carers; supervisors need to be aware of this issue, which can often be overlooked and remain unaddressed. Compassion fatigue may arise when there is an unusually high number of nursing home deaths over a short period of time, or when a resident's dying phase seems unusually protracted. Partnering with colleagues may help to reduce this distressing phenomenon, and sympathetic, empathic management is fundamental.

5.3   Pain: A Vital Sign

The Royal Commission into Aged Care Quality and Safety recognises the need for increased emphasis on pain management as an essential component of nursing home care. 'Residential aged care is a place where people will and do die, and that experience should be as free from pain and fear as possible . . .' (Commonwealth of Australia, 2019, p. 117).

Noting the difficulties of accurately assessing pain, particularly in the context of a resident's cognitive impairment, it is estimated that 'up to *ninety-three per cent*' (Goucke, 2018, p. 19) (italics added) of nursing home residents are in pain which is unrecognised and therefore untreated. 'Even if pain is not always visible, it is always real. It is possible to experience pain even without tissue injury' (Goucke, 2018, p.19).

'Pain should be considered the "fifth vital sign" after temperature, pulse, blood pressure and respiration rate....Persistent pain should be considered a disease in its own right not just a symptom of disease' (Australia & New Zealand Society for Geriatric Medicine, 2016). Another well-published clinical fact is that *pain can kill*. 'Pain can... have a major impact on morbidity and mortality. ... it can mean the difference between life and death' (Liebeskind, 1991, p. 3). Contrary to the common assumption, 'pain won't kill you', this decades-old research remains relevant today, urging increased referrals to a palliative care physician or pain specialist for the many residents with chronic painful conditions.

Comprehensive pain management requires collaboration between nursing staff, GPs, pharmacists, therapists and other experts. Drug dose and frequency deserve particular attention; for example, when the drug for chronic pain is ordered *prn (pro re nata)*, meaning 'when necessary', misunderstandings often ensue. Unfortunately, a judgement about 'when necessary' is too often left to the whim of the nurse holding the drug cupboard keys. Use of a widely recommended pain assessment tool is an important guide to effective pain management; when the pain is 'constant' the analgesia needs to be 'constant' – that is, given regularly. That means *prn* dosing should be reserved for *intermittent* pain, whereas most residents' painful comorbidities are associated with *constant* pain. It remains a sad fact, however, that many residents with persistent, chronic, painful conditions only receive intermittent, ad hoc analgesic relief. The analgesic order needs to be consonant with the resident's painful diagnoses and symptoms. In this respect, the ubiquitous 'prn Panadol' seems incommensurate, if not entirely inadequate.

A multidisciplinary approach is recommended, involving families in pain assessment and treatment where possible. When pain management is discussed within a team, a holistic plan is more likely to emerge. Comprehensive assessment indicates the need for broader, deeper questions, rather than a mere 'Are you in pain?' or 'Do you think he/she is in pain?'

Pain is a subjective experience, defying the all-too-common assumptions formed by nurses and other carers: 'I think she's in pain' or 'I don't think he's in pain'. A palliative approach to pain management assumes its complexity, all elements informing a plan based on thorough assessment, clearly articulated goals, multidisciplinary discussion, regular evaluation and review.

Pain management education and increased staff awareness may prevent differing opinions such as the following:

'I think he's in pain. He's crying'.

'I don't think he's in pain: he just gets a bit teary when his wife leaves'.

'I think her dementia masks her pain, so she doesn't need painkillers'.

'I think her constant screaming is due to pain: it's not just her dementia'.

Pain is not a matter of *opinion*. Pain assessment belongs to the highest order of clinical judgement: the pain experience needs to be understood in a multifaceted way. What social issues are impacting the resident's daily life? Have psychological and spiritual needs been assessed? How has this person responded to painful conditions in the past? Biography is an essential component of pain assessment. The professional satisfaction arising from good pain management, not to mention the life-changing benefits for the resident, cannot be overstated.

Reliable tools are the key to thorough assessment. The Abbey Pain Scale is recommended for use with residents whose cognitive impairment compromises their description of pain, especially for those unable to give an accurate response to the pervasive 'Any pain today?' or 'How's the pain?' questions. The Abbey Pain Scale is 'quick to administer and assesses non-verbal pain cues including behaviour along with physiological changes' (Goucke & Slatyer, 2021, p. 230). However, these authors emphasise that, in response to symptoms of concern, it 'does not discriminate between pain and other causes'. Repeating the assessment following analgesia is said to provide a guide as to the cause, and the use of other pain assessment tools is recommended for residents with no cognitive impairment.

It is worth noting here the often-neglected link between physical pain and existential suffering. Most carers in this context agree that residents seldom initiate reports of suffering, leaving it to staff to conduct regular and reliable assessments, ensuring the best possible attention is given to the alleviation of pain in all of its manifestations. The common question 'are you in pain?' is no substitute for a professional assessment using an appropriate, reliable rating scale.

While pain management in nursing homes may not attract the highest effectiveness rating, neither is it always without criticism in acute care. One would expect all patients, regardless of their age, in a hospital palliative care unit, to have optimum pain management. Contrary to this assumption, an analysis of older people in the last three days of life showed they received fewer analgesics than younger patients at a similar stage of illness and with comparable symptoms (Rashidi et al., 2011). Moreover, they were unable to articulate their distress. Older people are often less able to communicate their level of pain than a younger more articulate person, which points to the need for stringent assessment procedures, especially where the older person lacks language to describe their pain. The urgency of addressing this issue is pertinent for the majority of nursing home residents.

One example, of many, is included here to emphasise the need for urgent attention to investigating all instances of pain. A glance at the list of comorbidities included in a (typical) list of a resident's diagnoses reveals the high incidence of *osteoarthritis* (although accurate data is difficult to access) deserving an albeit brief mention here because of the pain associated with its end stage (stage four). In the absence of joint cartilage, the 'bone on bone' syndrome causes unrelenting pain, even when the resident is at rest (Australian Institute of Health and Welfare, 2020). Other research reveals the increasing intensity of arthritic pain in the three months prior to death. Thorough, regular assessment of every resident with this painful, terminal condition would surely lead to an apposite response, including meticulous use of *regular analgesia,* adjusting the dose as the symptoms increase.

Professional carers are not always aware of the 'hidden' nature of pain, nor of the reluctance/inability of many residents to self-report severe suffering, leading to underuse of opioids when, in other settings, they would be more widely used. Drawing on the vast literature of Dame Cicely Saunders, comprehensive pain assessment would emphasise the narrative, biographical nature of pain and suffering; aspects not always appreciated by modern medicine and not readily amenable to medication (Lucas, 2012). Saunders also makes a categorical statement about the relationship of opioids to addiction, speaking of

> '... the unceasing need to influence the education of physicians and the nursing profession and, indeed, to enlighten the general public of the fact that morphine and other opioids properly used do not lead to drug dependence and addiction. These fears often deny relief and have been shown to be unfounded' (Saunders, 1996, p. 321).

Saunders understood that the pain experience is total, encompassing social, psychological and spiritual aspects. Although her research focused on people dying of cancer, the concept of *total pain* resonates strongly when considering nursing home residents with their multiple comorbidities. Decades later, it appears the advice from a renowned world authority on this issue is too readily ignored.

5.3.1 Learning from the Literature

Leo Tolstoy's 'The Death of Ivan Ilyich' (Tolstoy, 1960), first published in 1886, provides unique insights into contemporary pain management. Part of Ilyich's torment was located in the failure of his friends, family and doctors to recognise that he was dying, and to acknowledge the existential reality of his pain. What seemed 'true' for this fictional character may be all too readily observed in the anguish experienced by those nursing home residents whose pain is not adequately assessed or addressed. Thorough analysis involves looking beyond what is immediately observable. In the words of one patient: 'I look for someone to look as if they are trying to understand me' (Saunders, 1965).

5.3.2 Learning from the Patient

Another example of looking 'beyond the obvious' comes from Dr. Archie Cochrane (of the Cochrane Collaboration), recording an event at a prisoner of war camp when a young Soviet prisoner was admitted to his ward, who had 'gross bilateral cavitation and a severe pleural rub'. In Cochrane's words:

> I thought the latter was the cause of the pain and the screaming. I had no morphia, just aspirin... I felt desperate. I knew very little Russian. .. I finally instinctively sat down on the bed and took him in my arms, and the screaming stopped almost at once. He died peacefully in my arms a few hours later. It was not the pleurisy that caused the screaming but loneliness. It was a wonderful education about the care of the dying. I was ashamed of my misdiagnosis and kept the story secret (Wiffen, 2003, p.75).

Cochrane's story serves as a sharp reminder of the multi-faceted nature of pain, and the amelioration that comes through a tender, tactile response. Education about the biopsychosocial model of pain would encourage staff to address residents' needs beyond the immediately observable. It also indicates that, notwithstanding the well-proven effects of most analgesia, other non-drug responses may also be very effective.

Another example of total pain concerns a resident who complained to her family each time they visited of 'unbearable pain all over her' (Goucke, 2018, p. 51). She said that 'moving makes everything worse', so she spent most of her time hunched in her chair, too scared to report her pain for fear she would be 'sent off' elsewhere. This example is a reminder for staff to look further than the physical manifestations of a resident's posture, to ask the more probing questions of why they are reluctant to move, and with gentle encouragement to uncover their fear of reporting pain.

Examples cited in this section on 'pain' merely skim the surface of the volume of literature and research into this vast subject, pointing to the need for increased regular education sessions on pain management to be translated into practice. Taking seriously and re-emphasising the estimation (above) that *ninety-three percent* of residents may be experiencing pain (Goucke, 2018), it would be a welcome, timely, evidence-based response to find that *ninety-three percent* were receiving *regular* pain relief in response to continuous assessment. Residents may then enjoy the remainder of their lives in relative freedom from their hitherto unrelieved suffering.

5.4 Palliation: A Cloak of Care

5.4.1 *Designing the Cloak*

Metaphorically, palliative care (from the Latin '*palliare*' meaning to 'cloak', cited in 6.1) is likened to a cloak of warmth and protection for those who are dying. A cloak offers protection from dangerous assault by ineffective, futile, or inappropriate interventions; its purpose is to protect the inner layers but it is far more complex than a cosy, cotton-wool cocoon. 'In palliative care symptoms are "cloaked" with treatments whose primary aim is to promote comfort' (Twycross, 2003, p. 9). Going beyond symptom relief, palliative care embraces all aspects of the dying person's life, helping them to face their impending death. When applied correctly, the dying person remains free to accept death on its natural terms, and the cloak is large enough to cover family and carers; embracing them as true partners in one unit of care.

The following scenario concerns a family member who wanted direct involvement in decisions regarding the care of her older sister, a nursing home resident who was dying.

> *'Miss A', aged 102, too frail to remain at home, required high level nursing home care. Fiercely independent, she wanted her wishes respected. "Don't you ever let them send me to hospital," she told the nurse manager. Reluctantly at first, she welcomed her "baby sister," aged 85, to attend care planning meetings. Previously refusing pain medications, Miss A was persuaded by her younger sister to have confidence in all the symptom management offered, reinforced by relevant written information to guide decision making. Miss A's sister was most grateful to be involved in the care planning and to witness her older sibling's peaceful, seemingly pain free death. "I had no idea about palliative care before this," she said. "This is what I want when my time comes."*

From a different perspective, and a former era, Florence Nightingale has some words of wisdom to offer about end-of-life care in her chapter entitled 'Chattering Hopes and Advices':

> I really believe there is scarcely a greater worry which individual have to endure than the incurable hopes of their friends. There is no one practice against which I can speak more strongly from actual personal experience, wide and long, of its effects during sickness observed both upon others and upon myself. I would appeal most seriously to all friends, visitors, and attendants of the sick to leave off this practice of attempting to 'cheer' the sick by making light of their danger and by exaggerating their probabilities of recovery (Nightingale, 1969, p.237).

Misplaced emphasis on 'cheer' and 'making light of their danger' is all too common, as in the following:

> *The nurse accompanied the GP to the bedside of a resident whose condition was, according to the nurse's assessment, deteriorating daily. 'I hope it will soon be over', said the resident in a feeble voice. Looking a little awkward, if not embarrassed, the GP responded: 'O, don't be silly Dora, cheer up, you've still got plenty of life!'*

Away from the bedside, the nurse appealed to the GP for a referral to the palliative care team for 'Dora', to which he reluctantly agreed. In a follow-up conversation, the GP admitted to the nurse his reluctance to prescribe even low dose opioids for a very frail ninety-eight-year-old. 'This is out of my league', he confessed. 'And, I didn't know what to say when she talked about wanting it to be over'. The nurse ensured the resident's concerns were fully addressed by a referral to the palliative care team, whose response included a multidisciplinary team meeting to cover other concerns, including the family's needs and the GP's reticence.

5.4.2 Cloak in a Basket

'We always get out the palliative care basket when anyone is dying' the nurse proudly stated. What was in the basket? Lavender oil, a clock, a small satin cushion, body oils and lotions, taped music. It was assumed that every dying resident would be comforted by these items, including the limited music selection. It was unclear in the conversation with this nurse as to *when* or at what point the palliative care basket would be brought to the resident, and the same basket was used for everyone. While some aspects of this care may be applauded, it seems at odds with palliative care's focus on the uniqueness of each person and their family, where attention is given to their carefully assessed preferences and idiosyncrasies, as well as expert symptom management. The basket's contents may well have provided a welcome cloak for some; for others an unacceptable, unwanted imposition, if not an affront to their senses.

5.4.3 A Cultural Cloak

'Cultural differences in end-of-life care matter a great deal and merit our most stringent scrutiny... failure to take into account cultural differences in end-of-life care can result in preventable and culpable moral harm' (Johnstone, 2012, p. 187). Such failures represent the opposite of a cloak of care. 'It fundamentally involves being "wronged"' (Johnstone, 2012, p.187). Careful assessment, including discussion with family, engaging a qualified interpreter where needed, would ascertain the distinctive cultural needs of every resident prior to their death. For example, the meaning of words can be misinterpreted, particularly for residents and relatives whose first language is not English. It may be more instructive for all parties to utilise questions rather than assumptions. 'What do you know about your situation at present?' 'What would you like to know?' Rather than waiting until the person is close to death, full discussion and documentation soon after admission would identify the resident's cultural preferences, ensuring the cloak is fit for purpose.

A cultural cloak is needed where there is a language barrier between the resident and staff. For example, it is not uncommon for a person with dementia whose first language is not English, to revert to their primary language as the disease advances. A health care worker with skills in dementia, language and culture may offer advice on specific aspects of care and communication, ensuring the cloak is adjusted accordingly.

5.4.4 A Legal Cloak

One account of poor pain management, uncovered by research into nurses' knowledge of the law at end of life, found that nurses were concerned about legal or administrative disciplinary repercussions when a patient died following administration of pain medication (Willmott et al., 2020). The following comments epitomise the confusion.

> 'Did you hear about 'Jenny'? She gave the man in room three the maximum morphine dose and he died thirty minutes later!'

> 'I'm not giving the woman in room fifteen any opioids. I don't want to be the last person to give a drug before her death'.

Aside from the impersonal references to residents by their room numbers, the comments above display a poor understanding of pain relief, including lack of confidence. Interviews with nurses revealed that pain and other symptom relief was often compromised by their hesitation to 'give the last dose', and their failure to understand the doctrine of double effect. It is clear from this, and other examples, that nurses need increased confidence in the law: particularly the protections in place for giving appropriate medication at the end of life (Willmott et al., 2020). The focus is on the goal of care, in this case, administering opioids for the purpose of relieving pain, carefully documenting the rationale in the resident's care plan. The law of double effect is not well understood, which addresses the specific issue of such legal protection: another reminder for nurses to practise from knowledge rather than opinion.

> The doctrine of double effect is an ethical principle dating back to the thirteenth century that explains how the bad consequences of an action can be considered ethically justified if the original intent was for good intention (Wholihan & Olson, 2017, p. 205).

In other words, administering legally prescribed opioids to a resident who is in pain, albeit when close to death, is legitimate, particularly when substantiated by clear documentation regarding the goal of care, for example, to alleviate pain and/or suffering. A legal cloak, well understood, provides reassurance and confidence rather than ambivalence or fear.

Fortunately, we have long passed the era when unreasonable, not to mention undignified, measures were used for an older person close to death, described in this account. 'And when they cannot swallow any more someone rams a tube down their gullet, or up their rectum, and fills them full of vitaminised pap, so as not to be accused of murder' (Beckett, 1951, p. 118).

This description, albeit from a former era, portrays in graphic detail some actions arising from misunderstanding the law!

5.4.5 A Comforting Cloak

The prescription for 'comfort measures' has many different connotations, as in this scenario:

> 'Mrs. Petroska' was transferred to the nursing home from acute care where it was determined 'nothing more can be done, so just give her comfort measures'. Rather than signifying a hopeless situation indicating absolutely no professional care options for Mrs. Petroska, the comfort measures occupied a substantial entry in the care plan: pain relief medication, her favourite music, flowers at the bedside, snacks of her choice and other creative comforting gestures. Both family and staff were constantly reminded that 'comfort measures' do not denote a withdrawal of care. Rather, they constitute the best of end-of-life care.

Emotional and social care deserve a place in the documented plan – essential components of a comforting cloak, particularly at the end of life. Lack of social

5.4 Palliation: A Cloak of Care

contact can negatively impact death's timing; enabling social relationships can increase happiness and longevity.

While end-of-life care does not lend itself to prescriptive principles or easy answers, there is a danger in providing false comfort – the antithesis of a well-fitting cloak. A first-person account is given by Wolterstorff (1987) who pleads with his comforters:

> Death is awful, demonic. If you think your task as comforter is to tell me that really, all things considered, it is not so bad, you do not sit with me in my grief, but place yourself off in the distance away from me. Over there, you are of no help. What I need to hear from you is that you recognise how painful it is. I need to hear from you that you are with me in my desperation. To comfort me, you have to come close. Come sit beside me on my mourning bench (Wolterstorff, 1987, p. 34).

The 'mourning bench' is no place for a carer wishing to pathologise or problematise suffering and death. It is the legitimate place for the carer who wants to 'come close'. The mourning bench may be a frightening place for someone who prefers to remain behind the desk in attitude or physical posture. It is the place for open communication. The mourning bench is no place for the carer who feels impotent without tasks to perform or a curative salve to offer. The vacant place on the mourning bench is for the companion who does not resort to pious platitudes or sentimental clichés; it is the very special place for the very ordinary action of one human being sitting beside another.

The notion of false comfort is starkly exposed by de Beauvoir (1969) as she takes the reader through a moment-by-moment description of the difficulties surrounding her mother's death from cancer. She describes with stark realism and graphic detail how the last hours were fraught with anxiety and a fight against death; her mother experiencing frightening spasms and difficulty in breathing. In the face of this, the staff were full of reassurance telling Simone and her sister that the death would be calm and peaceful. Describing her mother's final moments, at the end of agonising weeks of pain and distress, the author takes issue with the statements of those in charge of her mother's care. 'The doctors said she would go out like a candle: it wasn't like that, it wasn't like that at all,' said my sister, sobbing. 'But, madame', replied the nurse, 'I assure you it was a very easy death' (de Beauvoir, 1969, p. 78).

False consolation is the antithesis of a comforting cloak.

5.4.6 A Cloak for Suffering

Evidence continues to mount as some nursing home staff, albeit well-intentioned, attempt to soften suffering and deny death's finality. Skilled communication is required to describe death's subtleties without alarming families, or without feeling the need to dissect every detail. While gerontic nurses may not presume to know the 'right answer', the least they can do is offer an honest account of the situation, conveyed with compassion and a close, caring personal presence. When health professionals state, 'We'll make sure your mother will not suffer', what reassurance is offered as to how this promise can be achieved? As one writer puts it, 'Once avoiding suffering becomes the *primary purpose* of society, it too easily mutates into a license for *eliminating the sufferer. . .*' (Smith, 2001, p.222).

Palliative care does not claim to eliminate all suffering; it is, therefore, important to be frank and truthful when advocating its benefits. Broad understanding reveals there is some suffering for which no drug is totally effective. Similarly, documented goals of care, which include 'pain free', are unrealistic and are better described as 'minimised pain' or 'alleviation of pain' or 'mitigation of pain'. Suffering is a complex phenomenon, including far more than physical pain; it should not be reduced to a symptom a health professional can eliminate or cover up or accurately measure.

Howarth notes the emphasis within hospice care on the relationship between physical pain and other symptoms. 'In other words, if physical pain can be controlled, this will reduce emotional and spiritual suffering and, therefore, improve quality of life for dying people' (Howarth, 2007, p.141). Timing is of the essence. When palliative care is left until the very end of life, opportunities are missed for optimum pain and symptom management, engagement with families, and responding to residents' choices. The 'cloak' then fails to provide proper protection.

5.4.7 An Ill-Fitting Cloak

A cloak needs to fit well the person it is designed to protect. An ill-fitting cloak assumes that measurement is not needed, so a standard size is used. For example, a well-meaning carer wants to 'tell the truth' to all residents who are dying, including those whose cultural and religious belief would be violated by candid talk of death. This cloak is not made to fit the individual; it takes no account of each unique resident and family. When the cloak does not fit, suffering is compounded. While some residents and families will defiantly remove the cloak, or demand a different size, many will continue to 'wear it' for fear of causing offence or being charged with ingratitude.

The following scenario reveals one of the many types of 'ill-fitting' cloaks offered to older people:

> *'Mr Clark', a 78-year-old nursing home resident with Lewy Bodies Dementia (DLB) was regarded as a 'challenge.' 'Behaviour management' dominated his care plan. His cloak was not protective; the inner layers of this complex form of dementia were not adequately assessed and the 'treatment cloak' of antipsychotic medication was not only ill-fitting, it was dangerously inappropriate. Symptom management for Mr Clark's other serious medical problems was not addressed, and his suffering was not ameliorated. A protective cloak would have included pain management for the peripheral neuropathy associated with his diabetes and for the osteoarthritis pain, which had escalated in recent months. The care plan was not discussed with his family nor was advice sought from the palliative care team or a clinician with DLB expertise.*

A close-fitting cloak has been compromised in some situations in response to the COVID-19 pandemic, giving rise to situations where 'risk management' has replaced holistic interdisciplinary palliative care and leaving residents to suffer from isolation through very restrictive social distancing procedures and no (metaphorical) cloak on offer.

To 'cover' suffering as much as possible, the source needs to be 'uncovered' so the cloak fits well, leaving no parts exposed.

5.4.8 A Psychiatric Cloak

The psychiatrist's role in palliative care for older people in nursing homes does not always receive the attention it deserves. For example, the distinction between psychosis and dementia is not well understood, so residents are denied the protective cloak of comprehensive psychiatric care. Instead, many are inappropriately prescribed antipsychotic drugs in the absence of any 'psychosis' diagnosis. The added burden of delirium and depression is discussed in Chap. 3, emphasised here also, with the strong recommendation for health professionals to arrange referral for specialist assessment. In other words, to ensure the 'cloak' is 'tailor made'.

Aged care staff are also urged to consider the mental health of older people who are dying, and the benefits of psychiatry in identifying issues of 'loss, grief, anxiety, depression, hopelessness, suicidal ideation, personality change and confusion' (Aziz & Saeed, 2019, p. 37). The link between palliative care and psychiatry is well researched, with geriatric psychiatrists offering a biopsychosociospiritual approach. Such a cloak envelops the resident with reassurance and responsive care.

5.4.9 A Cloak of Sedation

Palliative care sometimes involves discussion about sedation at the end of life, to investigate whether an additional layer of protection is needed for severe, complex, unresolved symptoms.

> Palliative sedation is a *medical treatment* applied *when necessary* to relieve intense suffering: it offers *individualized* relief from pain and suffering (caused by conditions such as severe agitation) *as the situation may warrant.* It is *not directed at causing death or ending the patient's life* (Smith, 2001, p.102) (italics in the original).

Palliative sedation may be provided intermittently or continuously, usually at the direction of a palliative care physician. Although not widely used in residential aged care, staff need to be aware of palliative sedation's advantages in certain circumstances, offering another 'protective layer' or 'cloak' when all other measures have failed.

5.4.10 Frailty in Need of a Cloak

In a former era, 'frailty' would not have been considered a legitimate diagnosis; such reference to the syndrome was regarded as clinically unprofessional. As the population ages and more older people experience a much longer illness trajectory from diagnosis to death, frailty surfaces as a serious issue needing a comprehensive diagnosis and care plan. Frailty is

> ... a clinical syndrome in which three or more of the following criteria are verified: unintentional loss of weight, patient-reported feelings of tiredness, slowness walking, muscular weakness and low levels of physical activity. In general, therefore, the frail elderly are the reflection of a clinical reality that is very frequent, and severe stages lead inevitably to death (Pialous et al., 2013, p.75).

Applying Pialous et al.'s definition (above), some arguments can be refuted: 'You don't die of frailty!' or 'He's very frail, but it's not terminal'. Discussing various definitions and concepts, other writers state that frailty 'is not simply present or absent but a spectrum that encompasses cognitive, functional, psychosocial and somatic domains' (Hamker et al., 2020, p.262). Another report found frailty is 'not only a physical and psychosocial concern, but is a condition affecting the whole human being in relation to the well-being of self and others' (MacKinlay et al., 2020). This report, including participants' statements, shows 'the spiritual and emotional aspects of frailty' deserve increased attention.

Palliative care for people with frailty shows that it should be tailored to meet the needs of the patient and the family (Stow et al., 2019). Another survey revealed that families lacked knowledge about frailty, had unrealistic expectations, and were largely ignorant of palliative care's benefits (Harasym et al., 2020). With appropriate palliative care services, including use of one of the many assessment tools ('frailty scales'), evidence suggests that in some instances frailty can be reversed, particularly when issues of nutrition and/or exercise are addressed.

'Frailty' in nursing home residents is a subject worthy of much more research and attention; with an accompanying 'cloak' to match.

5.4.11 Cloak or Cover-Up?

'As a society, we are not very sophisticated when we talk about serious illness and death. We talk of fighting, of battling against, of staying positive and of not giving up' (Philip, 2018). Talk about 'not giving up' or 'there's nothing more that can be done' leads to serious misunderstandings about palliative care's benefits. To equate palliative care with 'no more treatment' is to misunderstand the range of care responses, ensuring maximum pain control and other symptom relief, including psychological and social factors. Rather than being confined to imminent death or 'the end of the line', palliative care aims to maximise choices and options well before the end of life. This is not to suggest that palliative care overcomes death;

rather, the outcomes and benefits include less distress for families, and in some cases, improved survival rates.

Negative views may serve as a 'cover-up' when palliative care is discussed in the context of euthanasia. Rather than emphasising its benefits in 'cloaking' the person with expert symptom management, it is assumed death is the only means of comfort. While opinions vary, some scholars believe euthanasia is no substitute for quality palliative care and Palliative Care Australia (in its 2019 position statement) upholds the view that voluntary assisted dying 'is not part of palliative care practice'.

A cloak does not cover up, or cover over, or hide, like a mask; true care involves looking into the face of the person and responding. A cloak of clear communication allows for joy and sadness; laughter and crying; doubts and certainties; words and silence. When the face-to-face encounter is neglected, such as talking about the dying person in the third person, speaking *about them at handover* rather than *with them at the bedside*, a cover-up occurs. The resident is exposed rather than protected.

5.4.12 A Cloak of Hope

Even in a seemingly 'hopeless' situation, a protective layer can be offered, as it was for 'Joan'.

> *'Joan's family had assumed she was 'no longer there', due to her end stage dementia, but they did not wish to see her suffer. They had no idea she was eligible for palliative care. It seemed a hopeless situation as Joan could not explain her distress or calculate her pain level. A well-informed nurse suggested referral to the palliative care service for advice. Joan's family were assured that, even within the seemingly 'hopeless' context of dementia, hope could flourish. Not hope for a magic cure, but hope kept alive by thorough assessment, relief of distressing symptoms, finely tuned analgesia and other appropriate comfort measures until she died.*

Hope, in this context, 'is not to be equated with some esoteric intangible fantasy; neither is it isolated from impeccable clinical care. While hope may not be scientifically observable it can, nevertheless, be grounded in human care' (Hudson, 2006, p. 241).

This chapter is written to foster hope for residents, families and carers that, while being mindful of the reality of suffering and death, a palliative approach offers reassurance through multi-disciplinary evidence-based care. Nursing home staff, acting unilaterally, sometimes base their care on 'hoping for the best'. Rather than the illusory protection of wishful thinking, a cloak of hope envelops the resident and family with a well-fitting garment, personally measured and suited to their unique needs.

The time has come for every nursing home in Australia to become a 'palliative care unit', and for staff to be educated and paid accordingly.

References

Anstey, S., Powell, T., Coles, B., Hale, R., & Gould, D. (2016). Education and training to enhance end-of-life care for nursing home staff: systematic literature review. *BMJ Supportive & Palliative Care, 6*(3), 353–361.

Australia & New Zealand Society for Geriatric Medicine. (2016). Pain in older people. *Australasian Journal on Ageing, 35*(4), 293.

Australian Bureau of Statistics. (2018). Causes of death, Australia, 2017. Australian government.

Australian Institute of Health and Welfare. (2020). *Osteoarthritis.*

Australian Medical Association. (2015). *Palliative approach in residential aged care*. https://ama.com.au/position-statement/palliative-approach-residential-aged-care-2015

Aziz, V. M., & Saeed, R. (2019). Palliative care for older people: The psychiatrist's role. *BJPsych Advances, 25*(1), 37–46.

Barber, J., & Murphy, K. (2011). Challenges that specialist palliative care nurses encounter when caring for patients with advanced dementia. *International Journal of Palliative Nursing, 17*(12), 587–591.

Beckett, S. (1951). *Malone dies*. Penguin Books.

Brito-Pons, G., & Librada-Flores, S. (2018). Compassion in palliative care: a review. *Current Opinion in Supportive and Palliative Care, 12*(4), 472–479.

Carter, A. (2020). To what extent does clinically assisted nutrition and hydration have a role in the care of dying people? *Journal of Palliative Care, 35*(4), 209–216.

Chapman, M., Johnston, N., & Lovell, C. (2018). Avoiding costly hospitalisation at end of life: Findings from a specialist palliative care pilot in residential care for older adults. *BMJ Supportive & Palliative Care, 8*, 102–109.

Coatzee, S., & Klopper, H. (2018). Compassion fatigue within nursing practice: A concept analysis. *Nursing and Health Sciences, 12*(2), 235–243.

Collingridge Moore, D., Payne, S., Van den Block, L., Ling, J., & Froggatt, K. (2020). Strategies for the implementation of palliative care education and organizational interventions in long-term care facilities: A scoping review. *Palliative Medicine, 34*(5), 558–570.

Commonwealth of Australia. (2019). *Royal Commission into Aged Care Quality and Safety Interim Report: Neglect Volume 2.*

Commonwealth of Australia. (2021). *Royal Commission into Aged Care Quality and Safety Final Report: Care, Dignity and Respect, Volume 3A The new system.*

Commonwealth of Australia. (2019). *Royal Commission into Aged Care Quality and Safety Interim Report: Neglect Volume*, 1.

Damarell, R. A., Morgan, D. D., Tieman, J. J., & Healey, D. (2020). Bolstering general practitioner palliative care: A critical review of support provided by Australian guidelines for life-limiting chronic conditions. *Healthcare, 8*(4), 553.

Deerain, M. (2019). What will emerging technologies mean for the future of palliative care? *Palliative Care Australia.*

Dementia Australia Policy Team. (2019). Australia must improve palliative care for people with dementia. *CareSearch.*

Elsner, A. (2021). After COVID-19: The way we die from now on. *Cambridge Quarterly of Healthcare Ethics, 2021*(30), 69–72.

Erny-Albrecht, K. (2019, 16 January 2019). Understanding palliative care and why it matters. *CareSearch.*

Fan, L., Lukin, B., Zhao, J., Sun, J., Dingle, K., Purtill, R., Tapp, S., & Hou, X.-Y. (2018). Cost analysis of improving emergency care for aged care residents under a Hospital in the Nursing Home program in Australia. *PLoS One, 13*(7), e0199879.

Ferrell, B. (2019). Triggering palliative care in late-stage dementia [Viewpoints]. *Medscape Nurses.*

Forbat, L., Chapman, M., Lovell, C., Liu, W.-M., & Johnston, N. (2018). Improving specialist palliative care in residential care for older people: A checklist to guide practice. *BMJ Supportive & Palliative Care, 8*(3), 347–353.

References

Fox, B. M. (2020). *Looking behind the fear of becoming a burden*. HEC Forum.
Froggatt, K. A., Moore, D. C., Van den Block, L., Ling, J., Payne, S. A., Arrue, B., Baranska, I., Deliens, L., Engels, Y., & Finne-Soveri, H. (2020). Palliative care implementation in long-term care facilities: European Association for Palliative Care White Paper. *Journal of the American Medical Directors Association*.
Gabbard, J., Johnson, D., Russell, G., Spencer, S., Williamson, J. D., McLouth, L. E., Ferris, K. G., Sink, K., Brenes, G., & Yang, M. (2020). Prognostic awareness, disease and palliative understanding among caregivers of patients with dementia. *American Journal of Hospice and Palliative Medicine®, 37*(9), 683–691.
Gerber, K., Lemmon, C., Williams, S., Watt, J., Panayiotou, A., Batchelor, F., Hayes, B., & Brijnath, B. (2020). 'There for me': A qualitative study of family communication and decision-making in end-of-life care for older people. *Progress in Palliative Care, 28*, 1–8.
Gonella, S., Basso, I., De Marinis, M. G., Campagna, S., & Di Giulio, P. (2019). Good end-of-life care in nursing home according to the family carers' perspective: A systematic review of qualitative findings. *Palliative Medicine, 33*(6), 589–606.
Goucke, C. (Ed.). (2018). *Pain in residential aged care facilities: Management strategies* (2nd ed.). The Australian Pain Society.
Goucke, R., & Slatyer, S. (2021). Pain assessment and management. In C. Vafeas & S. Slatyer (Eds.), *Gerontological nursing: A holistic approach to the care of older people* (pp. 223–239). Elsevier.
Halifax, E., Bui, N. M., Hunt, L. J., & Stephens, C. E. (2020). Transitioning to life in a nursing home: The potential role of palliative care. *Journal of Palliative Care*, 0825859720904802.
Hamker, M., van den Bos, R., & Rostoft, S. (2020). Frailty and palliative care. *BMJ supportive & palliative care, 10*, 262–264.
Harasym, P., Brisbin, S. A., Sinnarajah, A., Venturato, L., Quail, P., Kaasalainen, S., Straus, S., Sussman, T., Virk, N., & Holroyd-Leduc, J. (2020). Barriers and facilitators to optimal supportive end-of-life palliative care in long-term care facilities: A qualitative descriptive study of community-based and specialist palliative care physicians' experiences, perceptions and perspectives. *BMJ Open, 10*, e937466. https://doi.org/10.1136/bmjopen-2020-037466
Harwood, R. H., & Enguell, H. (2019). End-of-life care for frail older people. *BMJ Supportive & Palliative Care*, 1–16.
Hashimie, J., Schultz, S. K., & Stewart, J. T. (2020). Palliative Care for Dementia: 2020 update. *Clinics in Geriatric Medicine, 36*(2), 329–339.
Hickman, S. E., Parks, M., Unroe, K. T., Ott, M., & Ersek, M. (2020). The role of the palliative care registered nurse in the nursing facility setting. *Journal of Hospice & Palliative Nursing, 22*(2), 152–158.
Howarth, G. (2007). *Death and dying: A sociological introduction*. Polity Press.
Hudson, P., Collins, A., Boughey, M., & Philip, J. (2021). Reframing palliative care to improve the quality of life of people diagnosed with a serious illness. *Medical Journal of Australia(Perspective)*, 1–9. https://doi.org/10.5694/mja2.51307
Hudson, P., Morrison, R., Schulz, R., Brody, A., Dahlin, C., Kelly, K., & Meier, D. (2020). Improving support for family caregivers of people with a serious illness in the United States: Strategic agenda and call to action. *Palliative Medicine Reports, 1*(1), 6–17.
Hudson, P., Toye, C., & Kristjanson, L. (2006). Would people with Parkinson's disease benefit from palliative care? *Palliative Medicine, 20*(2), 87–94.
Hudson, R. (2006). Nurturing hope at the end of life. *Ageing International, 31*(3), 241–252.
Hudson, R., & Richmond, J. (1998). The meaning of death in residential aged care. In J. Parker & S. Aranda (Eds.), *Palliative care: Explorations and challenges* (pp. 292–302). Maclennan+Petty.
Iyer, A. S., Curtis, J. R., & Meier, D. E. (2020). Proactive integration of geriatrics and palliative care principles into practice for chronic obstructive pulmonary disease. *JAMA Internal Medicine*. https://doi.org/10.1001/jamainternmed.2020.1088
Johnson, S., & Bott, M. (2016). Communication with residents and families in nursing homes at the end of life. *Journal of Hospice & Palliative Nursing, 18*(2), 124–130.

Johnstone, M.-J. (2012). Bioethics, cultural differences and the problem of moral disagreements in end-of-life care: A terror management theory. *Journal of Medicine and Philosophy, 37*, 181–200.

Kaasalainen, S. (2020). Current issues with implementing a palliative approach in long-term care: Where do we go from here? *Palliative Medicine, 34*(5), 555–557.

Kis-Rigo, A., Collins, A., Panozzo, S., & Philip, J. (2021). Negative media portrayal of palliative care: A content analysis of print media prior to the passage of voluntary assisted dying legislation in Victoria. *Internal Medicine Journal, 51*(8), 1336–1339.

Latta, L., & MacLeod, R. (2018). Palliative care education: An overview. In R. MacLeod & L. Van den Block (Eds.), *Textbook of palliative care* (pp. 1–21). Springer. https://doi.org/10.1007/978-3-319-31738-0_95-1

Liebeskind, J. C. (1991). Paincankill. *Pain, 44*(1), 3–4.

Liu, A., Koerner, J., Lam, L., Johnston, N., Samara, J., Chapman, M., & Forbat, L. (2019). Inproved quality of death and dying in care homes: A palliative care stepped wedge randomized control trial in Australia. *Journal of the American Geriatrics Society.* (4 November 2019).

Lucas, V. (2012). The death of Ivan Ilyich and the concept of 'total pain'. *Clinical Medicine, 12*(6), 601.

MacKinlay, E., Burns, R., & Mordike, S. (2020). Finding meaning in the lived experience of frailty: Final report.

Mallinson, R. (2015). Guidance at the juncture of palliation and old age. *The Gerontologist, 55*(3), 503–505.

Mataqi, M., & Aslanpour, Z. (2020). Factors influencing palliative care in advanced dementia: A systematic review. *BMJ Supportive and Palliative Care, 10*(2), 145–156.

Meilaender, G. (1991). I want to burden my loved ones. *First Things, 201*(25), 12–16.

Nightingale, F. (1969). *Notes on nursing: What it is, and what it is not.* Dover Publications, inc. (First published by Appleton and Company in 1860).

Palliative Care Victoria. (2018). *The process of dying: What to expect and how to help.* P. C. Victoria.

Persson, H. Å., Sandgren, A., Fürst, C.-J., Ahlström, G., & Behm, L. (2018). Early and late signs that precede dying among older persons in nursing homes: The multidisciplinary team's perspective. *BMC Geriatrics, 18*(1), 134.

Peskin, S. (2017). The symptoms of dying. *The New York Times.* (Section 4), 4.

Philip, J. (2018). *The language of living and dying.* Health & wellbeing, Pursuit.

Philip, J., Crawford, G., Brand, C., Gold, M., Miller, B., Hudson, P., & Lau, R. (2018). A conceptual model: Redesigning how we providepalliative care for patients with chronic obstructive-pulmonary disease. *Palliative & Supportive Care, 16*, 452–460.

Pialous, T., Goyard, J., & Hermet, R. (2013). When frailty should mean palliative care. *Journal of Nursing Education and Practice, 3*(7), 75–84.

Pollock, K., & Seymour, J. (2018). Reappraising 'the good death'for populations in the age of ageing. *Age and Ageing, 47*(3), 328–330.

Randall, F., & Downie, R. (2006). *The philosophy of palliative care: Critique and reconstruction.* Oxford University Press.

Rashidi, N. M., Zordan, R. D., Flynn, E., & Philip, J. A. (2011). The care of the very old in the last three days of life. *Journal of Palliative Medicine, 14*(12), 1339–1344.

Reigada, C., Arantzamendi, M., & Centeno, C. (2020). Palliative care in its own discourse: A focused ethnography of professional messaging in palliative care. *BMC Palliative Care, 19*(1), 88. https://doi.org/10.1186/s12904-020-00582-5

Saunders, C. (1965). Watch with Me. *Nursing Times.* (November 26).

Saunders, C. (1996). Hospice. *Mortality, 1*(3), 317–321.

Sellick, P. (2007). The power of death in a secular society. *Quadrant, 51*(5), 44.

Shaw, R. (2019). *Family fulfils end of life wishes of mother with advanced dementia.* https://momentsthatmatter.org.au/written-stories/family-fulfils-end-of-life-wishes-of-mother-with-advanced-dementia/

References

Sinclair, S., Beamer, K., Hack, T. F., McClement, S., Raffin Bouchal, S., Chochinov, H. M., & Hagen, N. A. (2017). Sympathy, empathy, and compassion: A grounded theory study of palliative care patients' understandings, experiences, and preferences. *Palliative Medicine, 31*(5), 437–447.

Slife, B. D. (2004). Taking practice seriously: Toward a relational ontology. *Journal of Theoretical and Philosophical Psychology, 24*(2), 157.

Smith, W. (2001). Culture of death: The assault on medical ethics in America.

Stolberg, M. (2007). " Cura palliativa". The concept of palliative care in pre-modern medicine (c. 1500–1850). *Medizinhistorisches Journal, 42*(1), 7–29.

Stow, D., Spiers, G., Matthews, F. E., & Hanratty, B. (2019). What is the evidence that people with frailty have needs for palliative care at the end of life? A systematic review and narrative synthesis. *Palliative Medicine, 33*(4), 399–414.

Tolstoy, L. (1960). *The death of Ivan Ilyich (R. Edmonds, Trans.).* Penguin.

Turner, S., Pratts, S., & Hutchinson, S. (2019). *The impact of a death and dying simulation on nursing students levels of empathy, attitude towards caregiving, and fear of death.* Brenau University.

Twycross, R. (2003). *Introducing palliative care.* The Radcliffe Medical Press.

Verberkt, C., van den Beuken-van Everdingen, M., & Schols, J. (2020). Effect of sustained-release morphine for refractory breathlessness in chronic obstructive pulmonary disease on health status: A randomized clinical trial. *JAMA Internal Med, 180*(10), 1306–1314.

White, N., Kupeli, N., & Vickerstaff, V. (2017). How accurate is the 'surprise question' at identifying patients at the end of life? A systematic review and meta-analysis. *BMC Medicine, 15*(139). https://doi.org/10.1186/s12916-017-0907-4

Wholihan, D., & Olson, E. (2017). The doctrine of double effect: A review for the bedside nurse providing end-of-life care. *Journal of Hospice & Palliative Nursing, 19*(3), 205–211.

Wiffen, P. (2003). The Cochrane collaboration: Pain, palliative and supportive care. *Palliative Medicine, 17*, 75–77.

Williams, S., Hwang, K., Watt, J., Batchelor, F., Gerber, K., Hayes, B., & Brijnath, B. (2020). How are older people's care preferences documented towards the end of life? *Collegian, 27*(3), 313–318.

Williamson, E. P. (2019). Self-care isn't selfish. Nurse.com. https://resources.nurse.com/self-care-isnt-selfish-2020

Willmott, L., White, B., Yates, P., Mitchell, G., Currow, D. C., Gerber, K., & Piper, D. (2020). Nurses' knowledge of law at the end of life and implications for practice: A qualitative study. *Palliative Medicine, 34*(4), 524–532.

Chapter 6
Death and Dying

6.1 Preparing for Death

6.1.1 Drawing on Ancient and Contemporary Literature

If anything pulls one out of ordinary time, it is death. A cartoon character says: 'The results of a study released today confirm that living is the number one cause of death' and for Tolstoy's character, Ivan Ilyich:

> Caius is a man, men are mortal, therefore Caius is mortal. . Caius was certainly mortal, and it was right for him to die; but for me, little Vanya, Ivan Ilyich, with all my thoughts and emotions – it's a different matter altogether. It cannot be thought that I should die. That would be too terrible (Tolstoy, 1960, p. 137).

Not content to stand in awe of death's *mystery*, we try to analyse, grasp, control and triumph over it or deny its existence – 'let's get on with living' – to which CS Lewis responds: 'It is hard to have patience with people who say 'There is no death' or 'Death doesn't matter'. There is death. . . . She died. She is dead. Is the word so difficult to learn?' (Lewis, 1961, p. 16).

Silence on the subject is evident in contemporary nursing homes, where almost every person has one or more incurable medical conditions resulting in death, up to one third dying within 6 weeks of admission (Tjernberg & Bökberg, 2020). While most other chapters in this book describe ways in which *living in a nursing home* may be enhanced, the subject of death 'so difficult to learn' and most often left until last is here brought to the fore.

Education is fundamental for giving prominence to this issue, namely, the inclusion of the subject of death and dying in study programmes. 'This will obviously

---

While (some of) the stories are based on factual situations, real names and other details have been altered to protect the identity of the persons concerned. Resemblance to any particular person is therefore purely coincidental.

require a broader cultural change, so death itself will not be considered as something to ignore, or fear, but accepted as part of life itself, something to talk about, and a crucial theme for every healthcare professional's training' (Testoni et al., 2020, p. 11).

Language is important when speaking about death within the nursing home, especially considering cultural diversity; a pivotal point when communicating with families. Cultural change in the broader sense is best achieved through staff learning to speak frankly, albeit with empathy, about death and dying. However, such discussions are often resisted, and euphemisms are often preferred to plain language (Omori et al., 2020).

6.1.2 The Conversation

Whose responsibility is it to initiate a discussion about death? The GP? Palliative care staff? The manager? The chaplain? While some staff find such conversations difficult, if not impossible, palliative care specialists suggest it is more distressing for families *not to have the conversation* (O'Connor & Allison, 2020). Concomitantly, there is a paucity of research indicating how death is regarded by the doctors caring for this population, and how their various attitudes impact each resident's dying process.

Some writers have identified the benefits of clear communication between staff and residents, emphasising interpersonal relationships and their impact on the way a good death is defined (Steinhauser & Tulsky, 2015). Apart from this research, another writer, two decades ago, said, 'Very little is known from the literature about how older people die in aged care facilities or about how a death in these institutions effects a spouse, other relatives, or other residents or staff. Such research is eagerly awaited' (Stevens et al., 2000, p. 182). More than 20 years on, are we any further advanced?

When conversations about death are initiated by staff, they do not necessarily give prominence to matters of religious, spiritual or cultural care at the end of life (Hudson, 2014, p. 186). Rather than talking about death as part of life, death's reality is a subject often prompting passivity, avoidance, derision or an inability to 'face' it. Staff not necessarily equipped to discuss the topic with families may instead offer false comfort.

'Oh, your dad will probably outlive you!'

'Rather than dwelling on negative outcomes think of the positives!'

'We hope you mum will be here for years!'

'We prefer to think about living rather than dying'.

Knowing, intuitively, about death's inevitability, there is no place for guilt, shame, awkwardness, reticence, alarm or secrecy on the part of staff. Open discussion with residents and families/representatives should become a normal part of

6.1 Preparing for Death

preadmission and admission processes. This is not to suggest the subject of death should be introduced as the first item on the agenda on the day of admission. Nor does it imply the subject is spoken about in a morbid fashion. Conversation with the resident may begin: 'What has the doctor explained to you about your condition/s?' Or 'What is important to you as you consider your future?' Or, explaining the necessity for accurate documentation: 'Have you and other family members discussed funeral arrangements for when the time comes?' When it is time for a more comprehensive discussion with a resident, the following questions may serve as prompts:

- *Do you have any thoughts about the end of your life, and what it will be like?*
- *Would you prefer death to 'take its toll' (in its own time) or would you rather have whatever 'life-saving measures' are available?*
- *Who would you like to be with you at the end?*
- *Do you worry about the dying process, being in pain, or death itself?*
- *Would you like to have more discussion with staff about death and dying?*
- *Would you prefer to discuss these matters with the chaplain (or another religious leader)?*
- *Does discussion of this topic make you feel uncomfortable or anxious?*

For a person with poor cognition or who lacks coherent speech, these questions would be directed to family or another carer. The discussion need not involve a protracted emphasis on death itself; a more positive focus would include advance care planning as preparation for decision-making, ensuring the resident's personal preferences are recorded. Cultural sensitivities must also be acknowledged, allowing for situations where death is not readily spoken about and discussion and decisions are deliberately deferred to others. With reference to the palliative care discussion in Chap. 5, a palliative care consultant may be required to facilitate conversation where there is reluctance to speak about the topic.

6.1.3 Euphemisms and Misunderstandings

In the absence of plain language, misunderstandings on the subject of death and dying may arise, such as the following:

> The agency night RN did not know the family of the resident who died unexpectedly at 3 am. She phoned the next of kin, trying to soften the message: 'I'm so sorry to ring you at this hour, but I need to tell you your mother has gone'. Prone to hearing reports of her mother's 'wanderings', and her own English being less than perfect, the daughter offered a polite, albeit sleepy response of thanks and with a courteous 'Good night' ended the call. The morning staff wondered why the daughter did not come in earlier than her usual 3 pm visit. It soon became clear. 'Where's mum? Have you found her yet?'

As this exchange (above) confirms, euphemisms may unintentionally invite confusion and misinterpretations. For example, to 'pass' has many meanings, not necessarily associated with death; 'pass the salt' may be a more common expression. 'Passed' may not immediately mean 'died' to the hearer; it may refer to 'another car passed us' or 'my son passed his exams'. 'Passed on' has connotations of message transmission rather than a definitive declaration of death. 'Passed away' has given ground to (merely) 'passed', adding to the confusion of many hearers. Other euphemisms, also subject to misinterpretation, include 'resting in peace', 'slipped away', 'gave up the ghost', 'succumbed', 'kicked the bucket', 'called home', 'gone to meet her Lord', or 'gone to a better place'.

To 'depart' is not necessarily confined to leaving one location for another, but to life's final end. Similarly, 'gone' may be understood as leaving the room or the nursing home, as in the following scenario.

> 'Nico' arrived at the nursing home for his weekly evening nursing shift. Tucked under his arm was the Greek newspaper he brought each week for 'Mr. Papandreou'. 'Where's Angelo?' he asked, puzzled by the unfamiliar resident in the other bed in the two-bed room. 'Oh, he's gone', said Jo, the other resident. 'Where to?' asked Nico. 'I don't know', said Jo. 'The night staff didn't say'. The night staff knew exactly where 'Mr. Papandreou' had gone – to the mortuary.

Others in the nursing home community need not be shielded from the truth when a resident dies; and confusing messages are to be avoided, even when the aim is to soften what is a stark reality. For example, informing a relative 'the journey is now over' may invite confusion: 'journey' has different meanings in certain cultures and circumstances. 'Gone over to the other side' can mean crossing the corridor to the room opposite. 'We've lost your mum' can infer the person has (merely) gone missing. Rather than direct language, phrases such as 'lost his fight', 'battled with dementia' are commonly used in the public domain in preference to the unambiguous terminology of death and dying. Professional communication within the nursing home requires confidence and clarity.

Whatever the language used; staff need to ensure the message is clear, delivered with compassion and sensitivity. Although in most circumstances the words 'death' or 'died' are preferred, for some families such plain language is countercultural or offensive. Conveying a message includes cultural awareness, ensuring the hearer has heard and understood, especially if the communication is by telephone or in unfamiliar language. Assumptions are to be avoided, together with thoughtless 'assurances' such as:

- 'Time heals'
- 'God must have wanted her even more than you'
- 'You'll develop other relationships'
- 'You're such a strong person: you'll be okay'
- 'Now, you'll have some time on your own, to take up new interests and enjoy life'

Nightingale refers to words of false consolation as 'chattering hopes' which, she says,

> '. . . may seem an odd heading. But I really believe there is scarcely a greater worry which invalids have to endure than the incurable hopes of their friends. There is no one practice against which I can speak more strongly from actual personal experience, wide and long, of its effects during sickness observed both upon others and upon myself. I would appeal most seriously to all friends, visitors, and attendants of the sick to leave off this practice of attempting to "cheer" the sick by making light of their danger and by exaggerating their probabilities of recovery' (Nightingale, 1969, p.9).

Nightingale's wisdom is an apt reminder to those who prefer to 'cheer' rather than speak plainly of death's finality. Rather than offering 'chattering hopes' or other false consolation, more respect would be shown by a question such as: 'has this come as a surprise?' or, 'is there someone else you would like to discuss this with?' or 'would you like me to call the social worker or chaplain?'

6.1.4 Preference and Choice

Many families struggle with the process of nursing home admission; some remaining ambivalent, if not fraught, by the effects of the decision, often a reminder that 'this is the last chapter'. Such was the case with the members of a devoted family, devastated by the realisation they could no longer support their mother at home.

> *Following their mother's uneasy settling in period, the two daughters and son acknowledged she was receiving high quality care, with expertise beyond their own. When it seemed their mother's death was imminent, a family meeting was called to discuss their mother's care. When one daughter reiterated her regret, sadness and disappointment that their mother would not die in her own home, the manager was inspired to ask: 'Would you like to take your mum home to die? We could arrange palliative care visits and, if the doctor agrees, her pain management would continue without any changes'. The family, greatly appreciative of the offer, requested twenty-four hours to think it over. Returning the following morning, the older daughter told the manager: 'We discussed this option carefully and after serious thought we all agree that we wouldn't want to disrupt the continuity of care, and we're more than happy with the care provided here. But we are enormously grateful for being given the choice'.*

While the offer of being discharged home to die may not occur often, neither would it be appropriate in many circumstances; this account highlights the importance of choice, even for a long-term resident.

Researchers have discovered low rates of residents' recorded preferences for place of death and other issues important to them at the end of life. Rather than waiting for a crisis in order to raise these matters, it is recommended such discussions be initiated earlier, enabling comprehensive documentation about the resident's wishes and care preferences (Williams et al., 2020, p. 313).

6.1.5 Dying Naturally

Western medicine promises technological and pharmacological interventions often with the aim of curing the presenting disease in favour of what would have, in a former era, been regarded as 'dying naturally'. Within this paradigm, 'death is often considered an adverse event rather than a natural end to life' and little attention is paid to preparation for death, with a focus on existential, spiritual matters (Barbato, 2009, p. 1). Expanding this theme, Barbato claims that frail, aged, dying people do not want therapeutic interventions or 'cocktails' from their doctor; rather, they are looking for a listening ear and a degree of compassion that communicates genuine care. This includes honest, open conversations not necessarily centred on life-prolonging measures or 'miracle cures'; older people want a clear, factual description of 'what the end will look like' in terms of a natural death.

However, such preferences prove to be very challenging for some health professionals. With so much emphasis on curative medicine and 'not giving up', it is not surprising that many doctors are unwilling to cease invasive treatments, reduce medications, withhold investigations, or appear to 'do nothing'. In this respect Barbato quotes a geriatrician who counselled other doctors: 'Don't just do something, sit there' (Barbato, 2009, p. 5).

Just 'sitting there' is wise counsel for all who are ambivalent about 'what to do' in the context of death and dying.

6.1.6 Premature Deaths

The subject of premature death is analysed by researchers (Ibrahim et al., 2017) who conclude:

> The incidence of premature and potentially preventable deaths of nursing home residents has increased over the past decade. A national policy framework and implementation plan for reducing harm in nursing homes is needed (Ibrahim et al., 2017, p. 447).

In the absence of a national policy on this matter, some nursing homes keep meticulous records relating to residents' deaths, while others fail to record any details beyond the basics. It is only through accurate record-keeping and regular analysis that premature, preventable deaths can be reduced.

6.1.7 What Is Known of the Dying Resident?

Documenting each resident's essential health-related details and family contacts is important, but what is known of the *person* approaching death? What would it take to note Barbato's (2009) counsel to sit beside the person? One example concerns a resident known only by her first name, room number, and clinical status, until an inquisitive staff member asked her about her past, with unexpected consequences.

> 'Mary was born just after World War 1. . . . She was the school high-jump champion and competed in the state finals. She used to ride to school on her horse with both her brothers on board. . . Mary trained as a nurse and worked in London during the blitzkrieg. She met the prime minister Robert Menzies once. And Dawn Fraser. She preferred Dawn. . .! Her favourite dance is the samba, which she is still pretty good at' (Pilotlight Australia, 2007).

This knowledge helped staff to personalise 'Mary's' care, to broaden their conversations and acknowledge her sense of humour, rather than regarding her as another uninteresting old woman at the end of her life.

To concentrate on clinical care at the expense of personal, relational care is to deny the deep bonds that develop between staff and residents; sometimes over many months, if not years. The reciprocity of such relationships is seldom described in the clinical care plan or 'progress notes'. Absence of such details represents a failure to acknowledge the impact of mutuality, friendship and attachments, not to mention the sharing of biographical details which may have a profound effect on both staff and resident in the face of the latter's death. Nursing homes have the unique capacity to enlighten the general public about death's variations, by telling these stories, some of which are deeply moving, if not instructive, about life's final chapter.

6.1.8 Staff Preparedness

Current nursing home staffing profiles indicate that most of the care and attention given to dying residents is from personal care assistants (however named). Many of these dedicated and often highly experienced staff members are familiar with death; others have articulated the moral dilemmas they feel in such circumstances, and some have never been confronted by death. Frustration is sometimes expressed when policies, hierarchical staffing structures and time pressures prevent them from responding appropriately to the residents' psychological, social and emotional needs prior to their death. Two main moral challenges have been cited: 'ensuring that residents don't die alone; and providing the appropriate care based on residents' wishes' (Wiersman et al., 2019, p. 1). Due acknowledgement is needed for those staff who are often more intimately acquainted with the residents' needs than is the person 'in charge'.

It should not be assumed that the care of dying residents more appropriately belongs to senior, more qualified staff: rather, it is a shared responsibility. When the nursing home is regarded as a community of carers – residents, families, staff, volunteers, other health professionals – no one is shielded from the presence of death, particularly when it is unexpected. Nevertheless, Nay's (1993) doctoral dissertation described death in the nursing home to be unnecessarily hushed up, if not denied. It remains an open question as to whether Nay's findings would be replicated today. Questions are also raised regarding the content and frequency of education sessions; whether they result in ensuring all staff are adequately prepared for their encounter with death and dying.

6.1.9 Sudden Death

Although, in the nursing home context, there is usually ample time to know of a resident's impending death, sudden death is not uncommon, as in this scenario.

> On her first day in the nursing home 'Joy' was found wandering in the corridor. 'I'll help you find your room', offered the kitchen hand. This intuitive staff member gently redirected Joy and waited, intending to call for a nurse as soon as Joy had settled herself on her bed. To her utter amazement, Joy uttered a deep sigh and died.

What questions are raised by such a sudden death? 'Was it worth all the paperwork?' 'Why weren't we warned?' What acknowledgement would the kitchen hand receive for her sensitive interaction with the resident? Would she need further support? While death on the resident's first day might be an unusual occurrence, it can also happen within the first few weeks after admission. Frank discussions with families may include the facts about unexpected death, as well as the average length of stay. Without wishing to alarm them, such discussion highlights the uncertain nature of death's timing, and the difficulty in making predictions. Unexpected death also poses a dilemma for some staff. When 'Alex' collapsed without warning, responses to the nurse's decision-making varied.

> 'Why was he sent to hospital?'
>
> 'We had no chance to say goodbye'.
>
> 'Why didn't someone go with him?'
>
> 'If only...'

Responses of recrimination, guilt, sadness and blame are not unexpected in such circumstances. In the absence of a plan, the nurse witnessing Alex's collapse made what she deemed to be the appropriate decision. In other circumstances, sudden death may be welcomed. 'Thank goodness he was saved the indignity of an ambulance trip and then to die in emergency'. While it is impossible to make the correct decision in every situation, many inappropriate actions can be avoided by having the resident's wishes clearly stated in writing, for example, regarding hospitalisation in response to sudden collapse. Carefully prepared documentation may mitigate the problem of one nurse having to make a hasty decision.

6.1.10 Dying: A Time for Joyous Celebration

Given the all-too-common prolonged period of decline, a dying resident may experience major mood swings: alternating deep despair and loss of communication with occasions for enjoyment and celebration, as in this account by a resident's daughter.

> 'Ruth' decided that on her dad's 'good days' she would play games with him. While experiencing inevitable slow physical decline, for the most part he remained mentally alert. 'I might have a few more weeks left in me, yet', he would say. Ruth purchased some board games and tried to retain an atmosphere of normality in his nursing home room. She arranged a 'cocktail party' at her father's bedside; inviting several of his close friends and family members. Ensuring he was given the role of host, the event was a huge success, to be repeated two or three more times when his condition permitted.

Evidence of such thoughtful planning may well be documented in the resident's care plan; equally as important as their medications and clinical care as in the next example.

> 'Mr. C' was dying from Parkinson's disease but retained the ability to articulate his needs. 'All I want is one last night with my darling wife'. Mrs. C, with early Alzheimer's disease, lived in an adjacent supported living centre. When asked, she responded enthusiastically, 'Oh, I'd love to spend one more night with him'. Arrangements were made for two beds to be placed side by side so Mr. and Mrs. C could clasp hands. Catering staff arranged a special dinner for two. Later that night, after discreetly knocking on the door of the room before checking on Mr. C's complex medical needs, night staff were delighted to see Mr. C

sleeping peacefully with his wife's hand firmly clasped in his own. When Mr. C died next morning, Mrs. C thanked the staff for a 'wonderful, lasting memory' (Hudson & O'Connor, 2007, p. 139) (paraphrased).

Creativity has its rightful place in the nursing home. Mr. C's story arose from an intuitive staff member responding in a practical but sensitive manner to a profoundly personal life and death issue. Similar stories are replicated in many nursing homes where thoughtful staff ask questions beyond the clinical, and respond accordingly.

6.2 Perceptions and Perspectives of Death

6.2.1 Part of Life

Tisdale provides some timely reminders of the way death is perceived now, compared to a former age.

> In Victorian times, children were kept away from anything regarding sex or birth, but they sat at deathbeds, witnessed deaths, and helped with the care of the body. Now children may watch the birth of a sibling and never see a dead body . . . many people reach the end of their own lives having never seen a dying person (Tisdale, 2019, p. 1).

Other writers agree that while medicalisation indicates advancement and triumphs over many diseases, which once resulted in early deaths, there is, concomitantly, a pervasive fear of death, with medical personnel regarding it as 'unnatural' or a failure.

Not so for older people themselves: a survey of the 'oldest old' found that for this cohort death was certainly a part of life. 'Most were ready to die, reflecting their concerns regarding quality of life, being a nuisance, having nothing to live for and having lived long enough' (Fleming & Farquhar, 2016, p. 1). Survey participants were largely unconcerned about death itself; rather, on the impact on those left behind (p.25). Little evidence was found of current advance care plans; and many participants acknowledged their views on dying had never been sought (p.21). Remaining 'in place' without transfer to a strange environment was also a high priority (p.22). For others, a clear choice was for palliative care, emphasising 'comfort', rather than aggressive treatment involving hospitalisation (Bergman & Laviana, 2016). Such insights are instructive for nursing home managers, medical personnel and staff: all the more important because they come from the residents themselves – those most intimately affected by death and dying.

6.2.2 The Dying Process

The doubling of life expectancy in the last 100 years has resulted in a vast increase in the average age of nursing home residents (85 years for females and 80 years for males). The dying process, in an earlier generation, could be predicted with a fair degree of accuracy in months if not weeks or days. For the most part, dying in the twenty-first century seems to be taking longer every year.

For some medical practitioners, dying is seen as an unfortunate 'condition' and death an enemy to be conquered. Technology provides for the newly dead to be resuscitated, for those whose organs have failed to be placed on life support and for curative treatments to be continued for as long as the person has breath. In contrast, the nursing home can be a place where genuine *care* is offered at the end of life, according to each resident's regularly assessed goals which, according to many advance care plans, seldom include hospitalisation or resuscitation.

Research into nursing home deaths is often impeded by lack of details; for example, the cause of death is not always accurately reported. Others claim, with some discernment, that mortality immediately or soon after admission is not confined to the resident's health status; the transition itself is 'deadly'. On the other hand, through careful governance, sensitive communication and awareness of risk factors, the mortality rate may be reduced (Ferrah et al., 2017).

(These and other issues relating to the dying process are discussed in more detail in Chap. 5 on 'palliative care'.)

6.2.3 Anticipating Death

Consider two different scenarios.

> 'Robyn' was passing 'Gwen's' door on her way off duty. Gwen called to her, 'Nurse', 'can you come here a minute?' Robyn went to Gwen's bedside to hear Gwen's urgent question: 'Nurse, I think I might be dying?' Robyn, momentarily flustered, replied: 'Oh, don't be silly Gwen. You'll probably outlive me!' And off she went, surprised to learn next morning that Gwen had died through the night.

> 'Dave' was passing Gwen's door when she called, 'Dave, can you come here a minute?' Dave went to Gwen's bedside. 'Dave, I think I might be dying?' Dave moved closer to the bed, sat down and took Gwen's hands in his. 'Yes, Gwen, we think your time is coming. Is there anything troubling you? Anything you want us to do?' 'Oh, no', said Gwen, 'I just wanted to know whether I had to pretend, so you wouldn't worry'.

Some health professionals believe that nursing home residents need to be protected from the truth of their prognosis and the reality of their mortality. Others indicate the need for more open discussion, citing residents' opinions which included their fear of dying in pain, their wish to plan for their death, as well as discussing their funerals. While this subject is yet to be more fully explored in the aged care literature, the two brief scenarios (above) indicate the need for staff to be prepared for such discussion, especially when prompted by the resident.

6.2.4 Suicide

An Australian study, the first and largest to examine suicide in nursing homes, asks what signs (if any) may alert staff to such a tragic event; noting that over half of residents who had committed suicide had a diagnosis of depression (Murphy et al.,

2018, p. 786). Given the paucity of research into this topic, when more is known and understood about depression's link to suicide, aged care providers may be alerted to issues of prevention, together with a definitive diagnosis by a qualified clinician when indicated. As support for psychological, spiritual and social issues increases, the desire for suicide may decrease. When palliative care becomes more readily available to nursing home residents, including prompt and effective pain relief, the desire for hastened death may lessen (Sprung et al., 2018, p. 197).

Ibrahim puts the spotlight on the harsh reality of suicide in Australian nursing homes (Ibrahim, 2018): 150 residents took their own lives between 2000 and 2013. Drawing on the assumption that if one lives in a nursing home, depression and misery are to be expected, Ibrahim recommends greater access to mental health services and specialist psychiatric teams. His research reveals the commonly held theme: '. . . if I get old, I want to make sure I kill myself before I get to residential care'.

There may, however, be some room for optimism, when nursing home residents are assured of timely, professional symptom management, and when every federal budget allocates significant funds to mental health in aged care.

Some decades ago, Nuland offered a stark reminder of the link between suicide and depression in older people, noting:

> With proper medication and therapy, most of them would be relieved of the cloud of oppressive despair that colors all seasons gray, would then realize that the edifice topples not quite so much as thought, and that hope of relief is less hopeless than it seemed (Nuland, 1994, p. 152).

In light of the above, it is disconcerting to note the suicide statistics (although their currency may be questioned), suggesting greater attention is needed towards this continuing phenomenon, including careful screening for risk factors. The following (not uncommon) exchange exemplifies the totally inadequate response to the report of a resident expressing a wish to kill herself.

> 'She's always so negative!' says one nurse at handover.
> 'She often talks about that stuff', says the care attendant.
> 'Yes, with me too, and I never know what to say', says her colleague.
> 'I can't understand her heavy accent, so I'm never sure I've heard right' says another.

How often, and by whom, does professional screening for depression occur in the nursing home? (Some researchers suggest it should be mandatory on admission.) What access do residents have to a specialist and/or an interpreter when English is not their common language? Are staff inappropriately called upon to interpret instead of accessing a qualified professional? What benefits accrue when residents have the opportunity to express their thoughts, feelings, anxieties and fears? Although some GPs have the necessary skills to respond to suicide ideation, others prefer to seek advice from a psychogeriatrician. Systemic changes would also include easier, timely access to mental health services, and increased emphasis on the communal aspect of nursing home life (Jain et al., 2020). While insightful managers provide ongoing staff education on suicide, others ignore the problem so that grief after the event remains unaddressed.

When the tragic nature of suicide is acknowledged and its effects on families and staff observed, healing and recovery are optimised. When met with silence, recovery is thwarted. Thoughtful, sensitive, well-informed responses are called for in every situation; prompting open discussion, supported by education from a skilled health professional. Some suicides are entirely unpredicted, with no apparent link to environmental factors; others may be directly related to the resident's feelings of isolation, abandonment, real or perceived neglect or other existential factors. In the latter case, rather than treating a resident's suicide as a private, isolated event, deeper analysis may lead to changes in the nursing home environment, so that it reflects a community of caring people, rather than a custodial institution for isolated individuals.

6.2.5 Grief, Loss and Bereavement

Grief refers to the subjective experience and emotional reaction to death. *Bereavement* refers to the objective experience of loss through death. Raphael (1994) describes the many *losses* attached to growing old; including loss of vision, hearing, sexual function, loss of limb, body part, or body function, loss of health and well-being and loss of brain function, loss of work, loss of relationships. Raphael's deeply human account of the 'anatomy of bereavement' ends with the comment, 'we need to learn to comfort and console others with compassion' (Raphael, 1994, p. 405).

Grief at the end of the life cycle cannot be homogenised or reduced to neat definition, as shown by this account.

> *'Pipa's' husband died several weeks after admission to the nursing home. An anxious woman, Pipa was reluctant to accept that staff would ever care for her husband adequately. Despite repeated explanations she completely failed to grasp the principles of palliative care: his death confirmed her worst fears and she cursed the day she ever allowed him to leave home.*
>
> *Six months later a nurse met Pipa in the street. Suddenly, Pipa was clutching at the nurse, face contorted with emotional pain, words spilling out all around. For several minutes, she harangued the nurse, going over and over her husband's final days, biting back angry words; talking fast, bitter, bitter. The nurse thought Pipa might strike her, but she used her words to attack.*
>
> *People waiting for the tram stared at the furious woman in black. More and more words poured out. She said she could never, never come back; could not even walk past the street of the nursing home. She ignored the nurse's gentle reassurances that the staff had been thinking of her and that her husband would always be remembered. Suddenly Pipa scurried off, driven by her despair, shopping bags crashing around her, bent low with guilt and deep, deep grief, trailing anger* (Richmond, J, 1990 unpublished and paraphrased conference presentation).

'Peggy's' story was more positive.

6.2 Perceptions and Perspectives of Death

'Peggy', a mentally alert resident aged 70, suffered multiple medical problems. With a marvellous sense of humour, she knew exactly what was meant when asked about her wishes for terminal care. 'If I'm too sick to answer for myself, take me to my own hospital: they've a huge file on me and they'll know what to do. But if they can't do anything, they can send me back here to die in my own bed'. A few weeks later, Peggy suffered a massive brain stem stroke, and was transferred unconscious to 'her' hospital with a comprehensive letter outlining her wishes. In view of this information, the admitting officer decided Peggy would not be treated with intrusive procedures; she would be returned to the nursing home for palliative care where she died several hours later. Having no immediate family, it remained the role of her trusted staff to wave her off, as though she'd be watching and admonishing: 'Make sure I'm decently covered and not treated like a bag of wheat. And don't forget to put some makeup on me – I don't want to look like a corpse!' Peggy's funeral attracted few mourners but cast a distinct impression on the funeral directors. Hearing the nurse manager's brief eulogy one director stated: 'This person must have been greatly loved. We thought it was going to be just another impersonal nursing home funeral'.

Some staff, in certain circumstances, choose to attend the deceased resident's funeral, as in the scenario above. However, death's impact on staff in long-term care has received scant attention. Similarly, concerning the well-being of family carers after a death, 'the bereavement support scorecard reveals a "fail", being 'haphazard, under-resourced, and lacking a sound evidence base' (Hudson et al., 2018). Clearly, the need for grief and bereavement support in nursing homes requires further attention. Particular consideration may be needed for personal care workers, volunteers, and also GPs, who sometimes develop a close 'attachment' to residents.

6.2.6 Grief Among Support Staff

The (not uncommon) phenomenon of support staff impacted by a resident's death is represented by the following examples.

The cleaner wondered who would now give her a tip for the races every Saturday. When mopping around Alberto's bed they would always discuss the horses. Sadly, the resident who had replaced him was unable to speak.

The hairdresser had developed an affectionate relationship with a male resident who would request a weekly hair trim, even though he had few hairs left. 'I always give him a nice shave and when I'm not too busy he enjoys a facial massage'. On the morning he died, the hairdresser wondered why he hadn't come for his appointment. Since no-one had bothered to tell her he'd died, she went looking for him, finding to her astonishment a stranger in his bed.

These comments from ancillary staff are testament to the many and varied relationships which develop in a long-term care environment. Other examples abound as residents' deaths affect the administrative assistant, the part-time receptionist, kitchen and cleaning staff, accounts staff, maintenance workers and others. Some managers may frown on any 'unprofessional' encounters between support staff and residents; others acknowledge the foibles of human nature which prompt the development of such connections. The key to acknowledging these relationships is to communicate the death of a resident, clearly, consistently, sensitively and unambiguously, to the whole nursing home community.

6.2.7 A Volunteer's Grief

After hearing of the death of the resident she'd visited for 2 years, a volunteer offered her reflections.

> In an effort to hold in the tears I clamped my jaw together, my steps hurried and stumbling as I made my way up the street to my home, feeling raw and exposed. Blundering up the stairs, I threw myself into my apartment and let myself sob, holding onto the back of a dining chair. The size of my grief shocked me. Yes, I was crying for Betty; that her last few years spent quietly helpless in a wheelchair parked in front of a television were in no way indicative of the life she had led. . . . I dearly wished that she had lived longer . . . so I could have shown her the wedding photos she was so looking forward to seeing. . . Mostly I wish she could have held the baby that I am now weeks away from meeting, and perhaps tell me why it was she never had children when she was so clearly taken with them. . . In writing this, I realise that it is me who feels lonely now, not Betty (Joosten, 2016, pp. 202-203).

The volunteer's role is not always recognised following a resident's death, with little opportunity for expressing their grief, or for administration to acknowledge their contribution to the resident's care. Much remains to be learned from the unique experiences of volunteers.

6.2.8 To Hospital Or?

As for many other areas of aged care, there is a paucity of research and consequent recommendations for improved practice related to residents' place of death. The value of careful planning for end-of-life care is exemplified in this scenario.

> In response to the night nurse's call for a doctor to prescribe pain medication for 'Jim', the locum medical officer swept into the nursing home in the small hours of the morning, making the quick pronouncement: 'If we write he has a blocked mandibular gland that should get him seen in emergency'. The perceptive nurse, who knew the resident well, questioned whether it was appropriate to send him to hospital. Consulting with 'Jim's' wife, and another carer who knew what Jim would want, it was clear that his care plan should be adhered to. 'Comfort measures if death is imminent' was the clear directive. Jim did not want to be hospitalised. Clear, purposeful discussion with the doctor resulted in an order for low dose opioid medication and other comfort measures. With his pain eased, Jim died a few hours later with his wife beside him, and in the care of his known, trusted staff.

The 'doctor knows best' dictum does not always apply in nursing homes; particularly when the doctor does not know the resident, or is unfamiliar with residential aged care. In this story (above), the locum medical officer was surprised to see such a carefully framed set of goals in the resident's care plan. 'I've never seen such clear documentation. Without this, I would have felt it my duty to send him to

6.2.9 Impending Death

As noted in the first part of this chapter (7.1), much can be learned from literature about impending death and its consequences.

> Now Earl was nodding off, his eyelids flickering open and shut as Jean stroked his leg through the covers. He asked her to stay with him a while longer. He seemed to doze. Then he opened his eyes. He smiled at her and said, "You can go home now." He'd never said that before, in all the many long days he'd spent at Linda Manor. He had always asked her to stay longer... Earl died in his room the following morning' (Kidder, 1993, p. 211).

This episode provides another reminder of death's ubiquity and variability. For Earl (above), it was 'right and proper' that he die alone, giving his wife 'permission' to leave. What was right and proper for Earl may not have been appropriate for another resident, who may have longed for his wife to stay. These stories serve to 'teach' carers there is no recipe for dying – those who are dying are the best teachers.

The next story confirms what happens all too readily when specific goals are not recorded.

> *'Keith', a 99-year-old resident with severe dyspnoea and advanced dementia was showing signs of distress from other painful symptoms. The night RN was from an agency and, acting on her professional judgement, decided the best plan was to have Keith transferred to hospital. The day staff were distressed, frustrated and angry that Keith died in the emergency department. In her defence, the RN stated she could find no reference to his advance care wishes or any plan of action for an exacerbation of his emphysema, and nothing to suggest he ought not to be hospitalised.*

In this situation, the regular nursing home staff knew that Keith's family would be horrified to learn he had died alone in the emergency department. However, no clear plans were documented, although Keith had experienced similar breathing difficulties before, which were managed well without hospitalisation. While it is impossible to prepare for every scenario pre-empting a resident's death, in Keith's case, predictions could well have been made and a plan carefully formulated. In his defence, the agency RN described his frustration: 'I hardly ever find clear up-to-date directions in a resident's care plan regarding hospitalisation or response to an emergency'.

In the wake of the COVID-19 pandemic, numerous accounts revealed the numbers of nursing home residents transferred to hospital without their consent and, in many cases, without the knowledge of their next of kin. While the pandemic could never have been predicted, every resident's care plan should have current documentation about wishes related to hospitalisation and/or impending death. The tragic outcome of such omissions has been the untold numbers of frail nursing home residents dying alone in an unfamiliar environment. It is impossible to calculate how many would have benefited from timely referral to palliative care; a life-saving decision in some circumstances, including the comforting presence of a close family member or friend.

6.2.10 Solitary Death

Citing statistics about solitary death or 'hidden death', Nuland notes the wisdom of advance directives which give clear advice about the person's hopes:

> '... a restoration of certainty that when the end is near, there will be at least this source of hope – that our last moments will be guided not by the bioengineers but by those who know who we are' (Nuland, 1994, p. 225).

One distinct advantage of dying in a nursing home is that it may be claimed with confidence, in most instances, that residents will be cared for by those 'who know who they are'. 'Alan' was one who benefited from such knowledge.

> *When reporting on the death overnight of 'Alan', a long-term resident, the EN, confided to the manager: 'I brought in some strawberries from my garden. I knew his swallowing was compromised and I had a feeling he would die last night, but I also knew how he loved strawberries. I crushed a couple and gently fed them to him by teaspoon. His eyes conveyed his thanks, before they closed, and he died several minutes later. And, I'm sure he preferred the sensation of strawberries rather than a furry swab stick in his mouth!'*

Personalised care of a dying person could not be better illustrated than in this scenario (above) recounted by a thoughtful nurse who enacted Nuland's (1994) source of hope: Alan's last moments were guided by a nurse who knew him well.

6.2.11 Dignity, Euthanasia and Futile Treatment

One of the arguments in favour of euthanasia is that it preserves dignity by eliminating suffering. Based on the premise that suffering equates to lack of dignity, it is assumed that euthanasia achieves 'death with dignity'. Proponents of this view often equate dignity with usefulness, mental capacity and the ability to communicate one's wishes. How, then, may dignity be pursued, maintained, prospered when the nursing home resident is lying mute, immobile, suffering from end-stage dementia, unable to make any decisions about their own life and death? Fortunately, many nursing home staff regard the person in this scenario with compassion, practising holistic care, maximising the person's dignity through expert nursing, including pain relief, and regular discussions with the family.

> The term "dignity" is generally used to signify that an entity has an inherent and inalienable right to be valued, to be treated with respect, and to be treated in an ethically just manner. Contemporary usage of the notion "dying with dignity" dates back to the 1970s and was

6.2 Perceptions and Perspectives of Death

originally developed in the context of a growing collective desire to avoid burdensome, life prolonging medical treatment (Johnstone, 2013).

Johnstone warns against giving weight to slogans about 'dignity', claiming some definitions can be harmful and/or ambiguous. The term is used here to signify the inherent dignity of all older persons, commanding the right to be treated humanely, with respect, with recognition of their human rights and the avoidance of unnecessary treatment. Arguing against euthanasia, in favour of the 'dignity' approach, Somerville says, 'This approach presents a stark contrast to the quick-fix solution of a lethal injection being seen as the best way to enhance a person's dignity' (Somerville, 2017). Rather than regarding dignity as a matter of individual rights or autonomy, dignity is better understood as *relational*: the source being the love and care of one human being to another. As carers, we *preserve* our own dignity when we *conserve* another's dignity.

Dr Harvey Chochinov, world renowned for his clinical and research skills, together with his deep humanity, shows what would be lost by accepting interventions which cut short the experiences at the end of life. 'The concept of dignity is afforded a high profile in end-of-life care... However, while patients may be *dying* for lack of dignity, the medical literature is relatively silent on how dying patients experience or understand the notion...' (Chochinov, 2012, pp. 5–6). To this end, dignity therapy encourages the dying person to live in the moment so that the final phase of life is a time of living, rather than simply a time of anticipating death. Noting Chochinov's comments, it may be instructive for carers to ask a dying person 'What does dignity mean to you?' Such an exchange may prove pivotal in decisions about futile care.

An early (1987), noteworthy Australian Government report on 'dying with dignity' states:

> The Committee has also been aware of the great pressure on doctors and the health care team to perform the miraculous, sometimes by using extraordinary means of resuscitation, even when inappropriate, for fear of criticism or legal action by relatives or the public for not trying hard enough (Parliament of Victoria, 1987, p.xiii).

Those who, in contemporary circumstances, would also have expectations of the 'miraculous' are clearly unaware of the legislation which allows for circumstances where 'futile treatment' would be inappropriate, if not harmful.

The subject of futile treatment also has application to palliative care (discussed in Chap. 5), one of whose principles states: '... that unless required by law, doctors are not obliged to initiate or continue treatments that will not offer a reasonable hope or benefit or improve the patient's quality of life' (Palliative Care Australia et al., 2017, p.5). It is not always evident to families (or to all aged care staff) that doctors are not obliged to offer treatment that will be of little or no benefit. Concomitantly, families may not be aware of the negative factors associated with unnecessarily prolonging life through hospitalisation, extra medications, tests and procedures, not to mention the 'indignity' of resuscitation measures which leave the person exposed, semi-naked, dependent on machinery and devoid of inter-personal communication.

A resident or family cannot lawfully *demand* treatment not clinically indicated. If the treatment cannot achieve the aim, the burdens may well outweigh the benefits and other measures are offered (Radbruch et al., 2020). If the goal is comfort, continued administration of some powerful drugs may cause *discomfort*. Providing treatment which is unwanted by the person who is dying may also mitigate against peaceful death. What is articulated in the care plan? Are the resident's wishes unambiguously registered in their ACP or ACD? If the treatment is not commensurate with the goal, it should not be pursued, particularly when the resident is close to death. Focussing on the goal puts 'futility' in context: ignoring the goal jeopardises the central tenets of end-of-life care planning and dignified death.

The challenge continues, to minimise suffering without killing the sufferer.

6.3 After the Death

Thoughtful nursing home managers apply a variety of skills in dealing with death's aftermath, from professional documentation and verbal reporting to compassionate, sensitive conversations with families and staff. Others, however, prefer not to raise the subject.

While many relatives and staff would rather focus on *living*, the thought of death may be at the forefront of others' minds. In contrast to the many rather perfunctory accounts of death, 'Adriana's' story is personal and particular, connoting a sensitive partnership with her family, and clearly documented in her care plan.

> *'Adriana is to be assisted to maintain maximum dignity including all aspects of grooming and clothing, at the special request of family. Adriana's son has purchased a variety of silk scarves, requesting a fresh one be worn each day until her death'.*

When Adriana died several weeks later, it was gratifying to see the written record demonstrating the fulfilment of goals, including mention of the scarves.

> *'Adriana died with the dignity and self-esteem always so important to her and her family. Her son was delighted to see Adriana wearing one of her new silk scarves at the time of her death'.*

Regardless of other goals formulated in the care plan, the silk scarf takes a prominent place in this nurse's summation of the end of Adriana's life, demonstrating the importance of family involvement in personalised care planning and the gerontic nurse's responsibility and satisfaction in following the plan through. Readers will, no doubt, know of many other instances where a resident's death is framed by such imaginative care.

6.3.1 Death's Effect on Other Residents

In long term care, deaths are to be expected. However, there is little evidence that the needs of other residents are considered in relation to such loss. While some nursing homes are careful to include supports, resources, specific forms of

6.3 After the Death

communication, acts of remembrance, public commemoration of the deceased, in others, death's impact is treated with little regard, or intentionally silenced. What meaning is given to death by those who are still living? Are residents encouraged to articulate their sorrow and grief, or to express their reactions to the deaths of other residents in their own way? Researchers have found that if cumulative losses are not met with an opportunity for emotional release, a sense of despair or apathy may result.

> The greatest risk of all, however, is that when the death of a resident is given little pause by those in authority, the living residents may also come to devalue their own existence in a similar way (Djivre et al., 2012, p. 513).

With more discussion and attention to this subject, residents may have confidence that, even in death, their lives will be valued and their absence noted. When most residents lived in shared rooms, sometimes forming friendships, death would seldom go unnoticed, as in the following account.

> 'Fanny' was the mentally alert resident in the two-bed room, 'keeping watch' over 'Jessie' in the bed opposite, who, due to end stage dementia, had no speech and seemingly no awareness of her surroundings. She also received no visitors. It was 3.a.m. and Flora woke to find the nurses busy at Jessie's bedside. 'Oh dear, she must have died while I nodded off to sleep. I tried to keep awake as I knew she was close to the end'. 'Would you like to say goodbye to her?' asked the perceptive nurse. Fanny was wheeled over to Jessie's bed and, drawing close, whispered her farewell together with a brief, heartfelt 'prayer for her soul'.

While many prefer the privacy of single rooms, others enjoy relationships such as the bond developed over 2 years between Fanny and Jessie.

6.3.2 Documenting the Death

> 'Respirations ceased at 12.30. Doctor and relatives notified'.

Although clinically accurate, this entry is devoid of the person's story. Is this all that can be said of a resident at the end of their life? Who was the person whose respirations ceased at this hour? Is there nothing more to be written, particularly by a nurse who has known this resident for some time? This death, as with every resident's death, represents the last chapter in the last phase of that person's (usually long) life, deserving due acknowledgement. Accurate, descriptive documentation also makes a valuable contribution to further research. More expansive entries (although not necessarily long), may convey a valuable record of the final phase of a unique life.

> 'Tom' died at 12.30 am. Palliative care objective achieved. Family wished to visit Tom and stayed until the funeral director arrived. Family have been able to express their grief and talk to staff about their feelings. They expressed gratitude for all that the staff had done for their father.

While brief in content, this entry captures something of the deceased person's uniqueness and the distinctive relationships developed over time.

6.3.3 Death Notification and Certification

Relatives' responses to death's reporting vary widely, as in this account:

> 'Milly' reported to the manager: 'I was absolutely furious when I went to sign the cremation certificate to see the cause of death as "bronchopneumonia". He did not die of pneumonia. He had secondary carcinoma from cancer of the prostate, pyelonephritis and dementia'. This 'cause of death' seemed to diminish her husband's life. Milly, an 84 yr. old retired GP was reflecting on the death of her 90 yr. old husband who suffered serious illness for many years. This summary affronted her. Wasn't his suffering worth more than such a brief, cursory note?

'Milly's' comment raises serious questions regarding death certificates, an issue taken up by those who are concerned to see greater accuracy in documentation. Similarly, death notification is largely inconsistent, as in the following:

> 'Mrs. Eden's' daughter had been a faithful, regular visitor to the nursing home to see her mother, wanting to spend as much time as possible with her. The time came for her to resume her regular part time employment, reducing the time available to visit. Realising there may now be fewer opportunities to see her mother before she died, the daughter asked staff to ensure she would be notified immediately of any change in her mother's condition, checking her twenty-four-hour contact details were readily accessible. Unfortunately, when Mrs. Eden died at 6 pm, the daughter was not contacted until 11 pm. When expressing her displeasure and asking why she had not been called earlier she was informed that staff had to wait for a doctor to certify Mrs. Eden's death before notifying the next of kin. Mrs. Eden's daughter knew this was incorrect, as she had access to the relevant guidelines stating that next of kin could be notified without first obtaining a doctor's certification of death.

'Mrs. Eden's' daughter, and only child, had wanted to be with her mother as close to the time of her death as possible, making sure her wishes were well documented. Now, she felt cheated, and desperately sad she was not given the opportunity to be at the bedside earlier. This scenario raises many issues, most notably the means by which nursing homes adhere to the relevant guidelines for death notification, as well as ensuring that family's wishes are respected wherever possible.

Managers play a key role in this matter, ensuring clear policies, procedures and guidelines are accessible, regularly updated, routinely followed and reinforced by regular education sessions. When time and attention is not given to such procedures, others may be left to make their own judgements about the disappointments and discrepancies resulting from wide variations in practice.

6.3.4 Verification of Death

Guidance notes for the verification of death vary between states in Australia, including who can verify death, how to identify a reportable death, coroner's involvement and how the body is removed. Clear guidelines are also needed for rural/remote areas where twenty-four-hour access to a doctor or funeral director is not always available. Other related matters include specific cultural protocols which need to be honoured and respected. Consistency is also important, preventing unhelpful comparisons between families recounting their experiences which may differ, without obvious justification.

6.3.5 *Preparation and Removal of the Body*

In formal records, a paucity of detail may be evident regarding removal of the person's body after death, omitting an important last phase in the resident's narrative. Practices and protocols vary; some prefer an open scenario, where the deceased resident's body is removed on a trolley, with face uncovered, in full view of other residents, with doors open to convey a sense of normality. Those who wish to do so place a flower or message onto the deceased resident's body, or walk alongside if appropriate. Others prefer a process where the body is completely covered, all rooms have their doors closed, and the body is removed with minimum observance by others. The following anecdote highlights the difference.

> *'Amy' was contemplating the final chapter of her rather lonely, isolated life, with no friends or family to mourn her death. Having witnessed the 'closed' practice of removal of deceased residents, Amy welcomed the new procedure introduced by management after discussion with the residents' and relatives' committee. Amy confided to the manager: 'Now I can die happy that at least someone will wave me off, and my body will be covered with a coloured quilt rather than an awful black shroud'.*

This first-hand account (above) illustrates the benefits of treating death as a unique, albeit sad, event within the nursing home, a subject worthy of being open to view. By contrast, deliberate attempts to 'hide' the death is seen by many as disrespectful.

Part of the conversation with relatives about death and dying may involve particular requests such as the following.

> *The Greek resident's wish was to have her body washed in red wine, 'representing the blood of Jesus'. The funeral director was happy to oblige, coming to the nursing home at 4 am with sponge and small amount of wine, liaising with the RN who agreed to add this brief procedure to her own preparation of the body. The woman, always dressed in black as a mark of her widowhood, wanted to be buried in a floral dress. The RN responded readily, acceding to this resident's unique, personal preferences.*

In the absence of close family, the RN who had noted this woman's funeral preferences on admission, was careful to record her wishes accurately. This story serves as another reminder for staff to ensure relevant details are accessible regarding care of the body, well before the death occurs, ensuring due respect is given to all procedures.

Sensitivity to post mortem procedures includes experienced staff acting as role models by 'pairing' with inexperienced staff in caring for the body of the deceased person. Proper respect is also upheld through the use of protocols regarding 'filling the bed'. One PCA was deeply offended by the lack of such conventions, 'Our manager won't even let the bed go cold before admitting the new resident'.

6.3.6 Death Review

Given the all too familiar time pressures in the nursing home, it is not always possible to review each death after the event. Where this does occur, it provides an important indicator of whether or not the goals were achieved, and the plan of care delivered as documented.

When time is given, perhaps once or twice per year, to reviewing all the deaths in preceding months, results can be proudly reported to the whole nursing home community. 'We've completed our review and found that in 95 per cent of resident deaths, their end-of-life wishes were achieved. Well done, staff!' Further scrutiny may uncover some areas deserving more attention; the aim is for increased satisfaction and confidence in the process.

Opportunities for families who wish to comment on a resident's death may assist nursing homes in reviewing the circumstances of the death, together with factors relating to satisfaction (or otherwise) with the care. Rather than the resident's death ending all communication with the nursing home, some families may welcome an ongoing relationship, particularly if the review influences improved practice. Firsthand experiences related by those intimately affected by death contribute to the ongoing life of the nursing home, adding a significant layer to achieving best practice.

6.4 Community Reflections

6.4.1 Death Denial

Phenomena such as the medicalisation and institutionalisation of death, the reluctance of some people to participate in advance care planning conversations, or the common use of euphemisms when talking about death and dying, have all been taken as proxy indicators of death denial. When death's reality remains unspoken, opportunities are missed for residents' wishes to be expressed. Four dominant themes emerged from researchers' discussions with hundreds of people about how they wish to end their lives. 'Dying in my sleep' was the predominant response. Others wanted a 'social death' including the presence of family, friends and pets. Some concerns centred on 'pain control', while others focused on 'choice' (Sanderson et al., 2019, pp. 1–2).

Care planning includes the opportunity for residents to state in writing their preferred options for end-of-life care. When such discussions are part of routine care, death is faced squarely, thoughtfully and compassionately, rather than its reality denied.

Evasion of death's reality is nowhere expressed more cogently nor explored with greater depth than in Ernest Becker's *Denial of death* (Becker, 1973). Space does not permit a more comprehensive discussion of the vitally important matters discussed in this classic text. Suffice it to say much can be gleaned by asking each resident and/or representative how they understand death or life's end. For some, religious belief will be instrumental to their views, for example, whether they believe in an 'after life'. For others, a fatalistic response includes the injunction: 'eat, drink and be merry' (including the many variations of the original biblical reference). While a theological discussion is not intended here, it will often suffice to ask the resident: 'Do you have any particular beliefs about death?'

6.4.2 Wish to Hasten Death

Nursing home staff are familiar with statements such as: 'I wish I were dead' or 'I hope the end comes soon'. This does not necessarily mean 'Please kill me'. One extensive survey showed 'Expressions of a desire to die may not constitute a specific request for suicide or hastened death' (Hudson et al., 2006, p. 694). Indeed, as this research showed, a desire for death in the morning may be reversed in the evening: 'I feel so much better than I did this morning!' An expressed desire for death may be an invitation for staff to try and ascertain the person's underlying feelings, which may be a response to physical, psychological and/or spiritual suffering, a sense that life's meaning is lost, fear of the dying process, lack of control or merely a wish to be heard.

If or when a resident expresses a wish to hasten their death, it is important to document the discussion in detail, and initiate a referral, for example, to a palliative care consultant, together with an assessment for depression. Given the dearth of research in this area, it may be advisable for nursing homes to develop a comprehensive end-of-life policy which reflects residents' experiences and desires. Inadequate attention to this matter may also exemplify the lack of conversations about death and dying which older people say they want, but which nursing home staff seldom offer. Documentation using the resident's own words is pivotal to these discussions, also allowing for change of mind, which is not uncommon. Requests for means to end their life need to be taken seriously and while staff are not required to act against their moral principles, they must avoid making judgments based on their personal beliefs. Contrasting decision-making styles are exemplified below.

Nurse A reported at handover that 'Mr. D' had consistently refused all meals for two days. Nurse A stated she had documented this and changed the care plan to include: 'Do not offer any more meals as Mr. D has requested all delivery of meals should cease'. While family members and several other staff members were distressed by this directive, Nurse A reinforced her decision: 'I believe it's his right to state his preferences and we should not stand in his way'.

On the fourth day of Mr. D's failure to eat or drink, Nurse B sought advice from the palliative care team, who visited Mr. D and after sensitive questioning, offered their recommendations. Having thoroughly checked his mental status, and confirmed his earlier desire to stop eating, the palliative care physician documented the decision that small serves of modified meals would continue to be offered and removed without judgment if refused. Full attention would be given to pain management, regular mouth care and checking of other symptoms. Family members were apprised of this, and reassured that Mr. D's wishes would be respected without compromising his holistic care. Mr. D accepted small amounts of his preferred snacks and drinks from staff and family over the next two weeks and he died peacefully.

Mr. D's story is typical of an evidence-based, professional, multidisciplinary response to a resident's wish to hasten death. The emphasis here is on the *team response,* following referral for expert advice, rather than individual nurses acting unilaterally. 'When I'm on I'll make sure he eats his meals!' Or, 'I respect his wishes so I'm going to remove all food and fluids'.

This issue often causes serious differences between staff or between staff and family. Acknowledging these differences and the serious issue of a resident refusing to eat and drink indicates the need for expert opinion rather than accepting individual staff members' unprofessional and uninformed opinions and actions.

6.4.3 In Touch with Death

The following account is a reminder of the consequences when open discussion about death and dying is not forthcoming.

> I stayed with Bill, my husband, as he lay dying. I feel so bad now that I didn't lie beside him on the bed and hold him in my arms! I'm reading about this now, the need to touch. . . I wish I'd been told that it's okay to lie on the bed and put my arms around him and hold him while he's dying instead of sitting on a chair. I wish someone had told me that my children could stay with their father and sit beside him, sit on the edge of the bed, touch him, hold his hand, talk to him, and stay until he dies. But there was no one. I wish someone had been there to tell me how to do it right – which I now know but didn't know then (Kuhl, 2002, p. 116).

Doing it 'right' does not mean it is staff members' responsibility to instruct family members how to approach death. A simple question such as 'How would you like to spend the time you have left?' may provide reassurance and the opportunity to explore options. The role of 'touch' cannot be over-emphasised, particularly when there are no more words from the dying person. Touch can 'say' something powerful in the presence of death and dying. Some families may welcome a prompt which reinforces its legitimacy.

> *Without waiting for such a prompt, 'Mrs T's daughter took her shoes off and climbed into bed with her mother. The kitchenhand serving the teas looked askance, registering her shock and disapproval at such an unashamedly and, in her view, inappropriately public sign of affection.*

'Mrs T's' daughter was sufficiently confident of her own actions, which were for the benefit of her mother's comfort, to disregard the kitchen hand's obvious disapproval.

6.4.4 Death as Part of Life

Whether belief is based on religion or humanism, the issue of trust and hope in the face of death is related to trust in life (Faber, 1984). The concept of death being part of life is, of course, not new, being attributed to Martin Luther [1483–1546]: 'In the midst of life we are in death' (Luther, 1956, p.128). A contemporary writer puts it this way.

> We are born to die. Not that death is the purpose of our being born, but we are born toward death, and in each of our lives the work of dying is already underway . . . Death is the most everyday of everyday things. It is not simply that thousands of people die every day, that thousands will die this day, although that too is true. Death is the warp and woof of existence in the ordinary, the quotidian, the way things are . . . (A)ll our protest notwithstanding, the mortality rate holds steady at 100 percent (Neuhaus, 2000, p. 15).

This writer's wisdom reassures us that such explanations and advice are not often sought and seldom warranted. 'A measure of reticence and silence is in order. There is a time simply to be present at death . . . without any felt urgencies about doing something about it . . .' (Neuhaus, 2000, p. 16).

That is not to say, however, that death needs no preparatory discussion or planning. One of the hallmarks of end-of-life care is to encourage dying persons and their families to make considered decisions on where, how and with whom their last days or hours are to be spent. Even within the so-called confines of a nursing home, creative options can be explored.

6.4.5 Humour and Death

A sense of humour takes its legitimate place, even in the presence of death. For one family, it proved a refreshing release for their pent-up emotions, to make these comments when called to the nursing home on Melbourne Cup Day.

> While the horses were running their fastest 'Barth' was completing his own race against life, facing the final barrier in his battle with rapid progression of a rare neurological disease. The family realised there would now be no favourable 'winning post'. With a mixture of frustration, humour and sadness they came to his bedside. 'Trust dad to interrupt our lives and call us to attention right in the middle of the Melbourne Cup!' No more laying of bets as to the finishing time for this race.

Death's timing always takes its own course, resulting in different reactions from families. Staff who knew 'Barth' and his family well were pleased to acknowledge the appropriate place of humour. Laughter can enjoy its rightful place, while also acknowledging the solemnity of death.

6.4.6 Acknowledging Family

Family members' care of residents is not always acknowledged, especially after a death. The following exchange indicates what seems to be an exception:

> *'I've always admired the way you cared for your husband so tenderly', said the nurse to 'Edith' when her husband died. 'Thank you so much' Edith responded. 'I often wondered if anyone cared whether I was here or not. I'm not looking for praise, but I'm so glad you noticed'.*

This brief, but significant, conversation illustrates the value of a spontaneous, heartfelt acknowledgement of a family member's actions. Such a comment, which takes only a few seconds, is often warmly received and in this instance made all the difference to a careworn, grieving woman following her husband's death. As indicated elsewhere in this book, each family's chosen involvement (or non-involvement) deserves a place in the care plan; with specific comments after the resident's death, where relevant.

6.4.7 A Good Death?

In typical descriptive fashion, borne of deep insights into contemporary life and death, Ivan Illich says: 'Modern medicine has brought the epoch of natural death to an end' (Ilyich, 1976, p. 198). Perhaps a 'good death' is best summarised by principles that recognise the dying person's right for comfort, control, dignity, palliative care, autonomy, independence, wishes respected, and time to say goodbye, at a time when these issues are critically important (Smith, 2000). A contemporary writer puts death in perspective, noting from her own experience as a physician how society's expectations have changed.

> A good death – an ideal death – is pre-planned, perfectly timed, excretion-free, speedy, neat and controlled. Birth is not like this. Life is not like this. And yet we think we have a right to ask it of death. We want a caesarean-section death. The only way we could come close to meeting all these criteria for a good death would be to put people down when they reach a predetermined age, before the chaos of illness sets in (Hitchcock, 2015, p. 64).

More emphasis on the way we accompany the dying person rather than trying to achieve an 'ideal death' may be welcomed by residents and their families. When dying is reduced to a series of biological, physiological stages, or problems to be solved, the psychological and spiritual elements are easily ignored. Expectations which ignore the 'chaos of illness' (Hitchcock, above) may also be misplaced. Death cannot be *controlled,* and ultimately, we need to maintain an attitude of deference (if not reverence) in its presence.

In contradistinction to a 'good death', an Australian Government minister expressed dismay at the number of negative scenarios or 'bad deaths' at the height of the COVID-19 pandemic. 'Many died undignified deaths, unable to say goodbye to loved ones, incoherent, dehydrated and confused. Families are traumatised, with a lifetime of good memories tainted by a bad death'. While being mindful of the totally unexpected rapid increase in the number of nursing home deaths due to the pandemic, it remains a source of profound regret that so many families, not to mention aged care staff, are left with harrowing memories.

On the other hand, some nursing homes were quick to develop procedures for responding to the unexpected crisis with professionalism, creativity and compassion, optimising a 'good death' wherever possible.

6.4.8 Cultural Differences

In a multicultural society, it is inevitable that a variety of factors may arise in the context of death and dying; hitherto not encountered by most nursing home staff. For example, in some societies, the 'tribe' or the community is what glues them together; so that when one member is dying, everyone is affected. In some communities, it is an alien concept for an older person to spend the end of their life in a nursing home; all decisions are customarily made by the extended family or community. For others, being cared for by a young staff member of the opposite sex is inappropriate and unacceptable. Staff cannot be expected to know all of these details, so specific beliefs and practices are best elicited from direct discussion with families or community leaders on an individual basis.

One writer highlights the culpable moral harm that may ensue when a person's cultural beliefs are not considered at the end of life. This is not merely a matter of acknowledging the dying person's wishes and preferences: it has to do with the meaning of what it is to be human. Failure to ascertain a person's cultural beliefs about death therefore amounts to serious 'preventable and culpable moral harm' (Johnstone, 2012, p. 186). A comprehensive cultural assessment includes specific questions relating to end-of-life care, noting particular requests to be met wherever possible. Otherwise, residents are likely to die in an alien environment while longing for their country of birth.

Rather than making assumptions about practices relating to death and dying, a well organised meeting (with interpreter, if necessary) with key stakeholders will ensure relevant, individualised cultural care. The time and attention given to such discussions will result in increased knowledge, trust and confidence for all concerned; with satisfying memories for those deeply affected by the death.

6.4.9 Aboriginal and Torres Strait Islander People

As noted elsewhere in the book, the number of residents from Aboriginal and Torres Strait Islander communities in 'general' nursing homes is small by comparison with other cultural groups. However, it means that careful attention is needed when a resident from one of their communities dies: especially where specific rituals and procedures are to be observed. Best practice would indicate making direct contact with a representative from their community so appropriate guidelines can be followed.

6.4.10 Notification of Dying

Night staff in nursing homes are well acquainted with 'after hours' deaths and what is entailed, particularly when they have not met the resident's relatives and may not be familiar with the family's circumstances. Is the person designated next of kin (NOK) the most appropriate person to receive a phone call at 3 am? For non-English speaking relatives, what language is needed to convey news of the death? In some instances, such as in 'Liliana's' story, it is more appropriate to wait until morning to make the call.

> 'Liliana' had been a faithful visitor to her husband, dying of dementia and other comorbidities, for the three years he'd lived in the nursing home. In her twice daily visits, Liliana would befriend other residents as well, for her husband was now beyond conversation. Liliana, aged 92, lived alone, a short walk from the nursing home, and with no regular contact with other family or friends. At the regular review meeting to discuss her husband's care, the question arose about notification of death, now that his condition was deteriorating. 'Oh', she said, hesitatingly, 'you'll probably think I'm awful, but would you please write down that I don't want to be called during the night?' Seeing the nurse was a little taken aback, Liliana explained. 'I get up early so I can do my chores, then I come in here to see Ern, and as you know, I stay all morning, then I go home for a rest before returning later in the afternoon. Then, at night, I'm exhausted. I take my 'sleeper' and by 8.30pm I don't hear a thing until morning. I never go out alone at night anyway. Every day when I say goodbye to Ern, I kiss him, say a prayer for him, and I know one of these goodbyes will, one day, be the last. So, please don't call me before 7am'.

Liliana's request was honoured, though not without some raised eyebrows and judgemental comments by staff: 'Why wasn't she notified immediately? You'd think she'd want to know as soon as he died!'

The manner in which a resident's death is conveyed needs careful thought and articulation; some staff find it helpful to write down before the call what words they will use to convey this message. Unmistakable clarity is required, allowing time for the person's response. Such an important phone call needs to be sensitive, unambiguous and professional. Other important points include (a) checking the correct phone number and name, (b) confirming the person's relationship to the deceased, (c) allowing time for questions and (d) providing the option of seeing the body before removal to the funeral home. Other reassurances relate to the important issue of the resident's belongings. In the following story, 'Connie' could sense the deterioration after her mother's hip fracture, but staff tried to reassure her that all was well.

> Sooner than anyone expected, her mother died – sadly, at a time Connie was not present . . . When she got the call, she rushed there. No staff member was on hand to greet or support her. She felt terribly alone . . . 'No one was saying anything to me. She had lived there for

nine years', Connie said in disbelief. To make matters worse she had been assured that her mother's room would be left as it was, until Connie had time to pack up her mother's belongings. But when she got there the next day, 'someone was in her room, and all of her belongings were in these four garbage bags. It was horrible... It was unbelievable' (Baker, 2007, p. 178).

This account provides a timely reminder of the need for organisation policies and staff education about death notification, care of the body, and care of the resident's personal belongings. Some may think that when the family have been notified that is the end of the matter. It may indeed be 'the end of the matter' for the nurse, but it is the beginning of another chapter for the family. While it may be of no consequence to the staff member, the handling of personal items may have great significance for others; for some, with lasting memories of garbage bags.

The way families are told of a resident's death may also influence lasting memories, not always positive. There are no 'right words' after a death. However, a sincere acknowledgement of the loss is conveyed by mentioning the person by name, or a simple gesture of a hand on a relative's arm: 'I'm so sorry to hear of George's death', or 'We're so sad that Emily has died', or 'How are you feeling after your mother's death?' Again, the unambiguous use of 'death' can often be more comforting than saying nothing.

6.4.11 *Communicating the Fact of Death*

It would seem appropriate for each nursing home to devise a formal protocol for responding to such a common occurrence; such as posting a clear, formal (though not impersonal) notice for all to read. Confusion arises when responses are inconsistent or unclear. 'It depends who's on the desk', said one staff member when asking about death notification. Another was puzzled by the fact that a notice was *not* posted after the death of a particular resident. 'We couldn't find anything nice to say about him', was the response. A clear, formal, routine notice obviates the need for sentimental embellishment, or comments about the deceased resident's character. One such revised protocol brought much comfort to this resident.

'Evelyn', a single woman institutionalised her entire life, and with no living relatives, enjoyed living in the nursing home where she had maximum independent movement with the aid of her electric wheelchair. Severely physically disabled but mentally alert, Evelyn was the 'nursing home gossip'. An avid reader of the notice board, she commented to the new manager: 'I see you put the names of any residents who die on the notice board. Nobody has ever done that before. Now we know who's died, rather than wondering why their room is empty. When I die, will you put up a notice about me?' she asked. Acutely aware that she had no relatives to mourn her death or even to post a death notice in the paper, this new protocol brought a sense of enormous satisfaction. Someone would notice she had died.

Equally as important as welcoming a new resident into the nursing home is the way they are farewelled after their death.

6.4.12 Reportable Deaths

Policies vary across states in Australia so that it is wise to check whether a particular death should be reported to the coroner. Such reports may include death which is (a) unexpected by the resident's doctor, (b) due to accident or injury and (c) a violent or unnatural death. Such deaths would prompt specific procedures for review.

6.4.13 Death as Loss of Community

Death is more than the extinction of an individual's life. Death means loss of relationships and therefore loss of community. One of the continuous themes of this book centres on the nature of persons in community: when a person dies, we experience the fracturing of that community. As Hertz (1960) asserts: 'When a man [*sic*] dies, society loses in him much more than a unit; it is stricken in the very principle of its life, in the faith it has in itself' (p. 78). This does not presume irreparable damage, for as a community is ruptured, so it can be healed. Callahan has this to say of death of the community.

> As the ultimate form of separation from the human community, those undergoing that passage must need the company and care of others, to keep them socially in the community until the last possible moment, to assure them that they will not be forgotten, that the death of their body will not be preceded by the death of their social self, pushed out of sight and out of mind by fearful medical workers or families (Callahan, 1990, p. 146).

Aged care workers have the potential and the privilege to preserve the whole community by ensuring residents do not suffer the death of their 'social selves' even when others would prefer to push them out of sight. Community is preserved not only by recognising the death as a loss to the community, but paradoxically, also representing a gain through the solidarity of shared grief.

The foregoing discussion on death and dying is designed to encourage aged care workers and others to focus on who we are in our reciprocal relationships. This is not a discussion about death in general but, as the stories are intended to signify, it is about death in the particular. Unless we are shattered, surprised or affected in some way by each death, then we have not understood death's meaning. To remain indifferent to death is to deny who we are as persons-in-relation, for every single death is also about life in the community – who we are together. 'The worst thing is not the sorrow or the loss or the heartbreak. Worse is to be encountered by death and not to be changed by the encounter' (Neuhaus, 2000, p. 16).

6.4.14 Planning for Death

Much of this chapter has alluded to the need for anticipating death and to plan accordingly. This frame of reference asks not what we should *do* about the *problems* of death and dying but what their *meaning* is. This starting point would then look for different resources. What resources are given to people who attend retirement seminars? Largely, I suggest, tips for financial planning and other, albeit important, legal matters. In addition, we may ask what will we need to sustain us as we face the inevitable prospect of life's end? If we see no meaning beyond decay, dependency and death, we will do all in our power to abolish it. If, on the other hand, we regard each life as unique and irreplaceable, we will support each other in our difference and diversity. Our personal future will not be constrained by cheerful optimism, or by frenetic attempts to keep old age and death at bay; we will see in our days being numbered a sure and certain sign of our finitude, and the opportunity to plan ahead with freedom and hope.

References

Baker, B. (2007). *Old age in a new age: The promise of transformative nursing homes*. Vanderbilt University Press.
Barbato, M. (2009). *The dying game*. International Conference on Ageing and spirituality, Aukland, NZ.
Becker, E. (1973). *The denial of death*. The Free Press.
Bergman, J., & Laviana, A. (2016). Opportunities to maximize value with integrated palliative care. *Journal of Multidisciplinary Healthcare, (9)*, 219–226.
Callahan, D. (1990). *What kind of life: the limits of medical progress*. Simon and Schuster.
Chochinov, H. (2012). *Dignity therapy: Final words for final days*. Oxford University Press.
Djivre, S., Levin, E., Schinke, R., & Porter, E. (2012). Five residents speak: The meaning of living with dying in a long-term care home. *Death Studies, 36*(6), 486–518.
Faber, H. (1984). *Striking sails: A pastoral-psychological view of growing older in our society* (K. R. Mitchell, Trans.). Abingdon Press.
Ferrah, N., Ibrahim, J., Kipsaina, C., & Bugeja, L. (2017). Death following recent admission into nursing home from community living: A systematic review into the transition process. 1–21. https://doi.org/10.1177/0898264316686575.
Fleming, J., & Farquhar, M. (2016). Death and the oldest old: Attitudes and preferences for end-of-life care: Qualitative research within a population-based cohort study. *PLoS One*, 1–25. https://doi.org/10.1371/journal.pone.0150686
Hertz, R. (1960). *Death and the right hand (R. a. C. Needham, Trans.)*. Cohen & West.
Hitchcock, K. (2015). Dear life: On caring for the elderly. *Quarterly Essay, 57*, 1–78.
Hudson, P., Hall, C., Boughey, A., & Roulston, A. (2018). Bereavement support standards and bereavement care pathway for quality palliative care. *Palliative & Supportive Care, 16*(4), 375–387. https://doi.org/10.1017/S1478951517000451
Hudson, P., Kristjanson, L., Ashby, M., Kelly, B., Schofield, P., Hudson, R., Aranda, S., O'Connor, M., & Street, A. (2006). Desire for hastened death in patients with advanced disease and the evidence base of clinical guidelines: A systematic review. *Palliative Medicine, 20*, 693–701.
Hudson, R. (2014). Palliative care for the older person: Cloak or cover-up? *Journal of Religion, Spirituality, & Aging, 26*(2–3), 186–200.
Hudson, R., & O'Connor, M. (2007). *Palliative care and aged care: A guide to practice*. Ausmed Publications.
Ibrahim, J. (2018). Spotlight on the harsh reality of suicide in Australian nursing homes. *Australian Nursing & Midwifery Journal, 25*(9), 4.
Ibrahim, J. E., Bugeja, L., Willoughby, M., Bevan, M., Kipsaina, C., Young, C., Pham, T., & Ranson, D. L. (2017). Premature deaths of nursing home residents: An epidemiological analysis. *Medical Journal of Australia, 206*(10), 442–447.
Ilyich, I. (1976). *Limits to medicine: The expropriation of health*. Marion Boyars.

Jain, B., Kennedy, B., Bugeja, L. C., & Ibrahim, J. E. (2020). Suicide among nursing home residents: Development of recommendations for prevention using a nominal group technique. *Journal of Aging & Social Policy, 32*(2), 157–171.

Johnstone, M.-J. (2012). Bioethics, cultural differences and the problem of moral disagreements in end-of-life care: A terror management theory. *Journal of Medicine and Philosophy, 37*, 181–200.

Johnstone, M.-J. (2013). 'Death with dignity' – Doubts and demands. *Australian Nursing and Midwifery Journal, 21*(4), 26.

Joosten, M. (2016). *A long time coming*. Scribe.

Kidder, T. (1993). *Old friends*. Houghton Mifflin Company.

Kuhl, D. (2002). *What dying people want*. ABC Books.

Lewis, C. (1961). *A grief observed*. Faber and Faber.

Luther, M. (1956). *Luther's works* (Vol. 13). Concordia Publishing House.

Murphy, B. J., Bugeja, L. C., Pilgrim, J. L., & Ibrahim, J. (2018). Suicide among nursing home residents in Australia: A national population-based retrospective analysis of medico-legal death investigation information. *International Journal of Geriatric Psychiatry, 33*(5), 786–796.

Neuhaus, R. (2000). Born toward dying. *First Things, 100*(February 2000), 15–22.

Nightingale, F. (1969). *Notes on nursing: what it is, and what it is not*. Dover Publications, inc. (First published by Appleton and Company in 1860).

Nuland, S. (1994). *How we die: Reflections on life's final chapter*. Alfred A Knopf.

O'Connor, M., & Allison, M. (2020). Facilitating a good death in residential aged care settings, with support from community palliative care. *Australian Nursing and Midwifery Journal, 26*(10), 41.

Omori, M., Jayasuriya, J., Scherer, S., Dow, B., Vaughan, M., & Savvas, S. (2020). The language of dying: Communication about end-of-life in residential aged care. *Death Studies*, 1–11. https://doi.org/https://doi.org/10.1080/07481187.2020.1762263.

Palliative Care Australia, Alzheimer's Australia, COTA Australia, Aged & Community Services Australia, Leading Aged Services Australia, Catholic Health Australia, & Aged Care guild. (2017). *Principles for palliative and end-of-life care in residential aged care*.

Parliament of Victoria. (1987). *Inquiry into options for dying with dignity: Second and final report* (19). Parliament of Victoria,.

Pilotlight Australia. (2007). *Dying to know: Bringing death to life*. Hardie Grant Books.

Radbruch, L., De Lima, L., Knaul, F., Wenk, R., Ali, Z., Bhatnaghar, S., Blanchard, C., Bruera, E., Buitrago, R., & Burla, C. (2020). Redefining palliative care–a new consensus-based definition. *Journal of Pain and Symptom Management, 60*(4), 754–764.

Raphael, B. (1994). *The anatomy of bereavement*. Jason Aronson, Incorporated.

Richmond, J (1990). Unpublished and paraphrased conference presentation.

Sanderson, C., Miller-Lewis, L., Rawlings, D., Parker, D., & Tieman, J. (2019). "I want to die in my sleep"-how people think about death, choice, and control: Findings from a massive open online course. *Annals of Palliative Medicine*, 1–9.

Smith, R. (2000). A good death. *British Medical Journal, 320*(7228), 129–130.

Somerville, M. (2017). The importance of stories in the euthanasia debate. *Mercatornet.*. https://mercatornet.com/the-importance-of-stories-in-the-euthanasia-debate/8888/

Sprung, C., Somerville, M., & Radbruch, L. (2018). Physician-assisted suicide and euthanasia: Emerging issues from a global perspective. *Journal of Palliative Care, 33*(4), 197–202.

Steinhauser, K., & Tulsky, J. (2015). Defining a 'good death'. In N. Cherny , M. Fallon, S. Kaasa, R. Portenoy, & C. D (Eds.), Oxford textbook of palliative medicine (5th ed.). Oxford University Press. https://doi.org/10.1093/med/9780199656097.003.0008.

Stevens, J., McFarlane, J., & Stirling, K. (2000). Ageing and dying. In A. Kellehear (Ed.), *Death & dying in Australia* (pp. 173–189). Oxford University Press.

Testoni, I., Wieser, M. A., Kapelis, D., Pompele, S., Bonaventura, M., & Crupi, R. (2020). Lack of truth-telling in palliative care and its effects among nurses and nursing students. *Behavioral Science, 10*(5), 88.

Tisdale, S. (2019). *Advice for future corpses (and those who love them): A practical perspective on death and dying*. Gallery Books.

Tjernberg, J., & Bökberg, C. (2020). Older persons' thoughts about death and dying and their experiences of care in end-of-life: A qualitative study. *BMC Nursing, 19*(1), 1–10.

Tolstoy, L. (1960). *The death of Ivan Ilyich (R. Edmonds, Trans.)*. Penguin.

Wiersman, E., Marcella, J., McAnulty, J., & Kelley, M. (2019). 'That just breaks my heart': Moral concerns of direct care workers providing palliative care in LTC homes. *Canadian Journal on Aging*, 1–13.

Williams, S., Hwang, K., Watt, J., Batchelor, F., Gerber, K., Hayes, B., & Brijnath, B. (2020). How are older people's care preferences documented towards the end of life? *Collegian, 27*(3), 313–318.

Chapter 7
Leadership

7.1 Leadership and Governance

7.1.1 What Makes a Good Leader?

Nursing home culture is influenced by the calibre of its leaders, embedded in the home's philosophy. Good leadership is framed by the setting of clear goals, high expectations, mutual respect, encouragement, support and recognition, rewarding good work and actively promoting a positive community profile. A career path to leadership, while uncommon in residential aged care, can be fostered by a culture of learning, continuous improvement, high staff morale and careful, selective staff recruitment.

Transformational leadership includes role modelling fairness and integrity, encouraging others to see past their own interests, motivating staff and creating a vision that inspires others. Leaders focus on clear communication, giving and receiving regular feedback and acknowledging staff members' personal career goals. Political skills are also needed, for managing organisational aims and communicating with the wider community. Honesty, integrity and fair-mindedness are essential qualities. Personal strengths such as the ability to cope with stress and capacity for maintaining positive interpersonal relationships are attributes for good leadership. It is an uncontested fact that the quality of management affects the quality of care, as well as relationships with families, shown by the following research:

> A well-run aged care home had managers who were collaborative, good communicators, empathetic and visible. Although managers need to spend a significant amount of their time in their office, relatives were reassured when they saw managers *"on the floor"* supervising

---

While (some of) the stories are based on factual situations, real names and other details have been altered to protect the identity of the persons concerned. Resemblance to any particular person is therefore purely coincidental.

© The Author(s), under exclusive license to Springer Nature Switzerland AG 2022
R. Hudson, *Ageing in a Nursing Home*, https://doi.org/10.1007/978-3-030-98267-6_7

staff. They also appreciated managers who collaborated with residents' families (Russell, 2017).

Leadership includes qualities of fairness, determination, dependability, innovation, openness and courage. Leaders are more credible when they follow through on promises, speak with conviction, experiment and take risks, treat others with dignity and respect and show support and appreciation (Kolzow, 2014). Communication is also enhanced when leaders sit down and talk with their employees, or join them 'on the floor', as Russell (Russell, 2017) suggests. Others have suggested managers to spend a whole day with the residents they are paid to care for, in order to increase their awareness of the oversight required, together with insight into the complexity of residents' needs.

It is interesting to note the preponderance of male managers, disproportionate in numbers to the largely female nursing and personal care staff. While no judgement is intended regarding this comparison, it may be an issue for discussion when managers are appointed, to explore attitudes of leadership in relation to this potential imbalance.

Nay shows deep awareness of residents' needs, concluding her analysis of 'the good, the bad and the downright ugly' aspects of aged care: 'If I could do one thing to improve care of older people it would be to invest in whole of organisation leadership development' (Nay, 2016). In other words, the best systems in the world will be ineffective without competent leadership. Several commentators cited in the Royal Commission into Aged Care Quality and Safety report (2021) have emphasised the need for skilled leadership, strong regulation and greater transparency.

7.1.2 Emotional Intelligence (EI)

What is the relationship between EI and nursing home leaders? 'EI is a self-development concept that is characterised as the ability to influence and motive others by being attuned to their emotional needs...' (Butler, 2021, p. 19). Effective leadership includes the ability to motivate and inspire staff, to be aware of their own emotions, to acknowledge mistakes and to be aware of their limitations and continually aim to improve their self-awareness. These are some of the criteria for emotional intelligence in nurse leadership.

As EI is an involving concept, aged care leaders may require assistance in updating their credentials. These include attributes such as empathy (discussed in more detail in Chap. 5), advanced social and communication skills and the capacity for developing authentic relationships with colleagues.

The outcomes of building EI within the leadership team include 'decreased burnout, chronic stress, lower staff turnover, increased job performance, positive work-life balance and job satisfaction, ultimately enabling the provision of superior patient care' (Butler, 2021, p. 20). An energised focus on leaders' EI will benefit staff, volunteers, residents and their families.

7.1.3 Board Governance

'The board has ultimate accountability for the quality and safe care provided to its clients. Quality, safe care is a core business and thus needs to be a key focus of the board' (Australian Institute of Health and Welfare, 2020). The board's role in the nursing home is to clearly set out its expectations and 'continually test whether those expectations are understood and brought to life . . . through its leadership practices, policies and structures' (Australian Institute of Health and Welfare, 2020). The authors of this report also highlight the many instances where poor practices have fallen far short of what is expected of boards of governance.

A competent and sensitive board understands its role is to *serve* the nursing home, to develop a culture of continuous improvement and to set the tone of the organisation, whether large or small. It may be assumed that board members are chosen for their interest in improving quality at all levels, principally in the care of residents. To this end, board membership should include representation from residents and/or relatives.

Independent governance remains an issue of continuing debate, with one report emphasising the urgent need for change at the top levels of government. '. . . (K)eeping the department in charge would merely continue the unacceptable status quo. The department has a weak track record on aged care. Its failure to adequately manage the COVID-19 crisis . . . provides the case example' (Duckett et al., 2021, p. 15). Ensuring government officials at the local level are appointed according to their interest in, if not their passion for, all aspects of aged care would be a step in the right direction.

7.1.4 Clinical Governance

Clinical governance is the means by which managers and staff assume responsibility for residents' safety and quality care, including the development of policies and procedures, regularly reviewed according to prescribed standards of care. Clinical governance includes organisation-wide systems and mechanisms for continuous improvement.

A case in point is the COVID-19 pandemic, which led to a variety of governance responses both negative and positive. While it is acknowledged that this serious crisis raised issues never before confronted in the wider community, nor in residential aged care, some responses were less than optimal. Many examples were cited of management's failure to communicate with residents and families regarding the presence of the infection and the plans to respond. On the other hand, examples of clear communication included daily updates with honest messages, inspiring confidence in the leadership, albeit in a climate of crisis. In such critical situations, it may be prudent to employ a healthcare worker with specific public communication skills based on contemporary knowledge and the capacity to inspire confidence. However, this latter suggestion may not be commensurate with some organisations where the main focus is defensiveness, the bank balance and the size of the profit margin.

7.1.5 Reforms Needed

The Commonwealth has key responsibility for aged care; the *Age Care Act 1997* determines the structure of that responsibility in terms of system planning and monitoring. 'Reviews and inquiries into the aged care system have pointed out that the system is complex and fragmented and attempts at reform have often added complications' (Commonwealth of Australia, 2019, p. 123). This report has painted a bleak picture of aged care in Australia, reflecting poor governance in many areas, presaging major reforms required to improve the entire system. No attempt is made here, to describe in detail the complexities of reform needed in nursing home governance, particularly in relation to financial matters. Rather, the suggestion is made that governing bodies need not remain separate from others involved in the nursing home community. When governance is enacted in partnership with other health professionals, staff and families, the whole community's needs will be better served.

One of the governance challenges is to determine how best to ensure residents' enjoyment of life matches their experience of care. With imagination and creativity, it may be possible to ensure that concentrating on the latter does not come at the expense of the former. Together with excellent clinical and personal care, the system should prioritise residents' rights, choices and dignity so they can *thrive* and not merely remain alive. While the Royal Commission report, quoted above, emphasises the many instances of excellent care identified by the inquiry, the disturbing details of their three-volume publication highlight the alarming levels of poor care, described and reflected in the title: *Neglect*. Reports of abuse (discussed in more detail in Chap. 8) lead to questions of the board's culpability and the results accruing when criteria for good governance are disregarded.

Good governance is challenged by difficulties in staff recruitment, low wages, heavy workloads and insufficient supervision by qualified nurses. Many would attest to the level of dedication and excellent care provided in some nursing homes by all levels of staff, even within such constraints. However, commitment and dedication need to be rewarded and encouraged by proactive leaders. Those charged with such oversight need also to have a degree of insight into the needs of the frail residents for whom they are responsible. Rather than relying on market forces where residents are regarded as (fee-paying) 'customers', nursing home governance needs to focus on the highly complex and uniquely personal needs of residents at the end of their lives.

According to the first page of the report quoted above (*Neglect*), much of contemporary aged care is abysmally deficient:

> This cruel and harmful system must be changed. We owe it to our parents, our grandparents, our partners, our friends. We owe it to strangers. We owe it to future generations. Older people deserve so much more. (Commonwealth of Australia, 2019, p. 1)

Leadership includes not only rewarding good care but also detailed knowledge of all incidents negatively impacting residents' daily lives, factors which are vitally important to a philosophy of continuous improvement. One of the recommendations discussed in various places in the Royal Commission into Aged Care Quality and Safety (Commonwealth of Australia, 2021) describes a government leadership model with a newly constituted authority such as an Inspector General of Aged Care, with power to investigate systemic issues. Such a system would, potentially, promote the much-needed overall improvement in Australian nursing homes. 'As an affluent world-leader, Australia cannot continue to provide such low standards of care to its vulnerable older citizens' (Duckett et al., 2021, p. 8). This report goes further in advocating the urgent need for reform. 'Aged care reform is more than a political challenge, it's a moral imperative' (Duckett et al., 2021, p. 3). A new 'rights-based' approach would require a profound shift by placing the needs of older people first and employing staff with the capacity to meet those needs.

7.1.6 Choosing the Right Staff

At present, there is little incentive for nurses to choose residential aged care as a professional pathway: absence of regulation, pay differentials, lack of support, high stress levels, inadequate medical oversight and rapid staff turnover have the opposite effect. On the other hand, some nursing homes provide vocational training, attracting nurses and other carers to enter this challenging environment and rewarding them accordingly.

What calibre of staff provides the foundation for a thriving nursing home community? What caring attitudes are essential? How is staff performance regularly appraised to ensure the residents are receiving optimum care and the staff enjoy a level of satisfaction? What happens when things go wrong or when there's a mismatch of leaders, staff and residents? Does a partnership philosophy translate into effective practice, with genuine benefits for the residents and their families? The answer to these questions lies with the manager's leadership skills, including recruitment of staff committed to achieving excellence.

Excellence in nursing home care impacts the lives of frail older persons in their last 'chapter'. What kind of person is more likely to make this difference, and how does this influence staff selection? Leading questions may be raised at interview: 'Tell me your reason for wanting to work in aged care'. 'What do you think are essential qualities for nursing home staff?' 'What do you consider important for end-of-life care?' These discussions become fertile ground for first impressions: both for the interviewer and the prospective employee. The nursing home's culture is readily apparent at this introductory meeting. What is the prospective employee's first impression? What will this new staff member contribute to resident's care? How will other staff respond? Will the new staff member be welcomed or ignored on their first day?

Most managers would agree that reference checking is vitally important; and much is to be gained by face-to-face interview, for both parties. Attention to these matters influences the calibre of the staff and ensures their confidence in management. Within an open, friendly atmosphere, new staff adjust quickly, feeling 'at home' in a community of care. For others, first impressions leave a 'bad taste'.

An important component of community building within the nursing home is the way new staff are acknowledged and welcomed. A regular newsletter or a prominent, up-to-date noticeboard conveys relevant information to the whole community, including residents, relatives and other visitors. Otherwise, new staff may be left floundering, finding their own way even to the tea room, as in the following example:

> *'Patricia' wanted to work as a nurse assistant, to test her vocation for tertiary nursing studies. Her first impressions were not positive. 'The receptionist told me to report to the first floor but I couldn't find anyone. After a while, someone (she didn't introduce herself) told me to follow her and without further instructions told me to wash the residents in rooms two and three.' Fortunately, Patricia had sufficient self-esteem and incentive to find her own way around this new environment, quickly learning to rely on her own criteria for care, while deeply disillusioned about the lack of leadership.*

Orientation programmes vary across nursing homes: however, their potential to influence staff (positively or negatively) cannot be overstated. A comprehensive induction session for nursing and non-nursing staff has inestimable value for building a sense of belonging in a new community, as well as upholding the standards of care. Similarly, when staff are not farewelled appropriately, they may assume their service has not been appreciated. Concomitantly, when staff are well cared for and their work acknowledged, their commitment is consolidated and residents' care enhanced. One provider stated that a by-product of managers' care for staff, including appropriate rewards, is longer service and a culture of loyalty.

According to the *Aged Care Act 1997*, aged care homes must 'maintain an adequate number of appropriately skilled staff to ensure that the care needs of recipients are met'. However, in the ensuing decades, discussion with residents' relatives found much concern about the loose interpretation of this statement. While acknowledging the dedication of staff who, they believed, were inadequately paid, these relatives showed some alarm that most resident's care is provided by unqualified staff whose training is variable: inadequate to meet the needs of residents with complex and multiple comorbidities (Russell, 2017, p. i). One of the recommendations in the Final Hearing of the Aged Care Royal Commission into Aged Care Quality and Safety was 'making Certificate III and IV qualifications mandatory for personal care workers' (COTA Australia, 2020, p. 7). Further to ensuring staff have minimum qualifications, training in dementia and 'rights-based care', minimum care hours, adequate nurse supervision and independent assessments are keys to reforming the workforce (Duckett et al., 2021, p. 18).

Embedded in the aged care standards (Australian Government Aged Care Quality and Safety Commission, 2019) is the notion that aged care workers should be 'skilled, kind and respectful'. Thus, an essential component of the interview process should include questions and discussions centred on the employee's understanding of kindness and respect towards older people, as well as evidence of their skills.

Acknowledging the strict, professional, regulatory framework does not obviate discussion of other human elements such as compassion, empathy, tenderness, joy and celebration. Another important component involving staff is ensuring these standards are maintained, by obtaining regular input and feedback. Employing staff with the necessary skills also entails programmes for upgrading skills and developing new competencies. The role of a human resource department (however named and sized) is important in holding employees to account and building a culture of excellence in the workplace.

First-hand accounts from RNs recruited from acute care to work for short periods in nursing homes during the COVID-19 pandemic are notable for their positive, rewarding, pleasantly surprising outcomes. Many enjoyed their engagement with residents, noting especially the significant improvement in the residents' medical conditions and emotional well-being as a direct result of increased care hours by nurses capable of assessing residents' complex needs. The noticeable contrast in residents' care was not lost on family members, who, to their profound disappointment, knew this episode of high-quality care was only temporary.

7.1.7 Caring for Staff

Research comparing staffing in Australian nursing homes with staffing in the USA finds Australia 'does not meet benchmarks internationally'. One 'unacceptably low' benchmark relates to care hours and continuity of care. Analysing these and other figures, Eagar finds that not enough attention is paid to clinical governance (Eagar et al., 2020). Managers and leaders have a responsibility to ensure staff are cared for, while understanding the serious constraints on staff numbers and residents' acuity levels.

Caring for older people at the end of their lives and responding to a number of deaths sometimes in a short period of time can be taxing, leaving some staff to consider ways of self-care and replenishment. Many will have their own rituals to assist their coping: for others, a professional counsellor, spiritual advisor or chaplain can assist in the process. Some nursing homes have embedded rituals, centred in a quiet space, to provide 'timeout' after a stressful period. Many employees attest to the stressful nature of their work, particularly given the lack of prescribed staffing numbers and qualified personnel. Others show their dedication and strong desire to influence residents' care, even within the constraints.

While it is not unusual for nursing home employees to claim they are undervalued, optimum staff support is conveyed by the following:

- Regular performance appraisals with written feedback
- Notifying all staff of critical events
- Managers meeting regularly with staff
- Well-chaired staff meetings, allowing each voice to be heard

- Staff wellness programmes, such as on-premises gym or gym membership discounts
- Public acknowledgement of staff members' service
- Acknowledging staff's emotional needs
- Welcoming new staff, with comprehensive induction
- Farewelling staff with due acknowledgement of service

Lack of staff support has often been reported. For example, the 2019 Royal Commission into Aged Care Quality and Safety heard the testimony of an aged care worker suffering from post-traumatic stress disorder (PTSD) after she had witnessed one resident killing another resident. The aged care worker also recounted physical assaults from residents with dementia and the inadequate responses by senior staff (Burke, 2019). It is difficult for staff who are mistreated to provide optimum care for residents.

What reforms are needed to attract and retain high-quality staff? Only if work conditions are significantly improved, including pay rates, training opportunities and genuine care by management will quality staff be attracted to nursing homes. Attractive work conditions will enhance longer-term employment rather than the current norm in many places, of a constant turnover. 'To enhance continuity of care, providers should be relying much less on independent contractors or casual staff, and much more on directly employed staff' (Duckett et al., 2021, p. 19). Benefits of such an emphasis include permanent employees taking professional pride in their choice of workplace, despite the poor pay rates.

Transforming working conditions, pay rates and staff support require action from a government committed to an appropriate level of funding for nursing homes. Without such support from the highest levels of government, leadership at the local level will always lack sufficient resources to provide the care they would regard as optimum. As in other areas, this subject suffers from a lack of comprehensive, contemporary research.

7.1.8 Acknowledging Night Staff

The sector would benefit from research into the reasons why some staff choose to work at night, gaining insights from their unique perspectives. The following example is instructive:

> The manager seldom exchanged personal words with this nurse who had for many years worked only at night. On this particular morning, 'Margaret' was late going off duty, resulting in an unexpected meeting with the manager in the foyer. What began as a mundane exchange of pleasantries became a frank discussion on death and dying, especially when the manager learned of the sudden death of Margaret's younger sister. This was the first death of a close family member and in spite of many years as a hospice nurse, for Margaret, this was a close encounter with death of a different kind. For the manager, whose previous conversations with Margaret were perfunctory at best; this brief discussion humanised the

relationship resulting in a deeper understanding between manager and nurse, including the reasons why Margaret preferred to work at night.

A chance greeting in the foyer between a nurse going home after a long night shift and a manager about to commence a new day of administrative duties is an encounter that transformed the relationship. 'She's actually very caring', the nurse told her colleagues.

Apart from such a chance encounter, managers' meetings with night staff are optimised through a comprehensive appraisal system where staff feel free to describe the 'highs and lows', make suggestions to optimise residents' care and improve their own working conditions where needed. A well-structured system allows for learning needs to be identified, professional development goals to be discussed and personal aspirations to be voiced. Staff may be praised for their work and under performance addressed wherever pertinent.

Regular performance appraisals provide a welcome opportunity for managers to get to know all of their staff and develop a continuing relationship. If staff know the appraisals are routinely scheduled and based on objective assessment including self-appraisal, there should be no cause for concern. It is clearly a time-consuming procedure but one which puts a 'human face' on manager/staff communications, leading to responses such as the following:

'Deborah' received the customary official letter following her performance review. Stopping the manager in the corridor she confided. 'Thank you for that letter. It made my day! I really felt so low that week; my self-esteem was at rock bottom and I felt I had nothing to live for. This is the first personal letter I've ever had from a manager. I have it pinned to my kitchen wall'.

This scenario presumes, of course, an organisation small enough for the night nurse to 'bump into' the manager, prompting such an exchange. Larger organisations may devolve performance appraisals and feedback to a team of senior managers. Whatever the system, staff have confirmed in a number of surveys that their self-esteem and work satisfaction are enhanced by a manager's acknowledgement.

7.2 Systems Guiding the Community

7.2.1 Accreditation and Quality Monitoring

The ACQSC (Aged Care Quality and Safety Commission) assesses the performance of residential care providers against the eight Aged Care Quality Standards in order for them to receive funding. For a commencing service, this occurs via desk audit, with initial accreditation for 12 months. For most nursing homes, accreditation is for a 30-year period unless the Commission considers the service poses an unacceptably high risk of non-compliance. The standards are assessed on a 'met or not' basis rather than a sliding scale. As at 30 June 2018, 96.9% of the 2669 re-accredited residential aged care services had been given a three-year accreditation. The most frequent areas of non-compliance across all types of regulatory activity over this period were human resource management, clinical care, information systems, medication administration and behavioural management (Egan, 2019).

The Commission uses three types of visits to monitor residential care services – site audits as part of the re-accreditation process, assessment contacts and review audits. Since 1 July 2018, unannounced audits apply to all nursing homes seeking re-accreditation. Assessment contacts can be made at any time – announced or unannounced. Government policy mandates each nursing home receives at least one unannounced assessment contact every year. Review audits are undertaken in cases of suspected non-compliance, including an assessment against all accreditation standards (Commonwealth of Australia, 2019, pp. 61–62). Research is yet to uncover the impact of reduced unannounced monitoring during the COVID-19 pandemic. Further detailed research is also needed into the quality, consistency and professional systems governing accreditation.

7.2.2 Policies and Procedures

This section highlights the benefits of formulating clear policies and procedures, regularly updated, and in conformity with the best contemporary standards of care. When the whole nursing home community is included in the development and transmission of such policies, a climate of confidence is more likely to ensue. Rather than a negative or ad hoc attitude, staff draw support from coherent documents, as in the following exchange:

> *RN comment to a colleague:* '*I think we should check the policy on transfers before sending the resident to hospital*'.
>
> '*Oh, policies!*' *replied her colleague.* '*I never look at them, they're so outdated. Anyway, I wouldn't know where to find them even if I wanted to*'.

When policies (presumably current) are not followed, actions based on staff's vagaries or opinions do not benefit the resident. On the other hand, when trustworthy policies guide practice, continuity and consistency of care is enhanced.

Systems vary across the aged care sector: some have multiple procedures for resident's care, documentation, administration and all other aspects of nursing home life. Others have few, relying on staff 'to do the right thing'. Reliable systems inspire staff confidence and, concomitantly, influence resident's care. As well as clinical practice guidelines, protocols and procedures for meal delivery, cleaning and administrative activities also influence safe, consistent practice, all key aspects of good governance.

There are, of course, many situations where best practice in following policies and procedures is exemplified, as well as others where improvement is needed. One of the challenges is to maintain regulatory compliance, particularly when inadequate funding compromises optimal care. For instance, the ACFI (Aged Care

Funding Instrument) has the capacity to ensure funding is matched to residents' needs. However, the Royal Commission Report (Commonwealth of Australia, 2019, pp. 76–78) suggests review is needed to provide for at 'least one registered nurse on site at all times' (p.76). Further, the Committee observed that various other reviews and inquiries point to the need for the regulatory system to be overhauled (p.77).

7.2.3 Communication: The Management Culture

'We never see her (or him)' is obviously not the best response from staff who are asked to describe their relationship with their manager. It requires little stretch of the imagination to picture the extremes from (a) the manager who never/seldom leaves the office, does not know the names of staff and is generally unapproachable to (b) the manager who spends a vast amount of time chatting with staff 'on their level' to the point of impeding the manager's own workload and leadership role. Somewhere in the middle are the following 'keys' for best practice:

- Flexible communication, from 'open-door policy' to private, planned discussions
- Staff development: encouraging staff to attend lectures, courses and study programmes
- Acknowledgement and reward systems for staff achievements
- Fostering extracurricular activities that aim to achieve a work/life balance for staff
- Flexible working arrangements to accommodate family needs wherever possible
- Comprehensive induction protocols for new staff to be 'paired' with an exemplar
- Appropriate recognition and rewards for long service, exemplary service or associated achievements
- Encouragement for staff to pursue postgraduate study in gerontology
- Development of specialisations, for example, advanced skills in diabetes, neurological conditions, dementia and pharmacology/medication management

Another important aspect of management culture is the influence of *racism* in some contexts. Policies should be avoided which perpetuate inequities by imposing Western values in an increasingly multicultural milieu. Recognition must be given to the fact that for some residents and families, particularly from diverse cultural backgrounds, nursing home care is set within an unfamiliar health system. Managers are influential in developing cultural differences amongst staff and residents, employing multicultural experts where appropriate, to guide an informed response.

Policies and procedures, formalised in well-publicised notices, strengthen the practice of 'calling out' racism wherever and whenever it is evident throughout the nursing home. There is no place for bias or bigotry; managers need to act swiftly in response to any reports of the same.

7.2.4 Staff Induction and Orientation

Comprehensive new staff induction takes time, requiring clear procedures, and delegation where necessary. Some of the basic issues include general courtesies and thoughtfulness, to prevent the following commonly heard remarks:

> 'Nobody told me where to put my belongings when I arrived'.
> 'Nobody told me about the lunch room'.
> 'Nobody introduced me to senior staff'.
> 'Nobody told me anything!'

These negative comments are countered in many places where thorough procedures govern the induction process. Rather than a 'depends who's on' attitude, clearly written guidelines including a checklist and a brief instructive video, buddying or mentoring, give rise to optimal induction experiences for every new staff member, promoting responses such as the following:

> 'The manager was very welcoming. He even took me to the kitchen and introduced me to the chef'.
> 'I was nervous entering an unfamiliar work place, but I was 'shown the ropes and my confidence was boosted 100%'.

Induction time, together with all recruitment and retention procedures, is time well spent: clearly an investment, often resulting in long-term employment. When followed by a clear orientation period, new staff will quickly feel part of the team, especially when due consideration is given to the needs of staff whose first language is not English.

7.2.5 Acknowledging Long Service

Some management structures allow for recognition of staff's long service, such as the following:

> When 'Bridget', the kitchen assistant, resigned after ten years' faithful service, about forty residents, family members and staff gathered in the lounge for a special afternoon tea. The manager asked if any residents would like to say a word of thanks or goodbye to Bridget. 'May's' speech was brief and to the point: 'I'll miss you, Bridget.' And, of course there were many silent tears and some not so silent. 'Madeleine' asked, quite shyly at first, 'I can't make a speech but I'd like to sing a song.' A call for silence was all that was required for Madeline to render her beautiful, poignant solo: 'When I grow too old to dream.' A fitting end to an important chapter in the life of the nursing home.

Human resource systems may allow for a certificate and/or choice of tangible gift for a staff member achieving a significant length of service, reinforced by management, staff meeting minutes and residents'/relatives' committees. Given the rapid expansion of the ageing population and the associated need for aged care workers into the future, it would be prudent for management to invest in innovative procedures (not necessarily affecting the budget) which attract long service. One example

7.2 Systems Guiding the Community

is an Aged Care Employee Day, honouring staff in general, and allowing for specific attribution related to long service.

7.2.6 Staff Departures

Managers play a significant role in the way staff departures are communicated not only for other staff but also for residents and families. Lack of information may lead to speculation, if not confusion, such as the following:

> *'Where's Sally? I haven't seen her around' asked the resident. There was an awkward pause in response, as Sally's position had been terminated.*

It is not appropriate for management to provide details of the reason for a staff member's employment termination. However, if a regular list of staff is maintained, with no other explanation apart from the date of their commencement and departure, then all staff notifications are treated equally and residents and relatives kept informed.

With or without management's protocols for staff departures, individuals develop their own responses, such as the following:

> *'Andrew' came to the administration office to deposit his 'leave form' at the same time 'Ding Xian' was completing his resignation form. A coincidental meeting and spontaneous exchange, signifying a genuine display of affection and mutual acknowledgement of professional expertise. 'You'll be sadly missed, mate!' 'So, will you; the place won't be the same without you.' They began to shake hands, then pausing for a split second, warmly embraced each other. As 'Andrew' deposited his resignation letter on the manager's desk and walked out the door, did she note a tear being wiped from his eye?*

Compared with some other healthcare contexts, nursing home staff often work closely together with varying degrees of intensity and time frames, with some significant implications for the much smaller male workforce. Due recognition between colleagues has the potential to favourably influence lasting memories. To that end, managers have a responsibility to adopt a framework for acknowledging staff departures, ensuring consistent practices. While there may not be time or funds for a 'farewell party' for every departing staff member, meaningful, consistent practices can be developed through wide consultation with the whole nursing home community.

7.2.7 Exit Interview

More than a formality, the exit interview provides the final means of communication with a staff member who is leaving for any reason. Goodbyes are formalised, an exit questionnaire may have been completed and time is set aside for any final comments from manager or staff member. As a means of 'closure', the exit interview provides formal acknowledgement of the period of service.

The exit interview may also be instructive for service improvement, particularly following honest, confidential feedback. Whatever the reasons for leaving, the exit interview is, potentially, a courteous, satisfying way of completing the relationship. In the absence of such regular protocols, scenarios such as the following are not uncommon:

> '*Sarah*' *was recognised by staff and residents as one of the most popular, caring, competent nurse assistants, so when she signified her intention to leave (after one year) in order to pursue medical studies, she was rather surprised at the perfunctory* '*system*' *for staff resignations.* '*Just pop your name tag and key on the notice board*', *said the receptionist.*

The way staff are introduced to the nursing home, regularly appraised, supported and farewelled, should be of paramount interest to those who lead and govern. Such leadership creates a strong community: demonstrating care 'from the top' down. In the absence of such formalities, the sentiment is reinforced: 'they couldn't care less!'

7.2.8 Accepting Gifts?

Residential aged care poses both opportunities and threats for relationships between staff, residents and families; inevitably some close relationships develop. The Australian nurses' code for professional conduct (Nursing & Midwifery Board of Australia, 2018) permits nurses to accept 'nominal gifts'; however, this is difficult to define categorically. Best practice would suggest each nursing home develop a policy about the giving and receiving of gifts, to ensure clear lines are drawn between what is acceptable and what is not. For example, a limit could be applied to their value and a prohibition relating to monetary gifts. Specific parameters should be developed regarding legal wills, especially those involving large monetary amounts. When policies are clearly understood by residents and staff, the possibility of exploitation is minimised, not to mention protracted legal procedures in some circumstances.

7.3 Education and Students

7.3.1 Routine Education Sessions

Well-educated staff translate their knowledge into effective action for the benefit of the residents and their families. Without contemporary gerontological support, residents may be at the mercy of outmoded or, at worst, unsafe practices. Well-designed education procedures counteract slavish, unquestioning attitudes: understanding *why* something needs to be done is invoked, rather than merely *what* needs to be done. Some nursing homes place a high priority on regular education sessions for all staff, clinical and non-clinical; others regard it as essential only for minimal

7.3 Education and Students

regulation compliance. Some have the advantage of an educator on their staff; others rely on external personnel. Clear policies on education may provide answers to some or all of the following questions:

- Are guest lecturers welcomed and treated with courtesy?
- Is there a clear understanding of learning objectives?
- Do managers and/or in-house educators attend, so they know what is being promoted?
- If the goal is to change practice, how will this be expedited as a result of the session?
- What happens to the evaluation forms?
- Are staff members paid to attend education sessions and/or reward in other ways?
- What constitutes mandatory and non-mandatory attendance?

Making time for education, albeit within a 'time poor' environment, has the potential for saving time, for example, in correcting unsafe practice, responding to incidents and promoting consistency. Education is not only the key to good practice; it is the essential ingredient for a human rights approach, that is, residents' rights to evidence-based care. Seen from this perspective, an education curriculum would include residents' and relatives' priorities and responses, following an invitation to contribute ideas for specific sessions.

The COVID-19 pandemic provided a timely reminder of the need for continuous education on *infection control,* regardless of significant outbreaks, bringing to sharp focus the need for greater medical, public health awareness. Considering the frailty and vulnerability of nursing home residents, it is not surprising to find they are more susceptible to such infections than the general public. The other significant factor brought to light by the pandemic is the need for staff training in the use of PPE coupled with a clear rationale for such infection control measures. The challenge is to ensure that infection control education is understood and translated into practice by a workforce largely unskilled in this area.

A qualified educator has the potential to change attitudes, enhance practice and inspire excellence.

7.3.2 Innovative Education Role Play

The following scenario highlights the value of innovative education sessions:

> The setting was an in-service education session involving several role-plays. Staff were asked to consider two different scenarios involving a nurse responding to a resident calling out at night. Volunteers were called to enact the role-play. 'Jennie', considered by some as one of the least sensitive nurses, offered to play the part of 'Aaliyah', the resident. Using improvised props, two chairs became her bed. She acted the part well, assuming the role of Aaliyah, a defenceless resident reacting to the towering figure of the night nurse summoned by her relentless calling out. The staff member playing the night nurse adopted the overbearing posture of one who had 'come to see what all this noise is about'. Aaliyah immedi-

ately stopped calling out, cowering under the blankets, only to resume her noise when the nurse left the room.

In the alternative scenario, another night nurse adopted a more friendly posture, gently assisting Aaliyah out of bed, wheeling her to the lounge, making her a cup of tea and engaging her in conversation (despite the resident's lack of English), after which she returned to bed and to sleep.

Staff were moved to silence, not only by the talent of the 'actors' but by the profound effect of the normally garrulous and brash nurse taking on the feelings of the resident. The 'actor' commented she had never realised what it must be like to be literally 'under' the authority of the nurse, particularly when complicated by a language barrier.

Many other examples reinforce the role of innovative education in transforming resident's care. Attention is also needed to the significant proportion of staff whose first language is not English, ensuring the education sessions are relevant to their needs.

7.3.3 Students

At least one nursing home has opened its doors to an adjacent community of university students, offering rent-free accommodation for several, in response to their contribution to residents' care. Entering into a contractual arrangement specifying hours and conditions, the students provided (non-nursing) assistance to selected residents. This was especially helpful at meal times; but also, the students befriended the residents, offering companionship to those who were lonely; helping to celebrate birthdays and other occasions and assisting residents on outings were safe to do so.

Is student's involvement considered a liability or an opportunity? What is the quality of the information they are given prior to and on their first day? How are they welcomed?

'Oh, my God. Here's the students. I'd forgotten they were coming today!'

On this occasion, the manager had not given serious thought to the arrival of several nursing students, arranged many weeks prior. Another manager gave the arrangement much more attention. He ensured the details were included as a special agenda item at the staff meeting, together with a notice in the nursing home newsletter. 'Please prepare to welcome the students next Monday, doing your best to ensure they have a positive experience'. Ideally, a student information book would be available, well indexed and commencing with warm, sincere words of welcome. Information may include the mission/vision statement, map, directions, car parking and advice regarding meals. A comprehensive statement of resident's care standards would be an important inclusion, together with an evaluation form for students to complete on leaving.

Students from any discipline are often confronted (if not affronted) by new experiences when they are formally, as part of their curriculum, 'placed' in a nursing home; their response is influenced by their orientation (or lack thereof). When used wisely, the orientation session is an instructive forum for hearing the students' attitudes towards older people. When asked to rank nursing home placement on a list of preferred health agencies, many students place this last (Hudson & Richmond, 2000, p.147). However, such a placement provides an opportunity for reversing negative perceptions. One student noted in her journal: 'Sat and massaged creams and oils into Fred's wasting limbs. He especially seemed to like the back massage. I think it made a difference. I hope I helped him to die peacefully' (Hudson & Richmond, 2000, p.148).

In the nursing home, students will find people with multiple comorbidities, each impacting on the others. Given the high numbers of residents with cognitive impairment, this is the environment for a student nurse or other health professionals to learn assessment and communication skills. Enhancing their learning experience, supported by the highest-quality clinical supervision and education, will help to shape a talented workforce of the future, potentially, the cutting edge for clinical experience. Furthermore, if the experience is positive, students may well direct their attention towards gerontology as a career path.

Raising the profile of 'teaching nursing homes' can be achieved through formal partnerships with universities, so that a nursing home or group of nursing homes can be profiled as places of excellence for student's experience from a variety of disciplines.

Clinical placement for medical students, allied health professionals, pharmacists, social workers, dieticians, dentists and others presents a challenge, namely, the availability of sufficient skilled mentors. When such placements are well organised and supported, the educational partnership has significant potential to influence future leaders in all these fields (Hockley & Kinley, 2016). For example, postgraduate students of psychiatry may, when their nursing home placement is well organised, learn much about behaviour management in dementia, as well as teaching staff new skills drawn from their own academic experience.

7.4 Involving Families

7.4.1 Who Is the Resident's 'Family'?

Given the long-term nature of nursing home care, and taking account of 'duty of care' principles, partnering with families is essential. The first step is to determine who constitutes the family for it may not necessarily be the next of kin. Who does this particular resident regard as family?

> The resident was asked about her family; the nurse assuming the resident's husband played a key role in her life. 'Oh, no!' she replied. 'Don't ask him: we haven't spoken for years! I regard my next-door neighbour, 'June', as my next-of-kin. She's the one who's been caring for me and she'd love to stay involved now that I'm in here. Here's her phone number'.

This brief scenario illustrates the value of thorough assessment, as not all families are loving or caring nor are they necessarily physically able. It also offers a definition of family as 'who the patient says it is' (Monroe & Oliviere, 2009, p. 1). A pre-admission discussion sets the scene, creating understanding and trust between staff and families, including clear parameters and responsibilities, and, where appropriate, opportunities for the family carer to remain involved in the resident's care.

Social support is often lacking in residential aged care and family caregivers' needs are not always addressed. With some notable exceptions, nursing homes are not known for their care of families, a situation in need of urgent review. Now, more than ever, given the serious understaffing of most nursing homes, families need to be drawn into the assessment, planning and decision-making relevant to the residents' needs. Unfortunately, rather than positive relationships and caring attitudes towards families, there is, in some settings, a propensity for treating them as 'the enemy'. Or, on the other hand, families are given inappropriate deference, as in the following 'rationale':

'Oh, we'd love to give her more pain control but the family won't let us!'

What does 'the family' mean in this statement? The oldest son, the youngest daughter? Who is the designated spokesperson for the family? What happens when there are conflicting opinions? Who has the legal responsibility for the resident's care: qualified staff or families? Education on these matters would assist aged care workers to be clear about their role.

Many decades of research have demonstrated that family caregivers may have significant needs which remain unmet (Hudson et al., 2004). This can be the case when a relative has cared for a frail person at home, before the transition to residential care. Acknowledgement of the carer's role, skills and experience can pave the way for a relationship of trust. Sometimes well-meaning 'advice' is offered to families, but with misplaced assumptions. While intending to be empathetic, the following comment takes no account of the guilt, remorse and even depression, caused by 'Herbert's' separation from his wife and the sudden change from his caregiving role:

'Herbert, I know you've been caring for 'Lois' day and night for several months. It's now time for you to take care of yourself and leave your wife's care to us. Why don't you take a well-earned break?'

This well-meaning 'advice' was no solace for Herbert. A wiser approach may have been as follows:

'Herbert, let's sit down and talk about what this admission means for you. We know you've been caring for Lois at home and doing an amazing job. Are there any aspects of her care you would like to continue; or are you happy to leave all the care to us? Would you like to be involved in regular meetings about her care? And, how is your own health?'

Some well-intentioned but unenlightened nurses counsel relatives: 'You take a break dear, don't bother coming in each day, especially now she doesn't know who you are'. One elderly gentleman responded to this suggestion by saying of his wife of 68 years: 'She mightn't know who *I* am but I know who *she* is' (Hudson, 2019, p. 24).

7.4 Involving Families

Greater recognition of role changes would assist family members to adapt from their full-time carer status to that of 'visitor'. Inclusion in care decisions and consultations is a helpful means for family members to stay connected. A simple gesture such as staff enquiring after families' own health also helps to bridge the gap. More formally, care is demonstrated through well-organised family meetings.

Partnership with families is more than a management device for prevention/solving conflict. True partnership is an invitation to a narrative, entering the story of each resident and their family, as well as participating in the wider story of the nursing home community and beyond. Contrasted with the short-term nature of most hospital admissions, the longer-term environment of the nursing home may be considered as a place where each older person and their family can become part of a community of care. Through the telling of true life stories, evidence abounds concerning the rich legacy of each unique life, including 'Martin's' (below):

> *'Daphne', aged 70, had prepared well for her own impending death from advanced cancer, confident that her much older husband Martin, would be well cared for in the nursing home. Martin was surprisingly calm after the initial shock of his wife's death. A staff member who attended the funeral found him 'taking charge' as host from his wheel chair, thanking people appropriately for their visits, accepting condolences and seeming to enjoy being the centre of attention.*
>
> *Daphne's daily visits to the nursing home had been described by one staff member: 'She comes barging in and before she even looks at Martin or says hello, she's delving into the wardrobe to see if we've put his clothes in the right place and criticising him if he's soiled his pyjamas. She was often observed to be reading the paper rather than communicating with Martin. Prior to his wife's death Martin was emotionally labile, demanding of staff, calling out, often with incoherent speech. After Daphne's death his speech was slower but always intelligible: 'Thank you for caring for me. Thank you for being so kind.' An astute EN commented: 'Now Martin is coming into his own – no more nagging wife! It's up to us now to encourage his independence and strengthen his confidence in making his own decisions'.*

This story highlights the inevitable ambivalence experienced by staff witnessing a lengthy episode of care, compounded by a seemingly fraught relationship between the resident and his wife, not to mention the latter's impending death. Some staff sided with Martin and others with Daphne. Some were outraged at the perceived lack of care shown by Daphne; others who were aware of her illness were more sympathetic. Some were offended by the couple's constant bickering; others acknowledged a difficult relationship now compounded by serious illness. Martin and Daphne were never going to represent a picture of an ideal partnership, and the conclusion (described in the scenario above) highlights the fact that after a family member's death, there may be relief, rather than grief.

7.4.2 Family Meetings

'A family meeting is a *procedure* . . ., and it requires no less skill than performing an operation' (O'Mahony, 2016, p. 131). Preparation for a family meeting includes a concise definition of its purpose, a printed agenda, a venue free from interruption, a comfortable, welcoming atmosphere and a competent chairperson who ensures the meeting starts and ends on time (suggested length is one hour) and includes the GP. The resident should be included where possible and the meeting closed in a safe manner; ensuring disagreements, conflicts or emotional responses are resolved or deferred for another time. Brief notes should be filed with a copy for the family. When meetings are incorporated as routine, family members' comments such as the following can be avoided:

> 'I've been called to a family meeting. Nobody else I know has received a similar summons. I wonder what mum has done wrong?!'
> 'We've never been invited to a family meeting in five years'.
> 'I've asked for a family meeting, but they never seem to have time'.

When it is not practicable for family, resident and staff member to meet in person, 'virtual' meetings are now readily accessible, via Zoom or similar platform. Similarly, videoconferencing offers a practical solution for rural/remote areas where a well-planned case conference or meeting can occur without the resident leaving the bed, allowing input from the multidisciplinary team and providing the rural GP with support- and evidence-based solutions to problems. Such meetings optimise care planning and review, for the resident's benefit, staff convenience and family satisfaction.

One of the many benefits of a family meeting, particularly soon after admission, is to identify conflicts or misunderstandings over the goals of care. Other benefits include clarifying the role of palliative care, checking concerns about medications (such as opioids) and providing a copy of the care plan, with explanations where needed.

Families are not all 'traditional': changes include the ageing of baby boomers, geographical separation from the nursing home, blended families, gender variations, cultural mores and ageing caregivers. Other factors relate to smaller families, working families and the recognition that contemporary families have less exposure to death and dying than in previous generations. Careful attention is needed to address family issues within cultures unfamiliar to management and staff and to offer alternative arrangements where there is no family.

Conducting a family meeting is not an essential component of aged care education or employment; most senior staff will require some guidance, perhaps from the regional specialist palliative care team. Being aware of family meeting guidelines means being alert to some of the pitfalls, enunciated by this EN:

> 'Oh yes, we have family meetings: they go on and on and on! Yesterday, it went for two and a half hours!'

This example clearly indicates a lack of knowledge and/or experience in conducting family meetings which should generally not exceed one hour. (If there are issues not covered, another meeting time may be scheduled.) A well-conducted family meeting, particularly when focused on changed treatments and options for future care or the crisis of impending death, is particularly helpful for all involved (Hudson et al., 2009). Those who question their value raise the issue of time: 'We don't have time for family meetings'. Research shows that, when properly planned and executed, family meetings *save* time. For example, time is often wasted

7.4 Involving Families

repeating the same information to several family members rather than conveying a consistent message to all who attend the meeting, with comprehensive documentation for others not present. Careful thought needs to be given to planning each meeting, including seating arrangements (where relevant), technicalities for Zoom or other electronic meetings, pre-set agenda, meeting procedures and processes, timing, use of interpreter and ensuring each person present has a voice. Such a well-planned meeting promotes person-centred care, allows for an interdisciplinary approach and fosters families' voices (Puurveen et al., 2019).

Misunderstandings can often be addressed or averted by meeting with the family, particularly given the increasing variety of customs and cultural factors within the community. Sharing of information also requires careful assessment. What do the family understand about palliative care, end-of-life care and residential aged care in general? What are their expectations? What more information would they like and in what form? Family meetings assist with communication during the settling in period, clarifying misunderstandings, explaining staffing arrangements and overcoming of barriers.

Poor communication has detrimental effects on residents, families and staff. In the absence of discussion, particularly for end-of-life care, residents are at heightened risk for aggressive treatment; families are unprepared for the death and experience moral distress regarding decision-making (Durepos et al., 2017).

Those who are better prepared have more positive bereavement outcomes following the resident's death. What kind of assessment is made of family caregivers' preparedness? Does anyone ask the question 'Has anyone had a conversation with you about what to expect?' Preparedness for death also includes planning for the future after the death. For some families, taking care of the physical realities does not necessarily mean they are emotionally prepared for the death nor for the bereavement to follow (Breen et al., 2018). Such was the case with one family member:

> *'Judith' was a social worker who took time off from work so she could visit her mother regularly and for long periods during her mother's dying phase. The manager made a routine 'how are you' enquiry, which resulted in Judith's embarrassed pause before confessing: 'I feel silly asking this, but I don't know what to do when mum breathes her last'. An experienced health professional, Judith needed similar support and advice given to others facing death of a family member. The manager was pleased to invite Judith into her office for a frank, open discussion, sensitively answering her 'silly' questions.*

Families sometimes make requests of management to collude with their internecine conflict: 'Please don't ask my sister to the meeting'. 'I don't want my son to know mum's dying'. An experienced leader will incorporate all views into a family meeting, concentrating on the goals of care while minimising conflict. Setting the ground rules and referring to another agency if the family is perceived as 'too difficult' are some of the principles for family meetings.

Allocating resources for training and planning, including 'learning on the job', assists inexperienced staff to gain new skills. One nurse said: 'I feel much more confident doing family meetings now; and I'm okay about people sitting in to observe, to learn what we've learnt' (Hudson et al., 2009) (paraphrased). Such was the experience of another nurse:

> *The senior RN had no experience of family meetings. New to the nursing home, she was intrigued with the routine practice of inviting the family of a recently admitted resident to a routine meeting. 'Does everyone have one of these meetings?' she asked. When it was explained that the routine was clearly outlined on the advertising brochure, she was persuaded. She sat in the meeting conducted by the manager, carefully noting the structured agenda, the strictly observed time frame, and the warm response of the family members. The manager suggested the RN take responsibility for the next scheduled family meeting. Clearly nervous, due to her lack of experience, the RN left the meeting with hands still shaking as she reported to her colleagues: 'I've done it! And it wasn't as bad as I expected; but I know it will be far easier next time'.*

The arguments in favour of family meetings far outweigh negative perceptions, particularly when staff inexperienced in the process are given the chance to 'learn on the job'.

7.4.3 Families' Needs

Now that older people are entering nursing homes in a much frailer state than a decade ago, their carers also exhibit changing profiles. What issues does this transition involve, particularly for the main carer(s), who may or may not be a spouse or other close relative?

Every transition is unique, calling for sensitive assessment and understanding. What were this carer's responsibilities? Are they happy/relieved/sad/frustrated or ambivalent about their changing role? What aspects of care (if any) would they like to continue once the resident is in the nursing home? Careful assessment is needed for each unique situation, such as the following:

> *'Florrie' felt 'torn' when given the responsibility of accepting the nursing home place for her husband of 65 years. 'Tom' had become increasingly difficult to care for with his challenging dementia complications. His worsening osteoarthritis also meant that physically he had become increasingly dependent on his wife. Florrie seldom left the house, willingly accepting the role of full-time carer. Now that he was to be cared for in the nursing home what would she do with her life? Take that long postponed overseas trip? How would she sleep at night, in the strange newness of being without her partner? Would she even bother to cook for herself? How often should she visit? Would he even recognise her? How well would they look after him? Would they know when he needed the toilet? Nobody enquired about her reactions, let alone the myriad unanswered, or as yet unformed questions.*

The only way to discern a carer's needs is to ask them, preferably in the pre-admission interview. At the very least, families should be given opportunities to voice their reactions to the resident's transfer.

It is not unusual for a carer to become depressed following their relative's transfer from home to nursing home. Some feel relief; others may experience a sense of emptiness: 'What's there to live for now she's gone (into someone else's care)?' Increasing understanding is needed for an elderly male who has cared for his wife at home for several years: a role reversal in many cases, involving learning new tasks such as laundry and cooking and the personal hygiene his wife can no longer

7.4 Involving Families

manage. After his wife's nursing home admission, he may feel in some measure 'compensated' when his caring role is acknowledged, as in the example below:

> 'Bertram', I know you've been a champion in persuading your wife to attend the toilet. Please share with us any 'tricks of the trade'. Bertram knew immediately what was being asked of him by the nurse developing the care plan. 'Oh, I know exactly what you mean. Well, my wife is an obsessive about washing her hands. So, when I thought she needed the toilet I'd say, "Come on love, let's wash our hands". While she was busy with the soap, I'd gently remove the incontinence pad and replace it with a new one'.

Rather than an attitude which says 'Leave it to us! We know best!', this brief scenario conveys a mutuality of care. It also represents an increasing phenomenon, when the main carer is a male unaccustomed to providing personal care. In some circumstances, it is appropriate for the nursing home to ensure carers such as 'Bertram' receive ongoing support, including involvement in the resident's care, if that is their wish.

After the resident's admission, family caregivers do not necessarily cease their caregiving role or experience relief. What support is most appropriate, especially for those admitting 'burnout'? Is there grief over lost or changed relationships, particularly where the resident can no longer function as a 'listening ear' and a sexual partner or where marriage or partnership may seem to have lost its meaning? Friendships may also be lost or under strain, through reluctance to impose an added burden, feelings of inadequacy or being threatened by confronting situations. Friends may feel helpless and guilty: 'I wish I could do more but he won't accept my help'. There is often no end in sight, compared with a hospital admission when the length of time is more or less predetermined, or in a short illness, or when death is expected within days or weeks. Family caregivers may struggle to acknowledge their own feelings and motives: 'I don't know why I'm doing this. We haven't been a close couple and he doesn't even know me any more'. Or 'I doubt my own motives; that I'm doing this out of guilt, or duty or, the opposite, so I'll earn the praise and gratitude of the family'.

Relationships between families and professional caregivers may also be fraught: 'Some nurses assume I know all there is to know about dementia'. 'Some talk down to me as if I know nothing'. 'If only they would give me some information material so I can understand his disease better'. A sense of humour or healthy balance of feelings may also be lost when the informal carer is overwhelmed by other concerns.

A different kind of dynamic is evident when the caregivers are themselves health professionals. Two such experienced family members have written of their experience caring for their elderly mother. They found they were not listened to and were certainly not integrated into the care team. They described their need for constant vigilance, particularly when their mother was hospitalised. 'We caught and corrected major mistakes: failure to follow-up abnormal test results, multiple medication errors, undertreatment of pain, poor fall prevention, and inappropriate assessment and placement for rehabilitation'. They identified the urgent need for family-centred care, improved communication, enhanced information systems and improved discharge planning. They felt strongly that their mother's health would have been compromised if not endangered if they had not been constantly vigilant.

'In a dysfunctional health care system, the family is, and must be, the ultimate failsafe mechanism' (Kaiser & Kaiser, 2017, p. 46).

Individualised family assessment may uncover a wide variety of family needs: whereas one person may wish maximum privacy for her visit, another may prefer visiting in the communal lounge. It ought not to be assumed that staff know what families need; they should be asked.

Family caregiver's support includes practical advice on aspects of care they may wish to share. Some know instinctively what is needed; others may need prompting. For example, a nurse may suggest the family visitor brush the resident's hair, massage their hands and feet, offer a snack, play a game, watch a movie, listen to music or take the resident for a walk (with aid if necessary).

Family matters or 'families matter' is an essential component of comprehensive nursing home life (Keast, 2016). In one instance, family matters were treated seriously by arranging a series of focus groups to establish what was important to them about residents' care. The response was not about the quality of food, or the physical environment; rather, it was about the forming of relationships. Further discussion with staff reinforced the importance of families becoming familiar with staff and vice versa. This aspect of family/staff relationships is dependent upon management's focus. If the manager has the view that staff should be 'professionally distant' at all times, avoiding personal relationships with residents or families, continuity is compromised. 'We shift staff around every three weeks', said one manager, 'so the residents don't become too dependent on the same staff'. Other managers place such a high priority on task performance that taking time to converse with families is frowned upon. Such an attitude was lamented by one family member as in the following scenario:

> *'They don't seem to realise that I've been caring for my husband for the past fifteen years. I know his breathing patterns, I know when he's in pain, I know his dietary preferences, I know what surroundings annoy him, I know what sounds infuriate him, I know what gestures he responds to. When I tried to engage the staff in this kind of conversation I was dismissed.' 'You take care of yourself now. It's time to leave all the care to us.' 'If only they knew what assistance I could be; if only they realised that he would get better care if the staff worked with me rather than without me'.*

Making families feel welcomed, as well as involved in the resident's care if they wish, is an essential component of the nursing home community, including an open-door policy which invites families to visit any time of the day or night.

Some family visitors may wish to become volunteers, having their care acknowledged in a formalised way; others may simply wish to help other residents at meal times or sit and converse with them.

Care is needed not to place the burden of unrealistic expectation on family members. 'Oh, it's so great that you come and feed your husband every day. I don't know how we'd manage if you weren't able to come'. This leaves the relative with a narrow range of options if/when their circumstances change, such as the following:

> *I'm utterly exhausted walking to the nursing home at the set time every day. I'm terrified of being late, and sometimes I'm just too tired, physically or emotionally to be there at the*

expected time. The staff think I'll be there promptly at 12 midday every day to feed him; but they don't seem to understand that sometimes I just can't manage it.

It is clear that informal carers do not always receive the support they need or deserve. Family circumstances will vary from time to time, so the best way to develop and sustain responsive family care is to meet with them regularly.

7.4.4 When Problems Arise

Problems may arise when a family member believes they have the right to dictate aspects of care. 'I want my mum taken off this medication. It's not doing her any good'. Or 'If my husband has a seizure, he must be sent to hospital'. No family member has the legal right to demand such responses, particularly if it is clearly against the residents' best interest or out of step with the nursing home's policies and procedures. Such discrepancies in understanding a family's role may occur in the absence of previous dialogue, such as in a family meeting (described above). Problems may also be averted by comprehensive ACDs (advance care directives) where the resident's wishes are clearly stated and regularly reviewed.

It should not be assumed that every resident is supported by a family united in their views, harmonious in their attitude and sensitive to the care required for their relative. The direct opposite could be the case, as in the following scenario:

> *The manager contacted the next-of-kin when 'Mrs. Adams' was admitted to the nursing home, to invite family members to a meeting, which was standard practice. 'I wouldn't trust my sisters to know what's best for mum', the oldest of the three daughters asserted. The youngest daughter confided, 'My sisters never listen to me because I'm the youngest and they think they should be dictating mum's care'. The middle one said: 'I'm the nurse in the family but they totally ignore me.' The manager met with the three daughters, quickly becoming aware this was not going to be an easy meeting. 'We're a dysfunctional family,' said the middle daughter, 'and we each have strong views about what mum needs. You'll never get us to agree.' The manager decided that an independent arbiter was needed. 'We want to provide the best care possible for your mother, but we cannot resolve your family differences, so I strongly suggest we take the matter to the Guardianship and Administration Board, so we can be clear about who will act in your mother's best interests'.*
>
> *After a brief discussion, the daughters agreed with this strategy, and an independent guardian was appointed, following all due processes. The daughters were clearly relieved; their mutual hostility ameliorated by the decision.*

This clear, firm decision-making by the manager resulted in optimum care for Mrs. Adams, particularly when the resident needed urgent hospital admission. While the daughters would have had differing views about this move, they were confident about the guardian's role, pleasantly surprised by his personalised care, including his commitment to accompanying Mrs. Adams to the hospital and visiting her regularly.

Mrs. Adam's story provides one example of resolving what was clearly a considerable management challenge, saving much time and energy in responding to ongoing conflict. Communicating with families may, in different circumstances, compromise decision-making and resident's care, as in the following:

> *Discussion at handover focused on a very ill resident's pain management; one nurse challenging another, 'Why isn't she on stronger opioids?'*
> *'We wanted to contact the GP, requesting a medication review, but the family wouldn't let us. We also wanted to refer her to palliative care but the family said "no"'.*

This scenario raises other important questions about responsibility for the residents' care and staff confidence about decision-making based on a clear plan of care. While some would argue there is no time for including families in care planning, if such had been the case, arguably the conflict described above would have been averted. The focus here is on the *goals of care*, which, ideally, would have been discussed with the family, accompanied by information (both verbal and written) about appropriate pain relief. A legitimate plan of care takes precedence over family's 'directives'.

7.4.5 Which Family Member to Ask?

One often hears of 'difficult families' and the way they impede the general smooth running of the nursing home. 'We love the residents but it's the families that we have difficulties with'. This attitude suggests a lack of education about the benefits of getting to know residents' families. When clear lines of communication are not agreed at the outset, staff face a dilemma. 'Ask the family' is not as easy as it sounds. Which family member? Every member of the family? The youngest son or the older daughter? It may be best to ask the resident, where possible, as in the following situation:

> *'Lucy' was an articulate resident in her nineties. Physically frail with multiple medical problems, she responded firmly and with confidence when asked about her family. 'I don't trust any of them', she said. 'My neighbour in the next street, Bert, has been my closest friend for the last thirty years since my husband died. We consult each other on many day-to-day decisions, big and small, by phone when we can't meet. I have every confidence in him, and he's agreed I should have his name as my 'next-of-kin'. While he's not legally a relative at all, he's the one who should be consulted if I'm in no position to decide for myself'.*

Further to the discussion on communication (Chap. 2), this example (above) highlights the broad definition of family and, in many instances, prevents the conflict arising when family members have different opinions as to the resident's care. It also reinforces the point that time is saved and conflicts are circumvented when a family spokesperson is identified as soon as possible after the resident's admission, noting that the person may not necessarily be the next of kin. Unambiguous documentation is essential, to prevent scenarios like the following:

> *The night nurse had never met 'Mrs. Dhorman's' family, so when Mrs. Dhorman died unexpectedly at 3.a.m., she immediately tried to telephone the 'next-of-kin'. Unfortunately, the*

7.4 Involving Families

documentation did not clarify that Mr. Dhorman, although technically the next-of-kin, was deaf, had limited English, and declining cognition. The nurse's attempt at conveying news of his wife's death resulted in distress caused by communication failure. Mrs. Dhorman's daughter was similarly distressed. 'Why didn't you call ME!' she asked.

This scenario could have been avoided by comprehensive documentation regarding which family member to contact in an emergency: a further example of the need for clear policies and procedures.

Some nursing home staff believe, mistakenly, that they are required to ask the family's 'permission' on a number of issues. This can lead to a conflict with what is stated in the mutually agreed care plan. For example, the care plan may state that the resident wishes to remain in the nursing home rather than being sent to hospital, supported by other documents such as 'power of attorney'. Understanding the legal parameters of decision-making is important; otherwise confusion may arise, such as the following:

The night RN reported at morning handover: 'I rang Lucrezia's son to tell him that her condition was deteriorating, but we had clear orders from the GP regarding comfort measures, including controlling her pain, and that we believed she would be content to remain under our care'. Lucrezia's son protested loudly: 'I know my sisters don't want her to go to hospital but I'm her power of attorney and I say she needs the best care possible, so I say send her to hospital'. The RN felt she had no option than to 'obey' the son's directive.

When comprehensive documentation is lacking, such as copy of legal documents, staff are often unsure about responsibilities for decision-making. In this case, Lucrezia's son did not, in fact, have a certified document confirming his status, and the relevant papers were not in her file.

Families need to understand the boundaries of their decision-making capacity. Conflict can also arise when staff lack confidence in their own responsibilities for the residents' care; otherwise, confusion can arise, as shown in the following scenario:

'Mr. Crisp' was often screaming out in pain, particularly when encouraged to mobilise. 'I can't walk because my leg hurts too much' he would say. The GP suggested regular paracetamol, with the option of a 'prn' dose prior to his walk. When this was no longer effective, the GP ordered a stronger analgesic, to which Mr. Crisp's son responded: 'I don't want my father on these addictive medications, so you're not to give him any'. The GP instructed the nurse: 'Okay, if the son refuses, you can't give it'.

Is this response typical of a fear of litigation? Or, is it a clear misunderstanding of health professionals' responsibilities towards those in their care? The potential for such misunderstanding can be obviated by a family meeting including the GP (as outlined above), and involvement in care planning, together with relevant drug information. This procedure paves the way for a clear understanding of the nursing home's responsibilities, avoiding potential conflicts. It also emphasises the need for more readily available literature (and in multiple languages) to be provided to families regarding analgesics, especially opioids.

7.4.6 Family Carer's Ageing

The increasing age of older people entering nursing homes may signify that their families and/or other carers have played a significant role in their care prior to admission, illustrated in the following account:

> When Cara rang for the appointment to add her mother's name to the waiting list, her clear descriptions of her mother's needs, together with her precision about getting the address right were indicative of a person eager to find the best accommodation for her mother. Her careful attention to detail also suggested thoughtful planning and thorough investigation of nursing home options. "I've cared for her all of my life, but the doctor tells me I must now slow down. But I'm not just going to put her *anywhere!*' When Cara arrived for the appointment, the receptionist was confused. "There's a very elderly lady to see you,' she told the manager, 'I think it must be the resident herself'. Cara did not expect any confusion, as if it were the most natural thing in the world for an 85-year old to be negotiating nursing home accommodation for her mother of 105' (Hudson & Richmond, 2000, p.157).

This story provides another reminder for nursing homes to prepare for an increasingly ageing population, including prospective residents and their carer(s). While the merits are evident of an older person being cared at home for as long as possible, other issues may be at stake. When the carer is also ageing, significant health complications may be overlooked. In some instances, the elderly carer is in good health, acquainted with the person's needs, seeks appropriate advice and provides exemplary care. Other elderly carers may be unwell themselves or lack the resources to equip them for the task. They may be unwilling, or unable, to seek assistance. Comprehensive pre-admission details are important, including social worker assessment where relevant.

7.4.7 Recognising Family Suffering

Aged care nurses need to show compassion to the relatives of dying residents, recognising that many family members experience significant suffering while watching their relative's condition deteriorate, particularly when this means the loss of a partner of many years. Nurses can also assist with practical issues such as details of treatment options, enabling family members to make decisions when the resident is incapable of deciding for themselves (Lopez, 2007).

It is evident from Lopez's (2007) research and others' that family members appreciate being asked what input is appropriate for them, as soon as possible after the resident's admission. An open-ended question shows understanding that nursing home admission can be fraught with relief, distress, resistance, pleasure, frustration or a mix of feelings which may of course vary over time. Nurses may not presume therefore to know the effect of the admission on the relatives in question. Some family members may welcome a meeting to discuss their needs; others may not wish to take up the offer (Toye et al., 1996).

Family members may suffer physical exhaustion and other symptoms as a result of their caring activities. Palliative care researchers recommend a method of screening which identifies a family caregiver whose psychosocial functioning might be impacted as a result of their caregiving experience (Hudson et al., 2006). A thoughtful gerontic nurse would enquire of the main family member: 'Tell me how you're feeling now your wife is in the nursing home'. The answer might indicate whether referral to another health professional would be advisable.

7.4.8 False Comfort

An example of well-meaning 'comfort' is portrayed in the following account:

> *'Will' confided to his friend, the nurse, about his visits to his wife. He spoke about all his friends who offer 'words, words, words, which are no comfort at all but you have to politely listen. They tell you to forget the pain, to look after yourself, not to worry, not to think about my dear Pat dying' The nurse listened attentively, without rushing in with more words. Will continued: 'They try to offer comfort but it's empty advice. I've found that comfort comes from being realistic, by accepting the sadness, but retaining a sense of humour'. Noting Will's increasing emotional distress, the nurse asked what he would find most helpful. 'What I'd find most helpful would be someone to be there when I get home at night, just for an hour, even to sit in silence while I cook my meal, or to chat about normal things'. To the question of volunteers Will responded: 'Yes, I've been offered volunteers but who would accompany me home at eight or nine at night and sit with me for an hour, have a drink with me and a normal conversation?'* (paraphrased from unknown source)

The nurse knew Will was not expecting her to solve his problems. No more words of false comfort; he wanted someone to listen and to understand.

7.4.9 The Absent Family

The statistics are quite stark: although comprehensive figures are lacking, some estimate that around 40% of nursing home residents have no visitors (an estimate pre-COVID). More research is needed into factors contributing to this phenomenon, leading to increased understanding. Misunderstandings can occur when family members do not visit as staff expect they should. For some, relief of the caregiving burden is so great that a spouse or other carer is left depleted of energy following the nursing home admission. For others, the strain of seeing their relative cared for by strangers is too much to bear. Misunderstandings about dementia (discussed in Chap. 4) also mean that families are uncertain, feeling guilty or overwhelmed or very uncomfortable about visiting. Others are unashamedly happy never to go near the nursing home, particularly where relationships have been strained or in the absence of any affection or commitment towards the older person.

7.4.10 Happy Families?

Leo Tolstoy's novel *Anna Karenina* begins: 'Happy families are all alike; every unhappy family is unhappy in its own way'. Although various family factors are described above, no attempt is made to analyse or make judgements about the family variations encountered in a nursing home. Suffice it to say that managers and staff are called upon to create an environment where families feel welcome and have an avenue for expressing their concerns as well as their appreciation for the residents' care.

References

Australian Government Aged Care Quality and Safety Commission. (2019). *Aged care quality standards*.
Australian Institute of Health and Welfare. (2020). *Board governance in the aged care sector*.
Breen, J., Aoun, S., O'Connor, M., Howting, D., & Halkett, J. (2018). Family caregivers' preparations for death: A qualitative analysis. *Journal of Pain and Symptom Management, 55*(6), 1473–1479.
Burke, C. (2019). 'Impossible' to feed residents on $7 a day: Maggie Beer at the royal commission. *Aged Care Insite*.
Butler, J. (2021). Emotional intelligence in nursing leadership. *Australian Nursing and Midwifery Journal, 27*(5), 18–21.
Commonwealth of Australia. (2019). *Royal Commission into Aged Care Quality and Safety Interim Report: Neglect Volume 1*.
Commonwealth of Australia. (2021). *Royal Commission into aged care quality and safety final report: Care, dignity and respect, volume 3A the new system*.
COTA Australia. (2020). Aged Care Royal Commission - on the home stretch? *One Cota for older Australians*, Summer (December 2020–February 2021), 6–9.
Duckett, S., Stobart, A., & Swerissen, H. (2021). *The next steps for aged care: Forging a clear path after the Royal Commission*. The Grattan Institute.
Durepos, P., Kaasalainen, S., Sussman, T., Parker, D., Brazil, K., Mintzberk, S., & Te, A. (2017). Family care conferences in long-term care: Exploring content and processes in end-of-life communication. *Palliative & Supportive Care*, 1–12. https://doi.org/10.1017/S1478951517000773
Eagar, K., Westera, A., & Kobel, C. (2020). Australian residential aged care is understaffed. *Med J Aust, 212*(11), 507–508.
Egan, N. (2019). Residential occupancy down and wait times up, report shows. *Australian Ageing Agenda*.
Hockley, J., & Kinley, J. (2016). A practice development initiative supporting care home staff deliver high quality end-of-life care. *International Journal of Palliative Nursing, 22*(10), 474–481.
Hudson, R. (2019). Death and dying in dementia care: A good end? *Australian Journal of Dementia Care, 8*(1), 24–27.
Hudson, R., & Richmond, J. (2000). *Living, dying, caring: Life and death in a nursing home*. Ausmed Publications.
Hudson, P., Aranda, S., & Kristjanson, L. (2004). Meeting the supporting needs of family caregivers in palliative care; challenges for health professionals. *Palliative Medicine, 7*(1), 19–25.
Hudson, P., Hayman-White, K., Aranda, S., & Kristjanson, L. (2006). Predicting family caregiver psychosocial functioning in palliative care. *Journal of Palliative Care, 22*(3), 133–140.

Hudson, P., Thomas, T., Quinn, K., & Aranda, S. (2009). Family meetings in palliative care: Are they effective? *Palliative Medicine, 23*(2), 150–157.

Kaiser, R. M., & Kaiser, S. L. (2017). The insiders as outsiders: Professionals caring for an aging parent. *The Gerontologist, 57*(1), 46–53.

Keast, J. (2016). Family matters. *Australian Ageing Agenda*, 38–39. https://www.australianageingagenda.com.au/author/jackie/

Kolzow, D. R. (2014). Leading from within: Building organizational leadership capacity.

Lopez, R. P. (2007). Suffering and dying nursing home residents: Nurses' perceptions of the role of family members. *Journal of Hospice & Palliative Nursing, 9*(3), 141–149.

Monroe, B., & Oliviere, D. (2009). Communicating with family carers. *Family Carers in Palliative Care. A Guide for Health and Social Care Professionals*, 1–20.

Nay, R. (2016). The good, the bad and the downright ugly: Reflections on 10 years. *Residential Aged Care Communique, 11*(4), 6.

Nursing & Midwifery Board of Australia. (2018). *Code of professional conduct for nurses in Australia*.

O'Mahony, S. (2016). *The way we die now*. Head of Zeus.

Puurveen, G., Cooke, H., Gill, R., & Baumbusch, J. (2019). A seat at the table: The positioning of families during care conferences in nursing homes. *The Gerontologist, 59*(5), 835–844.

Russell, S. (2017). *Living well in an aged care home* (Research Report, Issue. https://www.agedcarematters.net.au/living-well-in-an-aged-care-home/

Toye, C., Percival, P., & Blackmore, A. (1996). Satisfaction with nursing home care of a relative: Does inviting greater input make a difference? *Collegian, 3*(2), 4–11.

Chapter 8
Community Expectations

8.1 Meeting the Standards

8.1.1 Required Standards

The Aged Care Quality Standards (Australian Government Aged Care Quality and Safety Commission, 2019) are listed under eight headings, with detailed requirements for each, including the following consumer outcomes:

1. Consumer dignity and choice
2. Ongoing assessment and planning with consumers
3. Personal care and clinical care
4. Services and supports for daily living
5. Organisation's service environment
6. Feedback and complaints
7. Human resources
8. Organisational governance

Unfortunately, the standards do not include mandated staffing levels for nursing homes, or a requirement for employing a RN. The Grattan Institute Report (Duckett et al., 2021) notes that the proposed reformed body responsible for aged care standards requires specialisation 'so that standards are set by experts that consider not only clinical, but non-clinical factors, such as lifestyle and rights-based criteria that enhance independence and dignity' (p.16).

Standard 1, 'Consumer dignity and choice', has received significant attention throughout the book, particularly in Chap. 6 where issues of death and dying are

---

While (some of) the stories are based on factual situations, real names and other details have been altered to protect the identity of the persons concerned. Resemblance to any particular person is therefore purely coincidental.

discussed. The subject of risk-taking pertains to residents' dignity and choice with greater emphasis needed on residents' rights (Duckett et al., 2021, p. 8). It is difficult to see how the next two standards, 'Ongoing assessment and planning with consumers' (No 2) and 'Personal care and clinical care' (No 3), can be achieved without professional staff at the helm. For example, Standard 2 is dependent on assessment and planning, with regular review of services, including the use of a behaviour support plan which describes all alternative strategies tried prior to using any restrictions for 'behaviours of concern'.

Standard 3 includes noting signs of a resident's health deterioration, referring to other health professionals, when necessary, particularly for infection control advice. Standard 4 'Services and supports for daily living' requires that organisations demonstrate safe and effective services including quality meals, a subject addressed in Chap. 3. Chapter 4 also addresses family involvement as well as a detailed list of clinical issues deserving a clear focus, together with the important role played by volunteers and other auxiliary staff.

Standard 5 'Organisation's service environment' is given particular attention in this chapter (8.1), with emphasis on mission statements and providing a welcoming environment. In Standard 6, 'Feedback and complaints' are discussed in the section 'When Things Go Wrong (9.2), including the important issue of elder abuse and management of the complaints process. Standard 7, 'Human resources', emphasises the need for a competent, qualified and well-trained workforce. However, the discussion in this chapter includes lack of research in equipping staff to respond effectively to residents' needs. 'Organisational governance' (Standard 8) is also addressed in this chapter, highlighting the importance of governors' expertise and aged care knowledge for selecting and managing the staff needed to meet the standards. Other references are made throughout the book to the importance of policies and benchmarks for achieving standards of care required by the community in order to maximise residents' 'rights'. Meeting and keeping the standards is for the benefit of everyone involved in the nursing home community, not merely to satisfy the regulators.

When new standards are introduced much depends on the way the facts are communicated. Would the average nursing home visitor be aware of these standards and their import? Would every staff member have a copy? Would residents be invited to comment? What education would be provided to ensure they are understood?

> All consumers of aged care must be given a copy of the charter and the provider must assist them in understanding their rights. They, or an authorised person, must sign a copy. (Aged Care Insight, 2019)

A spokesperson for the standards claims that, if implemented well, they will prove to be 'world class' in terms of transparency and improved care. On the other hand, when the standards are not promoted, with information for all relevant parties, they will be left to 'gather dust on the bookshelf', to the ultimate detriment of the residents for whose benefit they were introduced.

8.1.2 Mission Statements

The generally accepted ethos of a mission statement is that it covers the culture, core values and aspirations central to the organisation. Some nursing homes highlight specific religious or cultural issues in accordance with their affiliation; others focus on providing quality care to the residents and a respectful work environment for staff.

Mission statements are generally renewed every 2–3 years, following various levels of collaboration with residents, relatives and staff. Wide discussion increases the likelihood that the statement will be a true reflection of what is important to those who live and work in the nursing home and those who visit. A comprehensive mission statement provides a significant benchmark for regularly evaluating performance and outcomes at all levels of the home's care.

The Royal Commission's Report (Commonwealth of Australia, 2019) includes evidence that a nursing home's mission statement is not necessarily matched by the standard of care. Surveys reveal that the Australian communities are willing to pay more in taxes to achieve higher-quality care, as well as making their own co-contribution to fees (Ratcliffe, 2020). This evidence is a witness to the fact that the realities of daily resident's care often fall short of the wording on the wall.

8.1.3 Transparency

Transparency was considered an important aspect in evidence given to the Royal Commission, at least one facility cited as lacking transparency and accuracy in reporting details (Commonwealth of Australia, 2019b, p. 330). What do the board know? Although their governance does not include every detail of the home's daily operation, there should be avenues for reporting and feedback of specific matters. A reputable board will have an audit component, associated with risk and compliance. In the event of a crisis (such as the COVID-19 pandemic), the board would meet more regularly, inviting an external expert, ensuring additional resources were available for resident's safety, back-up plans for extra staffing and contingencies related to financial viability if necessary.

While formal compliance measures are important, governing agencies may pay little or no attention to the sometimes appalling day-to-day conditions, or to the increasing susceptibility for poor care, amounting to abuse. 'Our manager wouldn't recognise elder abuse if he tripped over it', said one staff member, 'and I have no idea who's on our board'.

On the other hand, where a nursing home has a community focus, broader issues are given priority, including publishing board members' names, and other factors such as the upholding of dignity and respect for every resident (Galloway, 2018), welcoming all comments, responding to those requiring remedial action and celebrating positive examples. Transparency implies nothing is covered up, for example, when complaints are published as they occur, together with management's responses. Whereas responses to regulations often prove to be *reactive*, transparency is *proactive*. Transparency is fostered by a culture of partnership and proactive attitudes rather than a mechanism for reactive defences and mere transactions.

8.1.4 Workplace Health and Safety (WHS)

Workplace health and safety (or human resource management) is included in an employer's general duty of care for their employees and other people in the workplace. Each state and territory have their own WHS legislation, details of which should be readily accessible by all who live and work and visit the nursing home. In some places, a representative is appointed to ensure the legislation is respected.

The following evidence of failure to comply with the legislation, not to mention the manager's duty of care, is provided (anonymously) by this nurse:

> Last week I had to work an evening shift 2.30pm to 11pm followed immediately by a night duty shift from 11pm to 7am. This happens regularly... I'm so tired that I can hardly think, let alone make critical clinical decisions... On the night duty shift I have 150-plus high care residents... and on the evening shift I have 75 -plus... I do not want to become a resident, ever, of any aged-care facility I have worked in.

These sentiments are echoed in the familiar mantra: 'I'd rather die than work in a nursing home'. On the other hand, some nursing homes rightly contend that the safety and security of residents and staff is a high priority. This leads to a more favourable response such as ensuring appropriate attention is given to workplace safety through regular staff education by a reputable registered training organisation.

WHS practitioners provide support to aged care managers in using associated technologies to ensure employees have the appropriate training and qualifications to ensure compliance with all standards. Forming a strong partnership, they can support corporate leadership in achieving competent and caring status as providers.

Compassionate, caring managers regard WHS as more than familiarity with the fire extinguishers; the physical, emotional, psychological health and safety of staff is their key concern. When such care is *received* by staff, residents benefit from the care *given* to them.

8.1.5 Prospective Residents

What may be gleaned about a nursing home's standards, from an initial visit? Prospective nursing home residents and their families/carers have every reason to be wary of the for-profit sector who can afford plush entrance lobbies with marble floors and glossy chandeliers, where an atmosphere of opulence does not necessarily reflect a high standard of care. Prospective residents and families need to make careful enquiries about how fees are spent and what proportion is allocated to direct care. While some in this sector are able to maintain all the standards of care within an attractive physical environment, others whose focus is first and foremost on the profit margin may fall short of achieving anything beyond the minimum.

What questions should prospective residents and/or their representatives ask when enquiring about residency? Some suggestions are the following:

Quality and standards
- What is the current accreditation rating?
- What is the staff ratio, for example, how many RNs on each shift?
- What level of qualifications do staff other than RNs have?
- What is the auditing procedure?

Residents' care
- How often is the care plan reviewed?
- What are the options for end-of-life care?
- What is the rate of pressure ulcers?
- What access to wound consultants, nutrition consultant and others?
- Is there provision of palliative care services?
- What are the criteria for transfer to the hospital?
- Who administers the medications and do they have the necessary pharmacology knowledge?

Families and visiting
- Are visitors welcomed at any time of the day or night?
- What prompts notification to families, for example, about a resident's changed condition?
- What are protocols/procedures for notification of a resident's death?
- What access/input do families have to the care plan?
- Are there routine meetings with families?

Medical and allied health services
- How many GPs visit, how often and what is their knowledge of gerontology?
- What access for a second opinion?
- What psychogeriatrician access for residents with dementia-related 'behaviours'?
- Can we buy in extra services?
- What access to after-hours medical attention?
- What access to physiotherapy, occupational therapy and other specialist therapies?
- What activities are provided?

Meals and snacks
- Is food cooked on site or brought in?
- What is the food budget per resident per day?
- What choice do residents have, including cultural preferences?
- What food and drinks are available between meals and overnight?
- Who monitors the residents' intake and associated weight?

The questions listed above are not exhaustive; prospective residents and families may have others. Being well prepared with a written list enables comparisons before making a decision, or adding a name to the waiting list. Where the nursing home has up-to-date comprehensive written information, additional questions may also be appropriate: is all information available in other languages, for those whose first language is not English? Are other queries welcomed? Would I feel 'at home' here, and would my family be welcomed? How many are on the waiting list? How long will I have to wait for a place?

Nursing home boards and managers have access to an abundance of contemporary evidence for promoting the highest possible quality of care. Family members or other carers seeking information are encouraged to check their options by researching whether the particular home is currently meeting all the standards. Visiting the home and requesting a 'tour' have the advantage of a 'snapshot' view of the atmosphere, together with readily visible signs such as cleanliness, residents' appearance and the nature of communication between staff and residents. Questions such as access to a resident's records, meetings with senior clinicians and the GP and involvement in daily care have the potential for a harmonious relationship, optimised by relevant information. Such knowledge is also obtained through observation: being present at meal time, drug round time, staff handover and evening and night time. Other signs such as call bells being answered, transparent complaints system and regular meetings with families provide essential information on which to judge the 'care'.

8.1.6 Resident Records: Beyond the Clinical

Acknowledging the importance of advancing technology's influence on record keeping, the Australian Medical Association (AMA) notes the following:

> Interoperability between My Health Record, My Aged Care, and clinical software systems would enable electronic health record sharing between the health and aged care systems, including sharing details of aged care assessments, care plans, advanced care directives, immunisation records, and past medical treatments for each older person. (Australian Medical Association, 2019)

Accepting the many advantages of these technological systems, they should not be expected to replace personalised, individualised resident's care planning and review. The purposes of documentation are the following:

(a) To serve as a communication tool for information exchange
(b) To ensure continuity of care
(c) To provide a legal record
(d) To aid research and education
(e) To meet funding requirements related to statutory information (Crofton & Witney, 2004, p.3)

Documentation represents more than basic clinical care: emotion and behavioural responses need also to be recorded in appropriate detail. To illustrate, the following story is told of 'Bertha' who suffered from dementia, lacking coherent speech:

> It was noted that Bertha was, unusually, refusing to eat. No physical cause was apparent. One insightful care attendant wondered if Bertha was missing her roommate who had been transferred to hospital. Opinions varied as to whether or not Bertha would understand why her roommate's bed was empty. The insightful carer sat beside Bertha and gently told her that 'Angelino' had been taken to hospital but would be returning soon. Bertha, albeit without speech, conveyed her pleasure at being given this explanation, and immediately accepted some food and drink.

The recording of these details by an observant care attendant provided vital information regarding Bertha's refusal of food and fluids, resulting in life-saving care.

The following is another example of comprehensive documentation that relates to spiritual care:

> 'Absalom's' behaviour was becoming 'unmanageable': screaming loudly, being physically aggressive towards staff, and reluctant to accept any care. An observant care attendant noted from Absalom's notes that he had been an assistant pastor in his local church. Surmising that he may respond to the chaplain's visit, she suggested this to the RN. Within minutes of the chaplain's visit, Absalom calmed down; he was observed to be quietly and reverently responding to the chaplain's prayers. The RN carefully documented this scenario, adding to Absalom's care plan his need for frequent visits from the chaplain.

Subsequently, the chaplain wrote a short, carefully worded prayer, framed it and posted it on Absalom's wall, with a note in the care plan inviting nurses to use the words with Absalom when he became agitated. Regardless of their own religious beliefs, nurses and other carers became accustomed to this 'procedure' as a vitally important part of Absalom's daily routine: testament to the place of spirituality in holistic care.

8.1.7 Quality: A Slippery, Subjective Notion

'Quality' is a word more appropriately used of objects or actions: 'I bought a high-quality sound system' or 'He spent some quality time with his daughter'. 'Quality' is not an appropriate term for a person's life. To say a resident has no quality of life is to make a judgement reflecting more the attitude of the 'beholder' rather than an objective measure applied to a living, breathing, human person.

Professor Ibrahim asks, in many of his publications: 'What is 'quality' in aged care?' (Ibrahim, 2018). His answers echo other views from around the world (Carnell & Paterson, 2017) which emphasise a homelike rather than institutionalised environment, with social connections and access to the outdoors. While these factors may appear incontrovertible, some readers will be aware of nursing homes which defy these basic principles, leaving at least one resident to complain: 'It's not at all like a home and there's no external view or access to outdoors'.

Continuing his discussion on quality, Ibrahim notes that friendships also play an important part, particularly when residents 'are allowed reciprocity with their caregivers' (Ibrahim, 2018). Such relationships are showcased in the following account:

> *The presence of fifty-six people at the funeral of a centenarian spoke wordlessly of her wide circle of friends. For friends they were — with only a handful of relatives. Leslie had married late in life and there were no children. She had lived her whole life within four km of the nursing home, and she was overjoyed to be offered a bed at the age of ninety-six. 'This will be like coming to my second home', she said. Leslie's capacity for making friends did not wither in this new community. She made her 'best friend ever' — one of the nurses — the one who made a special tray for her birthday breakfast, the one whose advice she sought on what to wear for an outing and the one with whom she spoke of her impending death. The rewards of this friendship were not Leslie's alone. This staff member was moved to tears when her name was mentioned at the funeral oration. It helped her to grieve and to 'finish off' what for her was one of the most significant relationships of her nursing career.*

As within any community, there were dissenting voices from those who felt threatened by this account of Leslie and her nurse. They were quick to criticise: 'She got too close. She overstepped the professional line'.

Does professionalism imply distance through impersonal contact? With more emphasis on holistic care, it seems well within the bounds of professionalism to include in a resident's care plan factors centred on the social component of their lives. These factors may well involve particular ways in which the resident relates to different members of staff. 'Friendships' may find a legitimate place in their care plan.

The point is also made although not always well understood that staff need to be both technically proficient and also 'good with people' (Ibrahim, 2018). This goes to the heart of staff selection (discussed in more detail in Chap. 7), where emphasis is given to staff members' capacity to interact well with older people. In other words, a carer may carry out all required tasks while failing to engage the residents on a personal level (Ibrahim, 2018). We may have progressed from an era when, for example, 'all beds must be made by 10am'; however, there seems to be a perpetual emphasis on the performance of tasks at the expense of personal relationships.

Food also ranks high when describing quality. For example, attention given to cultural variations which meet the needs of an increasingly multicultural society ought now to be reflected in nursing home meal times and food preferences. Traditional values, prohibition of some foods and celebration of certain feast days, together with religious sensitivities, are just some of the considerations needed to provide culturally appropriate food. (For more discussion on food, refer Chap. 3.)

For Ibrahim, all these factors can be summed up in four words: safety, dignity, respect and choice. 'Residents want and have a right to feel safe, valued, respected and able to express and exercise choice. Positive observation of these rights is essential for quality of life' (Ibrahim, 2018).

Another way of approaching 'quality' is to note the goals for each resident's care. What outcomes might be expected? How would the nursing home's philosophy or vision/mission statement be reflected in the resident's care plan? How would the care be measured by the residents and their families? How would the 'quality' criterion influence staff selection and ongoing appraisal? In other words, a nursing home may state confidently that they provide 'quality care' while unable to produce evidence in measurable outcomes of such a laudable aim.

8.2 When Things Go Wrong

The Shadow Assistant Minister for Health and Ageing, who had been a registered nurse, says it is time for 'proper, hard reform in aged care' (Kearney, 2021). Highlighting some of the harrowing stories reported by the Commonwealth of Australia, 2021b. Kearney describes an aged care system fundamentally flawed, unable to deliver 'uniformly safe and quality care, is unkind and uncaring towards older people and, in too many instances, it neglects them' (Kearney, 2021). She points to the many failures, including an underpaid and undervalued workforce. Another parliamentarian described the treatment of residents and staff in one nursing home as 'contemptible', calling for the senior management team to be dismissed. While some nursing homes have robust policies for addressing such failures, others seem to ignore them, leaving residents vulnerable, if not neglected.

8.2.1 Staffing Inadequacies

Some of these failures are related to inadequate staffing, described in graphic detail by a resident giving evidence before the Royal Commission:

> I feel I have been dehumanised, left as a carcass in an aged care abattoir ready to be processed like a slab of meat in a sausage processing factory at some future time... pain dominates my whole existence. Every second of every minute seems like an eternity. No one seems to get this. (James 2019)

This sobering, graphic description (one of many) adds to other first-hand accounts of the effects of staff shortages: others describe staff being overworked, overburdened and stressed. Compounding these factors is the issue of the perceived low pay which, according to some, is the reason the best people are not being attracted to work in this sector. Outlining the sophisticated jobs of the future requiring significant upskilling of today's workers, it is expected that

> ... only a tiny percentage of people in the post-industrial world will ever end up working in software engineering, biotechnology or advanced manufacturing... Many of the most important jobs of the future will require soft skills, not advanced algebra. (Gershon, 2017)

Gershon emphasises the need for increased respect and better pay for those generically dismissed as 'unskilled labour', appealing for greater emphasis on 'soft skills'. While acknowledging the need for technology-driven efficiency and the associated increased standard of living, he describes what is missing if all we see is technological progress. On a positive note he says: 'with routine physical and cognitive work out of the way, the jobs of the future could be opportunities for people to genuinely care for each other' (Gershon, 2017).

8.2.2 Making and Managing Complaints

It seems to be characteristic of most nursing home residents and their families that they are reluctant to file a complaint. Some, however, may be prompted by a trusted nurse, to 'report' an incident, as in the following:

> During his early morning 'rounds', checking on the residents in his care, 'Georgio', the care manager, greeted 'Dorrie', anticipating their customary brief but lively chat. Dorrie, a single, frail, 92-year-old, mentally alert woman with severe osteoarthritis and other major health problems, responded with tear-filled eyes rather than her usual bright smile. Some gentle coaxing elicited the reason: her shoulder was very painful. Slowly and hesitantly, in a barely audible whisper, the story emerged. 'It was that night nurse. You know . . . the one with the frizzy hair. . . I don't think she knows I can't hear too well . . . I didn't know what she wanted me to do . . . but . . . she just jerked me across the bed by my arm.' Further gentle probing evoked a look of fear that remained with Georgio for some time. With a fierce grip of his arm and a terrified look in her eyes Dorrie pleaded: 'You won't say anything will you?' Georgio had to work hard to reassure Dorrie she would not be punished for speaking out but that a thorough investigation was needed into her painful shoulder. Georgio reassured Dorrie he would, without naming the nurse, also take steps to ensure she was not treated roughly again.

Dorrie's story raises questions as to the number of similar incidents that go unreported, for fear of retribution. It also highlights the importance of a senior member of staff going regularly to each bedside with the deliberate impartial intent of checking residents' safety and well-being.

When a complaint cannot be resolved within the nursing home, residents or their families have access to the ACQSC (Aged Care Quality and Safety Commission):

> The complaints process seeks to resolve the concerns of the individual whose care is the subject of the complaint, but the process can also give an early warning of deficiencies in care and identify potential systemic failure. (Australian Government Aged Care Quality and Safety Commission, 2020, p. 48)

If the complaint is not resolved at this level, it can be referred to the Department of Health for potential compliance action. The ACQSC can also make referrals to other relevant regulatory bodies. The Australian government's Aged Care Complaints Commissioner Annual Report (2016–2017) states: 'The incidence of complaints in residential care have increased, accounting for 78 per cent of all aged care complaints received (Australian Government Aged Care Complaints Commissioner, 2017, p. 18)'. The most common complaints are listed as '(a) medication administration and management, (b) falls prevention and post fall management, (c) personal/oral hygiene' (p.19).

8.2 When Things Go Wrong

The complaints commissioner described 'multiple inquiries into aged care and elder abuse, fuelled by public concerns and media reports of shocking instances of poor care' (Australian Government Aged Care Complaints Commissioner, 2017, p. 5). The following report gives one example of improved care following the resolution of a complaint about communication:

> The resident had complained about poor hygiene care following episodes of incontinence, the lack of regular position changes, and food served lukewarm. The complaints commissioner met with the resident and her son, who stated they had regularly spoken with staff about these issues but they felt they were ignored. Following a meeting with management, it was agreed that a communication board with pictures of different care needs would be helpful to the resident whose speech was difficult to understand. The complaint was resolved to the satisfaction of all concerned. (Australian Government Aged Care Complaints Commissioner, 2017, p. 34)

Another complaint was raised by a family member about the nursing home's failure to act quickly enough in response to her father's deteriorating condition. Following the hearing of the complaint, the service provider apologised, developed training resources for staff and appointed a clinical care coordinator. The resident's daughter was pleased with the outcome (Australian Government Aged Care Complaints Commissioner, 2017, p. 35).

The commissioner highlights the need for aged care services to make the public aware of how complaints will be dealt with, such as placing relevant information on websites and in other publications. An example from one nursing home's advertising brochure is given below:

> *Nobody likes receiving complaints. While we try our best to give high quality care every day, sometimes things go wrong. It may be a staff member showing lack of consideration to a resident, a visitor feeling unwelcome, the quality of the food, or any number of other issues. We encourage residents themselves to speak out if they wish to raise any issues with their care. We encourage relatives and other visitors to raise any comments or complaints, preferably immediately after they arise. All complaints should be received and dealt with by management in a timely manner, to the satisfaction of the complainant. If you need to make a complaint, please use the official form you will find at Reception. Once the matter is in writing it will be dealt with as soon as possible.*
>
> *We wish to de-stigmatise the process of complaints, so that people feel free to speak their mind without any risk of reprisals. Our firm desire is to deal with any complaint thoroughly. We believe the benefits of this response will be in improved care. Our other desire is that all residents, families, visitors have confidence in the care provided.*

At the end of an article about improving the quality of residential aged care, Ratcliffe's recommendation includes 'the ability to lodge complaints with the knowledge that appropriate action will be taken' (Ratcliffe, 2020, p.4). Appropriate action, however, is not always forthcoming, prompting the response: 'I've sent three letters of complaint and I'm still waiting for an answer!' This comment reflects the common experience: 'It's no use making a complaint because nothing will change'.

From a positive perspective, when a well-designed complaints management system is in place, all who live, work and visit the home may be assured that the aim is to promote a culture of improvement. Within such an environment, complaints will be welcomed and brought to the attention not only of managers but of the board on whose governance residents' welfare rests.

Incident report forms provide a comprehensive framework, not only for observable accidents but for substandard care. The speed with which such reports are dealt with is one tangible sign of how seriously the issues are taken by management. On the other hand, when reports fail to attract a timely response, staff quickly lose interest in the system: 'What's the use of writing an incident report? They just sit on the manager's desk'.

Whether or not their fear is justified, residents and families may feel constrained by 'reporting' and therefore fail to make their concerns known to management. They may feel the vulnerability of a perceived imbalance of power between resident and service provider. A culture of acceptance is an indicator that management will listen and respond in a timely manner when complaints are received. However, not all nursing homes have a formalised complaints system, and if they do, it is not always widely advertised. A positive, rather than defensive, attitude to complaints will come from a management team who genuinely seeks opportunities for improvement.

An example of cultural differences in relation to complaints is described below:

> It's an all-Chinese nursing home, containing staff and residents who are culturally reluctant or scared to complain . . . [the air conditioning] was not fixed for years – senior management told us no-one had complained and they did not know, which we find hard to believe . . . residents were at times left sweltering in 30-plus degree heat . . . one resident's bathroom measured 40 degrees on a heatwave day. (Commonwealth of Australia, 2019c, p.6)

This episode highlights the need for a transparent system, together with readily accessible information about the availability of an external arbiter when complaints remain unanswered.

Other reports from relatives contrast a deficient complaints system with effective communication leading to positive outcomes, described below:

> Relatives described the current complaints system as ineffectual. They describe complaints escalating because managers of some aged care homes do not respond appropriately. Relatives appreciate managers who respond quickly and honestly to complaints – irrespective of whether complaints are from a resident or a relative – and welcome a genuine apology. Relatives are also pleased when managers work collaboratively with families and encourage feedback. (Russell, 2017, p.iii)

Repercussions from reluctance to complain are evident in the following:

> *The son of an elderly male resident was responding to the lack of care provided during his respite care. The resident himself was reluctant to complain and begged his son not to do so. The son's reasoning included weighing up the repercussions that may ensue; being quietly confident that as his father would not be remaining or returning to that nursing home, that he would not be labelled 'a complainer', perhaps compromising his father's care. In a carefully worded letter, the son listed, under clearly identified headings, his observations of his father's care, together with his reporting of instances when he was not there. The complaints included lack of attention to his pressure sores (developed while in acute care), meal trays placed out of reach, poor attention to hygiene, particularly following episodes of incontinence, for which the resident was acutely embarrassed. The resident, a retired university professor, had no cognitive deficits. He had multiple, complex health*

> *problems, due to seriously impaired mobility. Reluctant to call out or otherwise demand attention, he generally suffered in silence. His son chose to concentrate only on the most serious issues, making the judgement that poor food and lack of social options were not as serious as other matters affecting his father's overall health. A letter was lodged with management, followed by another two weeks later. No response was ever received. The son, with full time business and other family commitments, was emotionally exhausted by the process; and lacked the energy or motivation to take the complaint any 'higher'.*

This account raises the question of a person's ability or willingness to lodge a complaint, due to lack of knowledge, poor communication skills and/or lack of trust in the system. Sadly, the poor response to formal complaints, which may either be completely lacking or replete with defensive justifications, results in fewer residents or families willing to expend the energy on this process.

There are many reasons why families and friends, and perhaps even some staff, do not report what they perceive to be signs of abuse and/or neglect. They may fear they will not be believed or that raising the issue may cause conflict. Furthermore, many have a well-founded fear (based on their own or others' previous experiences) that they will be ostracised and/or that the care of the resident will be compromised. So, they prefer to 'turn a blind eye', further disempowering both residents and families by suppressing their right to complain. The following incident represents an incalculable number of similar scenarios:

> *'Cathleen' was bedridden and dying. She told her daughter that she had been given nothing to drink for several hours and could not reach the drink herself, or locate the call bell. Cathleen insisted her daughter promise not to report the matter 'otherwise they'll make it worse for me, if I complain'. Out of respect for her mother's wishes, and in recognition of her own similar fears of repercussions the secret suffering went with Cathleen to her grave.*

In response to such tragic stories, together with the Australian Law Reform Commission (ALRC)'s recommended actions (in 2017), the ACQSC issued its 'Serious Incident Response Scheme (SIRS)', effective from July 2021. The scheme focuses on residents' safety, health and well-being, with the aim of preventing serious incidents. It also strengthens providers' skills for dealing with such matters when they occur, for the benefit of residents (Bastian, 2020). Mandatory reporting of incidents affecting residents includes the use of unreasonable force, unlawful sexual conduct, unexpected death and neglect. Providers are required to ensure their systems for reporting and responding to incidents are compliant with the latest principles (Russell Kennedy Lawyers, 2021).

8.2.3 Why Staff Don't Speak Out

The reluctance of some nursing home staff to report incidents of poor care, or unprofessional attitudes by other staff, or clear breaches of policy or breach of formal standards is characterised by the following comment:

'I just come here to do my job to the best of my ability; it's not my business to interfere with the actions of others. Anyway, I've seen what happens to staff who register a complaint, and I want to keep my job.'

Where the 'turning a blind eye' attitude is entrenched, reform is difficult to achieve. Where complaints are treated fairly and openly, change is possible. When inappropriate staff behaviour continues unchecked, a culture of bullying may develop, described below:

An apparently competent EN was often given extra responsibilities, due to her willingness to take on additional roles. She was articulate, and, according to management, always obliging and responsive. Her nursing knowledge appeared superior to that of her colleagues; others working with her would feel compelled to justify their care of residents in some instances. Behind the scenes, however, she was constantly 'checking up' on colleagues and questioning their actions. When faced with the possibility that any of her co-workers were questioning her actions, she would threaten to report them. Hence, a culture of bullying was hidden for many months. When management were finally alerted to the nature of this EN's conduct, calling her to account, she promptly resigned. As befitting the 'airing' of this bullying, management provided all staff with appropriate education by an expert in this field. Staff who were previously intimidated were at last free to air their grievances.

In contrast to the response described above, some management reactions to bullying tend to 'blame the victim', dissuading others from reporting what often becomes an insidious workplace culture. Staff reluctance to discuss procedures for reporting highlights the challenge of transforming attitudes. When confronted by the evidence, managers may be persuaded to pursue all options for change, with an emphasis on education.

8.2.4 The Serial Complainant

While the serial or persistent complainant may not be common, it can be a troublesome and time-consuming problem for managers. A typical scenario is the complainant who remains dissatisfied regardless of mutually agreed procedures, for example, changing the timing of the resident's shower routine or making particular dietary requests. It could be that the requests are impossible to meet; in which case false expectations can cause distress. It is important to clarify what outcome the complainant is seeking and whether they would commit the complaint to writing. Sometimes the 'serial complainant' can be satisfied by the offer of regular meetings with a senior member of staff, regardless of any specific complaint. Once the complainant believes they have been heard, the complaints may cease. If not, referral to another agency may be warranted. The aim is to create a relationship of harmony and mutual respect, restoring trust through comprehensive, effective communication.

8.2.5 Complaints About the Complaint's Process

Some relatives have expressed dissatisfaction regarding the way complaints are received and dealt with: issues regarding lack of timely response, uncertainty as to who is responsible for resolving complaints, whether complaints should be forwarded to a government body rather than dealt with 'in house', management's defensive self-justifying responses and repercussions for the person(s) making the complaint, who may be labelled a 'complainer'.

Acknowledging the best intentions of a complaint's process, one government report found that the data 'may not be representative of the extent of problems within the aged care sector' (Commonwealth of Australia, 2021a, p. 146). The investigator claimed a significant climate of fear surrounds the complaints system, including the expectation that complainants would be treated poorly, accentuating the level of mistrust in the process.

Such views are antithetical to a comprehensive complaint's process whereby the complainant is heard without judgement. An 'open door' attitude ensures that residents, relatives, staff, volunteers and others are encouraged to contribute to a system clearly dedicated to continuous improvement. This involves policies and procedures relating to complaints being readily accessible throughout the nursing home.

8.2.6 Managing and Responding to Risk

It is unrealistic to expect that all resident-related risks can be eliminated: an overly protectionist stance may be taken in an effort to achieve this aim, thereby reducing residents' independence. For example, a resident may wish, for privacy reasons, to be left alone when in the bathroom or toilet, although requiring assistance to transfer from their bedroom. One nurse may consider it 'too risky' to leave the resident alone, while another may honour the resident's desire for privacy. Where there is doubt about what a resident may or may not do without supervision or assistance, it is important to have a discussion with the resident and family (if appropriate) so that a plan can be developed and recorded, outlining and weighing the relevant risk factors considered acceptable. For a comprehensive assessment, referral to a physiotherapist or occupational therapist would be instructive. All staff should adhere to the plan to avoid confusion arising from a scenario such as the following, described by the resident:

> 'Stephanie always insists on staying with me in the shower because she's afraid I'll fall, but all the other nurses know it is my preference to shower alone. They always make sure I have the buzzer within reach, and appreciate my strong desire to have as much privacy as possible.'

It should also be noted, of course, that in such circumstances, the plan may need to change should the resident's condition deteriorate or other factors necessitate the need for more supervision.

Discussion of various contrasting risks are noted by Nay who compares those generally applauded by society, with the overprotection of nursing home residents (Nay, 2002). Learning a little more about each resident's likes and dislikes, preferences for independent decision-making, and developing a comprehensive outline of acceptable risks, will lead to increased happiness for many residents. Nay concludes by stating it is impossible to eliminate all risks. 'Overzealous risk management may protect a physical body from bruising but it may also damage irreparably the already vulnerable human soul' (Nay, 2002). This is not to suggest a laissez-faire attitude to risks; managers need to be aware of the legal parameters and the relevant processes and guidelines for their organisation. This approach needs to be balanced, however, with the rights of residents to self-determination rather than always and in every circumstance placing 'safety' first (Duckett et al., 2021, p. 9). Individualised care planning is the key, where risk management for one resident may require a different approach from others, particularly if the plan arises from consultation with resident, family and relevant therapists.

Following comprehensive assessment, in some situations, there may be no other option than to apply some form of restraint, in order to prevent harm.

8.2.7 Physical Restraint

Differences of opinion relate to the need for high priority being given to restraint issues, some citing the lack of hard evidence that precautionary methods such as alarmed doors, physical restraints and other measures to prevent 'wanderers', 'absconders' or 'elopers' are in the residents' best interest. Questions are raised concerning the amount of harm resulting from limiting residents' freedom. While there are legitimate reasons for confining some residents, the widespread practice of restraint needs to be more fully investigated. What is the motive, what is the risk and what is the outcome?

The ACQSC (Commonwealth of Australia, 2019a) supported the use of restraints as a last resort and only after assessment by a medical officer or NP. Where managers aim for a restraint-free environment, it is necessary to involve staff in regular education and training as well as arrange for discussion with families. A restraint policy should be readily accessible and updated regularly.

> While in certain circumstances, physical restraint may be necessary to mitigate risks to a resident or others in an emergency, empirical evidence demonstrates that restrictive practices can cause death, as well as other serious physical and psychological consequences that may increase morbidity, or expedite the dying process. (Commonwealth of Australia, 2019a, p.5)

This government report lists almost 20 adverse impacts of physical restraint, including 'serious injury and mortality' (Commonwealth of Australia, 2019a, pp. 5–6), highlighting the need for careful, well-informed decision-making.

Ibrahim's evidence before the Commission referred to above focused on the inappropriate use of physical restraint, too quickly applied before a comprehensive assessment is made of the reasons a resident might be agitated or distressed (Commonwealth of Australia, 2019, p. 203). The key issue here is *assessment*, individualised, thorough, conducted by a person with relevant qualifications and discussed with others in the care team. Towards this end, it is recommended that 'a senior restraint practitioner' should be appointed to lead an education campaign to minimise its use' (Duckett et al., 2021, p. 9).

8.2.8 *Chemical Restraint*

Chemical restraint is defined as 'a practice where residents are given psychotropic drugs which affect their mental state in order to "control" their behaviour' (Westbury, 2019). Such drugs include antipsychotics, antidepressants, anti-epileptics and benzodiazepines (tranquilisers). Citing the widespread use of antipsychotic drugs 'such as risperidone and quetiapine' for management of behavioural symptoms of dementia, Westbury agrees with the extensive reviews claiming they are of minimal value for activities such as calling out and wandering (Westbury, 2019). Indeed, they have been shown to produce serious side effects such as stroke, pneumonia and death. Westbury acknowledges their use for 'severe agitation or aggression associated with a risk of harm, delusions, hallucinations or pre-existing mental illness'. Even within those constraints, antipsychotic drugs should be given only when 'non-drug strategies such as personalised activities have failed, at the lowest effective dose, and for the shortest period required'. The guidelines include documentation of the behaviour leading to the drug use, 'a trial of non-drug strategies such as activity programmes and dose reductions after six months' (Westbury, 2019). Others recommend that psychotropic medications be reviewed after three months and then reduced and/or ceased (Laver et al., 2016).

A pioneer in dementia cites 20 'behavioural issues' commonly given as justification for administering antipsychotics, when in fact, they are not only ineffective but may exacerbate the behaviour (Power et al., 2010, p.28). Similarly, another dementia researcher (Good, 2015) acknowledges the difficulty in accepting the situation where there is, at present, no effective medication, joining other colleagues in stating that antipsychotics are not the answer and that behavioural symptoms 'are typically a result of environmental factors'. Good (2015) describes the use of antipsychotics for symptoms other than those associated with psychosis, such as schizophrenia and bipolar disorder, as 'off-label', that is, 'when a medication is prescribed for something for which it is not approved by the FDA'. The article emphasises the rights of the person with dementia, such as the right *not to be sedated* and the freedom to enjoy their surroundings. 'We need to work together and spread the word to change this gross misunderstanding and inadvertent abuse of people with dementia' (Good, 2015).

Australian gerontologists have made a strong statement about the recommended use of pharmacotherapy only as second line of treatment for 'more severe and persistent BPSD (behavioural and psychological symptoms of dementia) not responding to *non-pharmacological measures alone* with close monitoring for side effects and regular review' (Australia & New Zealand Society for Geriatric Medicine, 2016). Informed consent should be obtained before these drugs are used, together with advice regarding their modest benefit and potential adverse effects. Importantly, these gerontologists emphasise that the drugs can be safely stopped with no detrimental effects.

A witness to the 2019 Royal Commission stated that some nursing homes 'were able to operate with minimal antipsychotic use, whereas others were giving these medications to over 40% of their residents' (Commonwealth of Australia, 2019, p. 205). Other research found that most staff were unaware of the serious adverse effects of the medication being administered.

In relation to the care of residents with dementia, Ibrahim asserts the following:

> We can't reduce the use of restraint with money alone: it will require a cultural shift in clinical and aged care practice. This includes having staff who understand the unique needs of a person with dementia and are trained to respond appropriately (Ibrahim, 2019, p.2).

Quality indicators and strict reporting mechanisms are needed to address the tendency in some nursing homes towards over medication (also discussed in Chap. 4 on 'dementia'). Rather than resorting to chemical agents (which are only effective in a small percentage of cases), education is needed to encourage staff and families to use non-drug strategies, such as activities or alternative lifestyle plans, to replace the use of these drugs.

The following points (amongst others) emerged from a symposium on the use of restraints (Gilbert, 2020):

- There are less aged care staff and fewer RNs than in 1990.
- 'Cottage model' residential care results in fewer psychotropic medications.
- Informed consent is needed for any form of restraint.
- Restraint use must include an explicit objective.
- Negative psychological and physical side effects include humiliation, confusion, anxiety, depression, cuts and bruising, anger, withdrawal and aggression, pressure injuries, muscle weakness, clots, incontinence and increased risk of falling.
- All aged care assessments should include the residents' trauma history.

A major report 'Restrictive practices in residential aged care in Australia' has found the use of restraint to be contentious (Commonwealth of Australia, 2019a). While noting the numerous media reports and enquiries highlighting the misuse of both physical and chemical restraints without consent, concerns were also raised about the impact on residents' liberty and dignity. Of particular concern is the possibility that restraint may constitute a 'civil or criminal offence, such as assault or false imprisonment', not to mention 'doubts about their effectiveness' (Commonwealth of Australia, 2019a, p. 1). 'In many cases, the agitation, discomfort and anxiety of the resident is only increased' (p.6).

These reports highlight the urgent need for nursing home managers to ensure all their practices are in accordance with contemporary evidence-based research, including an approach which protects the rights of both resident and staff. When a resident is restrained, oversight by a medical practitioner, NP or RN should identify any signs of distress and evaluate whether the restraint is proportionate to the need.

8.2.9 Alternatives to Restraint

Managers who aim for a 'restraint-free' environment for residents utilise evidence in seeking alternatives. If a resident is at risk of falling from the bed, the bed may be lowered as far as possible or a mattress placed on the floor, with an alarm to alert staff when necessary. Other alternatives include occupying the resident's attention by (a) giving them something to hold, such as a soft, familiar object; (b) keeping their hands busy, such as by folding clothing or turning magazine pages and (c) offering items of food requiring manual dexterity. A brief handover discussion may elicit other ideas from staff, particularly from care assistants who may be more familiar with a resident's likes and dislikes. Families may offer other creative alternatives. It is important to document what has been tried and its success or otherwise.

'A way to ensure the safety of residents is not compromised is to know what restraint free options are available' (Australian government, 2012, p. 2).

8.3 Negligence, Neglect and Abuse

The issue of negligence and its associated suffering came to the attention of a well-known writer several decades ago, who reminded readers as follows:

> We emerge discerning of little credit, we who are capable of ignoring the conditions which make muted people suffer. The dissatisfied dead cannot noise abroad the negligence they have experienced. (Hinton, 1967)

8.3.1 Negligence and Neglect

Nursing home residents often have several comorbidities, most of which are incurable and accompanied by varying degrees of suffering. Although they have complex needs, they do not necessarily receive the care they need, particularly from medical specialists. While it is difficult to access accurate statistics, Hinton's (1967) comments are apt. The 'dissatisfied dead' cannot articulate their suffering, leaving others to speak for them.

Evidence has emerged from a government's enquiry of the poor care, neglect and inadequate response from some managers. The following description is provided by a family member of a resident with dementia:

> Her toenails were allowed to grow for months without being trimmed, eventually cutting into the flesh of her toes. Her partial denture, which was meant to be removed and cleaned every night, was left in her mouth for weeks, causing several of her remaining teeth to rot. (McKinnon, 2019, p.17)

This is merely one example of dozens cited in the Royal Commission Reports referred to throughout the book. Promotion in such government reports may alert the general public to the tragic state of affairs, leading to the conclusion that residential aged care is 'in crisis' (McKinnon, 2019). Some years earlier, data from an extensive online survey revealed 'a systemic failure to ensure safe and adequate care to Australian aged care residents' (Australian Nursing and Midwifery Federation, 2016, p.5). Respondents found 'the lack of emotional and social care' for residents to be deeply disturbing, as well as the physical issues of neglect where residents were 'left wet, dirty, hungry, thirsty, dehydrated and in pain' (p.19). The level of boredom was also noted, with many respondents distressed about the dehumanising examples they encountered. Of most concern was the staffing levels. 'In some facilities on a pm or night shift there was one RN for 145-150 residents'; in others there was no RN, leaving a care assistant "in charge"' (Australian Nursing and Midwifery Federation, 2016, p. 21).

'The number of patients/residents assigned to a nurse has a direct impact on their ability to provide best practice care. For every patient added to a nurse's workload, the likelihood of dying increases by 7%' (Butler, 2018). A safe minimum level of staffing sets a 'floor' rather than a 'ceiling', the former potentially deadly dangerous.

Staff/resident ratios are higher in privately owned facilities, raising the question again, of the need for consistent, mandated staffing levels. Unfortunately, reports confirm a distinct lack of will by governments and industry to address these matters with serious, urgent intent. Tragically, in some instances, the profit motive reigns supreme, as in the following:

> Research shows nursing home residents are receiving two hours and 50 minutes of care per day from nurses and carers, well below the four hours and 18 minutes they should be getting... Last year, owners of aged care facilities pocketed over $1 billion in profits while cutting staff. It's time elderly Australians get the care they deserve... It's a national disgrace. It's a crisis that shames us. (Butler, 2018)

Another headline captures the essence of neglect: 'Aged care residents unwashed and unfed':

> From 2003 to 2016 there had been a 13% reduction in trained nursing staff working full time in aged care facilities... The urgent problem that needs to be addressed is not funding but the declining quality of care due to chronic understaffing of aged care facilities. (Butler, 2018)

Amongst many documented instances of negligence is the issue of dental neglect. One dentist visiting a nursing home reported his findings to the accreditation agency, with no effect. He had found, on inspecting inside the mouths of several residents, images which shocked him: from teeth caked with food, to cracked dentures, through to fungating mouth ulcers. Further evidence comes from a new staff member:

> A student nurse, working casual shifts as a care attendant, was astonished to find on her first day, a lack of residents' personal toiletries including tooth brushes, denture cleaning agents and tooth paste. She also noticed an absence of any reference to dental care in residents' care plans.

More residents are entering nursing homes with their natural teeth, compared with previous decades, when the vast majority would have artificial dentures following full teeth extraction. However, many residents who have retained their natural teeth suffer from lack of care. Insufficient attention is paid to the painful, distressing consequences of dental hygiene failure, especially for residents unable to articulate their discomfort. The link between this confronting reality and the increasing incidence of malnutrition is shockingly apparent: nothing short of abuse.

8.3.2 Elder Abuse Defined

Elder abuse is 'a single or repeated act or lack of appropriate action, occurring within any relationship where there is an expectation of trust, which causes harm or distress to an older person' (World Health Organisation, 2021).

This definition highlights two major points: 'that an older person has suffered injury, deprivation, or unnecessary danger, and that another person (or persons) in a relationship of trust was responsible for causing or failing to prevent the harm' (Pillemer et al., 2016, p.196). Although it is estimated that between 2% and 25% of older people experience abuse, 'comprehensive prevalence studies are lacking' (Joosten et al., 2017). Women are over-represented, and 'abuse in residential care can be perpetrated by any acquaintance or family member' (Joosten et al., 2017).

Physical abuse includes threatening, punching, slapping, shoving and restraining. *Psychological abuse* includes verbal abuse, name-calling, bullying and harassment, treating an older person like a child, threatening to withdraw affection, threatening to put them in a nursing home, stopping an older person from seeing family and passive-aggressive behaviour. *Financial abuse* includes incurring bills for which an older person is responsible, stealing money or goods, forcing someone to sign a will, contract or power of attorney document and abusing power of attorney arrangements. *Sexual abuse* includes rape and other unwanted sexual contacts. It may also include inappropriate touching and the use of sexually offensive language (Australian Law Reform Commission, 2017; National Ageing Research Institute Ltd, 2018).

Elder abuse includes substandard care (Commonwealth of Australia, 2019, pp. 6–7) evidenced as in the following:

- Inadequate wound care, leading to septicaemia and death
- Poor continence management
- High incidence of assaults by staff

- Use of physical restraint
- Overprescribing without consent
- Patchy, fragmented palliative care
- 22–50% of residents malnourished
- 75–81% are incontinent
- 1/3 have pressure injuries at end of life
- 61% on regular psychotropic agents
- 4013 notifications of alleged or suspected physical and/or sexual assaults in 2017–2018
- Complaints that go unanswered

While elder abuse is not always distinguished from *neglect*, the latter includes failing to provide someone with food or medical care, or family members failing to use social security payment for providing such items. Nursing home staff may also be guilty of neglect when such necessities are not provided. It is interesting to note that the government report, quoted above, is titled *Neglect*, amounting to several hundred pages in three volumes. The number of witnesses giving evidence is testament to the enormity of the problem.

The following must also be noted:

> The term "elder abuse", is problematic for people from Aboriginal and Torres Strait Islander backgrounds for whom the term elder does not necessarily refer to an older person but to a person who is respected for their knowledge of customs and lore. (National Ageing Research Institute Ltd, 2018, p.2)

This reference highlights the importance of taking cultural considerations into account when describing, reporting or investigating elder abuse. Elder abuse may also be associated with care dependency and the result of past abuse.

It is nothing short of astounding to learn that in the twenty-first century, nursing homes (where residents pay to receive full-time care) may also be places of abuse, including resident to resident, resident to staff and staff to resident. At a community forum, one speaker described finding her mother-in-law with extensive bruising as follows:

> [W]e saw heavy bruising on her face. When we got her home I undressed her and saw ... she was black from one end of her to the other ... My mother-in-law said that she had been left alone in the bathroom in the shower chair and had fallen onto the floor ... I wrote a detailed letter to the Director of Nursing and was told that 'old people bruise easily'. I suppose they do if you drop them onto concrete'. (Commonwealth of Australia, 2019c, p.5)

Commentary on neglect in nursing homes includes a description of resident's treatment as 'cruel and harmful' indicating a lack of leadership (Martin, 2019). Martin describes not only the inadequate funding but a lack of 'fundamental transparency' indicating the need for a renewed regulatory regime which ensures safety and quality of services.

It is clear from the evidence included here, a very small number of examples amongst hundreds of others, that education on elder abuse needs urgent ongoing attention from the highest level of nursing home management. Such education would include a focus on risk factors: cognitive impairment including dementia, social isolation and poor mental health and depression and psychological distress associated with other forms of cognitive decline, as well as transparent reporting mechanisms.

8.3.3 Deliberate Killing: The Ultimate Abuse

In 2016, a personal carer employed in a New South Wales nursing home was found guilty of killing two residents with dementia and diabetes in 2013 by deliberately injecting them with overdoses of insulin (Australian Associated Press, 2016). This horrendous account raises questions (a) about unqualified staff's unlawful access to drugs, (b) the need for thorough reference checking and (c) comprehensive interview procedures to establish as far as possible staff motivation in choosing this area of work. This particular personal carer (now serving life imprisonment) had advertised on his Facebook page: 'I hate old people'. Legal scrutiny of this abhorrent crime revealed a serious lack of oversight and comprehensive reference checking as this person moved from one nursing home to another. While reports of killing/murder are, thankfully, very few, it is highlighted here to show how, with stringent employment processes and ongoing supervision, such a horrific incident might have been prevented.

8.3.4 Resident-Resident Aggression

Resident-to-resident assault is one form of elder abuse which seldom receives the attention it deserves, 'ignored by staff as it is not identified as serious, often perceived as normative, and rarely reported, leaving residents at potential risk' (Ellis et al., 2021). This type of aggression is under-researched, and while it accounts for a small number of serious injuries or deaths, it is clearly a matter requiring management's careful attention. Researchers have identified that 90% of the identified cases involved a diagnosis of dementia including a history of aggression. Only 25% of the deaths resulted in a coronial inquest, and criminal charges were rarely laid. However, the emerging issue is a cause for concern. Resident-to-resident aggression is defined as follows:

> ... negative, aggressive and intrusive verbal, physical, sexual, and material interactions between long-term care residents that in a community setting would likely be unwelcome and potentially cause physical or psychological distress or harm to the recipient. (Murphy et al., 2017, p.2603)

Abuse perpetuated against residents by other residents poses a significant challenge for staff who are often at a loss to know how to respond. Education is called for, together with a clear mechanism for documentation and follow-up. On the latter, Joosten and colleagues found 'a lack of high-quality research evidence to support the effectiveness of most elder abuse interventions' (Joosten et al., 2017), highlighting the need for a response focusing on the perpetrator. Without such follow-up, they may be tempted to continue the abuse. Management intervention is warranted, based on meticulous assessment and documentation, with referral to relevant experts where indicated.

It is imperative that every act of aggression is thoroughly reported and acted upon.

Comprehensive investigation of such incidents may require assistance from a geriatrician or other specialist, to assess underlying causes and recommend an appropriate response.

8.3.5 Reporting Elder Abuse

Barriers to the reporting of elder abuse include threat of retaliation, embarrassment, shame and fear of the consequences. Staff may fail to report due to reluctance in acknowledging the abuse, lack of relevant knowledge, lack of reporting mechanisms such as protocols, fear of liability and lack of confidence in follow-up procedures.

According to legislation (Aged Care Act, 1997), the following offences are reportable:

- Unlawful sexual contact with a resident of an aged care home
- Unreasonable use of force on a resident of an aged care home
- Unlawful sexual contact (sexual abuse), intended to capture any sexual contact without consent of the resident
- Unreasonable use of force (physical abuse), intended to capture assaults ranging from deliberate and violent physical attacks on residents to the use of unwarranted physical force

Elder abuse is much more prevalent than is generally acknowledged, unfortunately described as 'routine abuse' by some commentators. It includes the common experience of residents not having enough time to eat their meals, failure to receive the necessary assistance, insufficient variety of fluids and lack of encouragement to drink. It also includes their call bells either out of reach or remaining unanswered, lack of personal relationships where a focus on tasks leaves no time for conversation and being left isolated and alone in their rooms. These examples do not necessarily evoke images of physical assault; they are, nevertheless, clearly indicative of abuse.

Some residents have experienced unlawful sexual contact, being any non-consensual inappropriate touching. Placing this in broader context, when two residents, without impaired cognition, consent to have sexual contact, this may be regarded as a private matter between two adults. However, rather than 'turning a blind eye', staff who witness non-consensual sexual contact between residents, or between staff and residents, have the serious obligation to report such matters, so they can be dealt with appropriately. Unreported incidents include the following:

- A resident with cognitive impairment is inappropriately touched or assaulted.
- A staff member reports witnessing a family member physically assaulting their relative.
- A staff member is seen to be applying an unreasonable level of force when directing a resident to the bathroom.
- A family member notices a staff member roughly treating a resident.
- A nursing assistant notices bruising when she is showering the resident.
- Food being withheld from a resident close to death as 'he's going to die anyway'.
- Staff poking fun at a resident's appearance, language or idiosyncrasies.

Regular education is recommended for all healthcare workers, addressing the reasons for failure to report, such as the following:

- Whistle-blowers tend to be punished rather than praised.
- Those who complain are likely to be shunned by colleagues.
- Making complaints against colleagues makes working relationships difficult.
- Lack of support from administration.
- Misunderstanding privacy laws.
- Confidentiality issues: 'It's the family's business, not ours'.
- Complacency and cynicism.

8.3.6 What Is the Planned Response to Elder Abuse?

A national action plan is warranted (Council of Attorneys-General, 2019, p.6.) with recommendations for enhancing our understanding, improving community awareness (p.15) and strengthening safeguards (p.29). Greater awareness is needed about the origins of elder abuse, often resulting from complex interactions/partnerships between the older person and the perpetrator. Many factors influence these relationships, not only their individual characteristics but also the social and cultural environment in which the abuse occurs:

> The Australian Government is committed to supporting measures to reduce the prevalence, severity and impact of elder abuse. Through the Attorney-General's Department the government is working collaboratively with all state and territory governments to develop nationally coordinated responses to the emerging problem of the abuse of older people. (Australian Government Attorney General's Department, 2019)

This government document acknowledges the complexity of elder abuse, especially when involving a trusted family member, friend or caregiver. The aim of the national plan is to provide ready access to information, enhance understanding, improve community awareness, strengthen service responses, increase coordination across jurisdictions and increase uptake of future planning tools such as powers of attorney. Further research will focus on 'strengthening the evidence base on the nature and prevalence of elder abuse in Australia' (Australian Government, n.d.).

However, sufficient evidence is readily available for policies and procedures to be enacted immediately following any instance of abuse.

8.3.7 What Else Can Be Done About Elder Abuse?

First and foremost is the raising of awareness. It should be noted, and widely broadcast in the nursing home's publicity network, that the World Elder Abuse Awareness Day is on 15 June every year. Encouraging wide participation in this event, through sharing resources and other material, helps to raise awareness. Specific needs of CALD (culturally and linguistically diverse) communities, particularly in rural Australia, need also to be understood and responded to. This may include notices in other languages, ensuring interpreters are available and seeking support from community agencies familiar with elder abuse in specific circumstances.

Elder abuse 'is largely preventable (unlike most diseases/conditions of old age)' (Pillemer et al., 2016, p.195), and the key factor in prevention is education: a key factor in empowering staff to recognise and report the signs. Without such education, elder abuse can easily be swept aside or incorrectly classified as 'none of our business'. Given the inadequate staffing levels, some other measures may help to prevent this problem. For example, the benefits of electronic surveillance as a risk management tool need to be further explored.

An increasing number of nursing homes are installing/trialling CCTV monitoring in order to provide greater transparency and offer stronger protection for residents. Cameras in communal areas and in some bedrooms (with the consent of the resident or next of kin) allow prompt response to any irregularities which may otherwise go unnoticed. Research has yet to uncover how many falls are prevented by continuous monitoring, together with analysis of other safety factors which may be revealed. On a more positive note, continuous monitoring may reveal whether a resident appears relaxed and free from pain, how residents are managing their meal, how they respond to visitors and whether they show signs of pleasure or anxiety when alone.

Researchers urge consideration from all perspectives related to elder abuse, including residents' privacy in relation to surveillance. Sawyer suggests that while 'the final decision should always be with the resident or their family', an opt-out system is a viable alternative. It needs to be understood that such surveillance is 'not about invading people's privacy, instead it is about ensuring transparency' (Sawyer, 2020). At least one Australian state has introduced a 12-month trial of CCTV cameras: this would seem a fair compromise pending a final decision, particularly when staff may be less than enthusiastic about perceived 'Big Brother' watching their every move.

8.3.8 Preventing Elder Abuse and Sexual Assault

The Australian Law Reform Commission Report advocated the establishment of a 'serious incident response scheme in aged care legislation', appropriate employment screening processes and national guidelines for community visitors (Australian Law Reform Commission, 2017, p.10). The outcome of these recommendations is yet to be seen.

According to other research, eliminating sexual assault in nursing homes is a major challenge which starts with acknowledging its existence and recognising its intensity. Sexual assault is considered the most hidden, as well as least acknowledged and reported, form of elder abuse, making it difficult to accurately estimate its prevalence (Smith, 2019). Societal attitudes towards older people are reflected in many nursing home populations, including ignorance about their sexuality:

> Negative stereotypes such as that older people aren't sexual beings, their greater dependency on others, potential divided loyalty to staff members or residents are unique barriers to reporting, detecting, and preventing sexual assault in nursing homes. Despite severe health consequences, efforts to prevent and address elder abuse remain inadequate. (Ibrahim et al., 2018)

Ibrahim and his colleagues also note that victims of sexual assault in nursing homes are often ignored by staff who are unwilling or reluctant to believe the accusations. Thus, reporting is uncommon and there is no follow-up. Lack of witnesses also makes accurate documentation difficult. Transformed care depends on timely, accurate reporting to appropriate authorities of any sexual assault involving a resident, staff member, volunteer or visitor. Sexual assault is a crime and must be treated accordingly.

Responsibility for preventing and/or responding to elder abuse and sexual assault lies with managers who provide regular training on reporting obligations, when to involve police and appropriate care for staff and/or residents following an assault.

The following account highlights the point about unwanted encounters not being recognised as abuse:

> A well-meaning (uninvited) pastor greeted 'Patty' from his superior height over Patty's diminutive frame in her wheel chair, with a pat on the head: 'How are we today? Have we been a good girl this week?' Unfortunately, Patty, aged 88, had experienced other infantilising, de-humanising encounters before this scenario was witnessed, reported and action taken. No-one liked to challenge the practice of this unofficial pastor who, in spite of his own poor health, continued to visit the nursing home every week, having done so for as long as anyone could remember. Nurses were reluctant to encroach on 'religious' territory and management were not sure how to raise the matter. However, pastoral visitors are not immune to discipline, and duty of care involves more than the physical.

When a report was finally made, the manager counselled the pastor about his unacceptable behaviour. The pastor acknowledged he was 'unwell' and 'not himself' and only continued the visits from a sense of obligation. He was relieved to be freed from his perceived 'duty' and, following acknowledgement for his (well meaning) pastoral care, was pleased to cease his visits. Of course, the situation also works the other way. Where pastoral care staff witness any dehumanising attitudes from regular staff, they should also have a mechanism for discussing the issue with management.

8.3.9 Investigative Reporting

A newspaper reporter cited evidence of 10 aged care homes which formally met the regulator's 44 accreditation standards, despite 'subsequent findings that they failed to provide a reasonable standard of care' (Burrell, 2017). Instances of neglect include an 89-year-old war veteran attacked by mice in his nursing home bed and a resident with dementia who died after falling into the nursing home fountain. Others died from heat exposure and from untreated urinary tract infections. The reporter noted that less than 1% of nursing homes had sanctions imposed for failing to comply with standards. This reporter joins the throng calling for staff-to-resident ratios to be introduced, similar to those at childcare centres, hospitals and schools (Burrell, 2017).

It is clear that this worsening phenomenon of negligence, neglect and abuse requires a renewed focus in every instance, such as rigorous research followed by immediate, meaningful remedial action. The first step is to recognise abuse when and where it occurs and act on it immediately, with the use of incident reports and thorough follow-up processes, including regular education. Only when the reporting and recording of abuse is prioritised with accurate, timely documentation at all levels will meaningful response be possible. The aim of such reporting is to reduce further risk of abuse, raising the confidence of all associated with the nursing home that their residents' continued safety is paramount.

8.4 Home Sweet Home: Expectations Met?

8.4.1 Residents/Relatives Committees

One of the most important nursing home partnerships is with residents' relatives and informal carers. A residents'/relatives' committee plays a pivotal role, having a significant effect on issues such as food, safety, décor, communication and other factors affecting residents' daily lives. In examining incident reports (with names deleted), a committee discussion may result in comprehensive systems reforms. When residents and relatives are involved in monitoring processes or measurements of the home's standards, other improvements may ensue. Such meetings, well chaired and accurately recorded, with minutes widely circulated, have proven benefits for the whole nursing home community. Residents and their families offer a unique perspective on aspects of care not necessarily noticed by staff, often including insightful recommendations and innovations. Many family members are highly qualified to offer such advice; their contributions should be welcomed rather than shunned. The benefits of residents raising their own voices, and being heard, also deserve greater attention. Nursing homes employing these processes reap the rewards in terms of commitment to excellence and communication through transparent systems, and ensuring expectations are met.

8.4.2 Liaison with Other Agencies

What partnerships arise or evolve between nursing homes and other agencies? Some of these matters are discussed in Chap. 2, where the importance of clear communication between nursing homes and hospitals is emphasised. Here, the topic is broadened to include other community groups. When a nursing home is considered a vital part of the community in which it is located, opportunities abound for effective, innovative relationships. Schools, kindergartens, day centres, sporting associations, elderly citizens groups, service clubs, multicultural centres, churches, volunteer associations and many others enjoy the advantages of strong inter-agency connections. In rural, regional areas, closer liaison is optimised through ongoing personal contacts between the nursing home and other community groups. Shared information is vital and communication is the key.

Managers may acknowledge the wisdom of such liaison while lamenting the lack of resources to extend residents' care beyond the nursing home. Raising the issue with volunteers and/or the residents'/relatives' committee may produce original ideas, including the establishment of a small group to discuss options which would benefit the nursing home and the wider community. Fostering interest from media groups also produces effective, positive liaison, with significant benefits for both parties.

8.4.3 A Homely Community

Although many would regard a nursing home as far removed, in both ethos and everyday realities, from their private home, some small gestures would help to close the gap, as in the following account:

> 'Alfred', a proud, dignified, well-educated gentleman, confused at times but fully alert at others, looked forward each morning to reading his newspaper as he ate his breakfast. This had been his custom at home, which he assumed would continue in his new environment. This request, albeit simple, proved beyond the capabilities of the nursing home. Whereas the newsagent diligently delivered his paper to the front desk early each morning, it was not always delivered to Alfred. When repeated requests were politely made by both Alfred and his wife, it still took several weeks for the task to be managed routinely. It became the responsibility of the kitchen hand. In strict fulfilment of her task, she would deliver the newspaper, still tightly wrapped in plastic to Alfred's bedside table, with his breakfast.

> *Requests to several staff members to assist Alfred access the paper went unanswered. 'It's not my job' was the usual reply, or 'I haven't got time for that', others would claim, defensively. It seemed utterly beyond staff imagination to have this simple task built into Alfred's care plan: a one-minute routine which would have assisted Alfred in maintaining his link to the wider world, not to mention feeling 'at home'.*

What are the outcome criteria for excellence in nursing homes? What makes them 'homely'? Would regular newspaper delivery (if requested) rank amongst such measures? There may be significant disparity between what the residents regard as 'homely' or 'quality care' and the way others measure such criteria. Residents are more concerned with communication and social issues rather than physical components of care (Nay, 1996). In a paper outlining the challenges residents face in communal living, Nay reiterates the findings from her 1993 doctoral study: *Benevolent Oppression: Lived Experiences of Nursing Home Life* (Nay, 1993). She challenges gerontic nurses to be active agents for change, to remove the barriers which prevent older persons from enjoying a more equitable life in nursing homes. As traditional divisions have been constructed, they can be replaced by negotiation and empowerment for change.

What is at the heart or centre of this change? In deciding what is the purpose of residential aged care, a more open view may embrace a hopeful, imaginative response. Social scientists look to the past for antecedents, to answer the question of how we should live. This view also encompasses the possibility of faith in the future, raising the further question of who determines the shape of any community. Jenson's wisdom is apt as follows:

> The chief political question is always, where is sovereignty located, where in the community is the decision about the community's future taken? Sovereignty is not necessarily located where the main pomp and circumstance is found nor even where the main power is found. Sovereignty is located where the decision is made about what sort of community the community will be. (Jenson, 1995, p.95)

Nursing home staff have the privilege and the responsibility of being involved in decision-making concerning their community, becoming catalysts for change and determining 'what sort of community the community will be'. Identifying where the nursing home's heart or centre is located will influence what sort of community it will be. On the other hand, when the community lacks cohesion, it will be destined to impermanence. Jenson (1995) sees the heart of the community as pivotal to its future. He also sees possibilities for those organisations where each person is regarded as 'an organ of community' (p. 36). That is to say, when each part of the organisation is considered as a change agent, uniqueness and diversity are equally valued. Belonging to a community is deeply rooted in mutuality, where each is considered pivotal to the whole. Jenson makes the strong statement about organisations that they may reflect a spirit of shared humanity when the centre is built on personal address and response. Belonging to such an organisation, he claims, is 'one of the ways we become human' (p. 36).

In this discussion on a homely community, the way we care for the most vulnerable people in our nursing homes is a witness to our humanity. Brown and Thompson

8.4 Home Sweet Home: Expectations Met? 273

(1994) state the issue succinctly, describing the needs of one very frail person in 'Martha House', exemplifying the central thrust of this discussion on community:

> It is a reminder of the Ghandian exhortation to consider our policy and its potential impact by looking into the face of the last person. If our policy does not ease the plight of this last person, it is wrong and should be discarded. The 'last person' in the residential aged care context is the one who is least likely to attract positive attention; the one to whom others are not necessarily drawn and may even be repelled by. (Brown & Thompson, 1994, p.132)

These authors (Brown & Thompson) urge us to adopt a different lens when looking at such a person:

> Does it suggest that quality of life for other elderly people would be different if they could be welcomed into a community of intention – where meekness, poverty and frailty were honoured; where one is listened to rather than dominated; where there was a capacity for intimacy; where life has meaning; where the gifts of each are recognised; where sorrow and joy could be freely expressed; where justice was lived as a reality; and, where one could dare hope that, in spite of great frailty and even death, one might find encouragement to continue on the journey towards wholeness? (Brown & Thompson, 1994, pp. 132–133)

What is the optimum environment in which a nursing home resident may 'journey towards wholeness'? This question, especially for the physically and mentally frail resident who may seem totally disintegrated, poses a challenge for carers to consider a positive framework of hope rather than a negative resignation of despair.

8.4.4 Size Matters

The propensity for nursing homes to grow larger has the effect of them becoming less homelike (Hampson, 2018). The average size of the 2672 nursing homes in Australia is 75 beds per facility, an apparent shift to a previous age where institutions were large:

> Large institutions for people with disability and mental illness, as well as orphaned children, were once commonplace. But now – influenced by the 1960s deinstitutionalisation movement – these have been closed down and replaced with smaller community-based services. In the case of aged care, Australia has gone the opposite way. (Hampson, 2018)

Research is showing that, particularly for residents with dementia, a smaller homelike model of care results in fewer hospitalisations with the cost of caring no higher than in the larger homes. It now seems that, contrary to the evidence, financial viability rather than quality of care is, for some, the driver for increased size. 'Around 45% of n/homes are operated by the private for-profit sector, 40% by religious and charitable governments and less than 1% by local governments' (Hampson, 2018).

A comprehensive analysis is needed to discern and compare the quality of care amongst various sized nursing homes. For example, what are the differences in outcome for residents in a large private for-profit home compared to the (very few) government-funded homes? The latter enjoy a substantially different staffing mix from the former, with a significantly higher resident/staff ratio, including a high proportion of RNs. Whether or not this amounts to a homelike environment is a perception yet to be more thoroughly analysed.

Other quality indicators which may be related to nursing home size include the number of residents' falls, fractures, hospitalisations, pressure injuries, weight loss, malnutrition, medication errors and other medical, psychological and social factors. The latter may be more difficult to quantify. Outcome differences may also be related to the type of provider, whether government, not-for-profit or for-profit, and the number of minutes per resident per day dedicated to direct care. Prospective residents and their families have every right to request such information, including relevant statistical data. Current wisdom and research suggests 'that residential aged care providers should be encouraged to adopt small household models, paired with appropriate relationship-based staffing arrangements' (Commonwealth of Australia, 2021b, p. 229).

8.4.5 Design Matters

Creative designs are evident in many nursing homes, purposefully built to create a sense of community. Some have within their grounds – or adjacent – a café or shop, a chapel, a hair salon, a school or kindergarten, gardens with safe, accessible paths and places for exercise. Community involvement is optimised by proximity to shopping centres, restaurants, libraries, music and sporting venues and theatres. For those residents unable to access external venues, spaces for small family gatherings including children, privacy for music preferences and conversations add to the options for individualised care. 'If we can't bring the residents to the community, we must bring the community to the residents by creating a space they will want to visit again and again' (Kaiser, 2019).

Dementia-friendly design is changing in response to research on lighting, floor covering, use of colours, clear signage and avoidance of extraneous noise. Rather than being built around a 'risk-free environment' where the focus is on management and secure containment, creative design expands the horizons, listens to the voices of those who live and visit the home and has at its heart, healthy, meaningful relationships. A total reversal of attitudes is needed, according to the following:

> For the culture at large, nursing homes still summon up notions of neglect and abandonment. For a society passionate about personal independence and self-sufficiency, nursing homes are too easily seen as habitations for diminishment and dependency. For a culture that prizes curative medicine and the dream of 'youthful ageing', nursing homes are places of intractable, therefore intolerable, frailty. For a culture with deep anti-institutional biases and often-romanticized versions of 'family' and the freedoms of 'home', nursing homes are especially suspect. They are 'total institutions' where personal freedom, privacy, and range of choice have little chance. It is no surprise, then, that our general cultural wisdom warns us to avoid the nursing home at all costs. (Collopy et al., 1991, p. 5)

Decades later, Collopy's 'cultural wisdom' prevails, creating a challenge for transforming institutions and 'facilities' into genuine homelike environments. Acknowledgement is made here of the many nursing homes who manage to achieve such transformation.

8.4.6 Home or Hospice

'Home' is a concept central to our sense of self. It is about belonging within an environment that is uniquely one's own, reflecting one's personality and providing security (Collier et al., 2015). Others (Phillips & Currow, 2017) ask whether reframing aged care facilities as hospices instead of homes would enable older people to be treated according to their unique needs. Given their increasing frailty, particularly when transferred from an acute hospital to nursing home, the question is raised as to whether the latter can realistically be regarded as their 'home'. Furthermore, due to their increasing medical acuity and decreasing ability to engage with the world around them, more older people are requiring intensive nursing care until they die. The authors of this analysis claim: 'Referring to it as a resident's "home" is outdated and may inadvertently restrict this vulnerable population's access to the expert nursing and medical care they require' (Phillips & Currow, 2017).

Those who argue for change suggest that nursing homes are looking less like homes and more like hotels, with lavish foyers, chandeliers and an impersonal atmosphere. In other words, the design needs to reflect the care being offered, that is, nursing care. This, of course, implies the nursing home will be staffed accordingly (a major issue accentuated throughout the book). Most people entering the nursing home will die there, requiring all of the expertise provided to those who are dying in a hospice (Phillips & Currow, 2017). However, nursing homes are not staffed according to the needs of hospice patients, although many of the residents' needs are similar. All of these factors signify the need for a different type of care than that provided in a 'home'. Renaming them as hospices would, according to some opinions, attract the necessary funding, and staffing levels commensurate with the needs of those who live and die there. This suggestion would benefit from wider community discussion.

Mounting evidence suggests the title 'homelike environment' is unsuitable for the complex clinical needs of many residents, particularly at the end of life when optimum care may only be sustained by daily medical oversight and skilled nursing. On the other hand, a renewed, transformed approach to nursing home care, embedded in legislation, may be instrumental in bringing the art and science of nursing 'home', that is, to the place where it has always belonged.

'Australia's future aged care system should be ambitious in what it seeks to achieve for older people and it should welcome accountability against that vision' (Commonwealth of Australia, 2021b, p. 135). Acknowledging the need for substantial change, incorporating the highest standards of medical, nursing, allied health and attendant care, it remains in the realm of possibilities for more providers to offer care commensurate with a 'homelike environment'. While nursing homes operating in isolation may find it difficult to achieve this aim, others benefit from partnerships, formal and informal, with the wider community, enhancing residents' care in a variety of ways. Transforming the care enables living and dying in a nursing home to be embraced with optimism and satisfaction rather than pessimism and fear.

When positive relationships are fostered, and evidence-based clinical care assured, residents may, indeed, enjoy living in their 'sweet home'.

8.4.7 Home Sweet Nursing Home

'Susan' faced the decision squarely. Although relatively young (aged 73 years) for a nursing home resident, her complex medical problems resulted in the need for increasing care for all her activities of daily living. She came with her social worker to look over the nursing home, deciding on the spot: 'This will be my home.' Alert of mind, and with a positive attitude, Susan wanted not only to receive care but to contribute to the life of the nursing home. She wanted to join the residents' committee, and to visit (in her wheel chair) any residents who were lonely. For three weeks Susan rode around the nursing home, learning many residents' names in a few days. Having struggled to maintain her independence at home, she said, 'I've had the best sleep I've had for three years' and after three weeks, 'I've done more things in these three weeks than I've done in years.' These 'things' included bus outings, pub lunches, art classes, music groups, cuddling pets, and film afternoons. Then, Susan's serious, complex, medical condition caught up with her, requiring hospitalisation for intractable symptoms. Visited by two nursing home staff members, two days before her death, Susan whispered 'All I want to do is come home'.

We wish we could bring you home, Susan.

References

Aged Care Act. (1997). https://www.legislation.gov.au/Details/C2017C00241
Aged Care Insight. (2019). Know your rights: Aged care quality standards and charter of aged care rights start this month. *Aged Care Insite*.
Australia & New Zealand Society for Geriatric Medicine. (2016). Position statement 26: Management of behavioural and psychological symptoms of dementia (BPSD). *Geriatric Medicine*, 1–8.
Australian Associated Press. (2016, 28 September 2016). NSW aged care worker guilty of murder and attempted murder of dementia patients. *The Guardian*.
Australian government. (2012). *Decision making tool: Supporting a restraint free environment in residential aged care*.
Australian Government. (n.d.). *Protecting the rights of older Australians*. https://www.ag.gov.au/rights-and-protections/protecting-rights-older-australians
Australian Government Aged Care Complaints Commissioner. (2017). *Annual Report 2016–17*.
Australian Government Aged Care Quality and Safety Commission. (2019). *Aged care quality standards*.
Australian Government Aged Care Quality and Safety Commission. (2020). *Making a complaint*.

References

Australian Government Attorney General's Department. (2019). *Protecting the rights of older Australians*. Australian Government.

Australian Law Reform Commission. (2017). *Elder abuse-a national legal response: Summary report*. Australian Law Reform Commission.

Australian Medical Association. (2019). *Innovation in aged care 2019 position statement*.

Australian Nursing and Midwifery Federation. (2016). *ANMF national aged care survey: Final report*.

Bastian, D. (2020, June 15). Abuse reporting rules to change with Serious Incident Response Scheme. *Aged Care Insite*.

Brown, C., & Thompson, K. (1994). A quality life: Searching for quality of life in residential service for elderly people. *Australian Journal on Ageing, 13*(3), 131–133.

Burrell, A. (2017). Aged-care checks for neglect 'failing'. *The Australian, 1*, 4.

Butler, A. (2018). Residents get barely half the care they need. *The Lamp, 75*(3), 9.

Carnell, K., & Paterson, R. (2017). *Review of national aged care quality regulatory processes*.

Collier, A., Phillips, J. L., & Iedema, R. (2015). The meaning of home at the end of life: A video-reflexive ethnography study. *Palliative Medicine, 29*(8), 695–702.

Collopy, B., Boyle, P., & Jennings, B. (1991). New directions in nursing home ethics [Special supplement]. Hastings Center Report, Special Supplement(March-April), 1–15.

Commonwealth of Australia. (2019). *Royal Commission into aged care quality and safety interim report: Neglect volume 1*.

Commonwealth of Australia. (2019a). *Restrictive practices in residential aged care in Australia*.

Commonwealth of Australia. (2019b). *Royal Commission into aged care quality and safety interim report: Neglect volume 2*.

Commonwealth of Australia. (2019c). *Royal Commission into aged care quality and safety volume 3, interim report: Neglect volume 3*.

Commonwealth of Australia. (2021a). *Royal Commission into aged care quality and safety final report: Care, dignity and respect, volume 2, the current system*.

Commonwealth of Australia. (2021b). *Royal Commission into aged care quality and safety final report: care, dignity and respect, volume 3A the new system*.

Council of Attorneys-General. (2019). National plan to respond to the abuse of older Australians (Elder Abuse) 2019–2023.

Crofton, C., & Witney, G. (2004). *Nursing documentation in aged care: A guide to practice*. Ausmed Publications.

Duckett, S., Stobart, A., & Swerissen, H. (2021). *The next steps for aged care: Forging a clear path after the Royal Commission*. The Grattan Institute.

Ellis, J., Ward, L., & Campbell, F. (2021). Managing resident to resident assault in residential aged care homes. *Australian Nursing & Midwifery Journal, 27*(2), 44.

Galloway, K. (2018). Aged care exemplifies the limits of markets. *Eureka Street, 28*(19).

Gershon, L. (2017). The future is emotional. *AEON*. https://aeon.co/essays/the-key-to-jobs-in-the-future-is-not-college-but-compassion

Gilbert, P. (2020). 10 things we learned at the 2020 aged care physical and chemical restraint symposium. *The Handover*, (7), 4.

Good, M. (2015). *Antipsychotics overused in dementia care for behavioral issues*. The Oakland Press.

Hampson, R. (2018). Australia's residential aged care facilities are getting bigger and less home-like. *The Conversation*.

Hinton, J. (1967). *Dying*. Penguin Books.

Ibrahim, J. (2018). What is quality in aged care. *The Conversation*.

Ibrahim, J. (2019). The aged care Royal Commission's 3 areas of immediate action are worthy, but won't fix a broken system. *The Conversation*.

Ibrahim, J., Smith, D., & Bugeja, L. (2018). It's hard to think about, but frail older women in nursing homes get sexually abused too. *The Conversation*.

James, E. (2019, 11 November, 2019). *'Tas aged care resident 'a slab of meat'*. Canberra Times.

Jenson, R. (1995). *Essays in theology of culture*. William B. Eerdmans Publishing Company.
Joosten, M., Vrantsidis, F., & Dow, B. (2017). Understanding elder abuse: A scoping study.
Kaiser, D. (2019). How technology and design can transform our aged care facilities into modern, fun places to live, work and visit. *Aged Care Insite*.
Kearney, G. (2021). The system is broken. *Nursing Review*, (2), 13.
Laver, K., Cumming, R., Dyer, S., Agar, M., Anstey, K., Beattie, E., Brodaty, H., & Broe, T. (2016). Clinical practice guidelines for dementia in Australia. *MJA, 204*(5), 1–3.
Martin, S. (2019, October 31). PM promises more funding after aged care system found to be 'harmful' and underfunded. *The Guardian*.
McKinnon, A. (2019, June). Aged care in crisis. *The Monthly*.
Murphy, B. J., Bugeja, L., Pilgrim, J. L., & Ibrahim, J. (2017). Deaths from resident-to-resident aggression in Australian nursing homes. *Journal of the American Geriatrics Society, 65*(12), 2603–2609.
National Ageing Research Institute Ltd. (2018). *Elder abuse community action plan for victoria*.
Nay, R. (1993). *Benevolent oppression: Lived experiences of nursing home life* [doctoral dissertation, University of New South Wales]. Sydney, Australia.
Nay, R. (1996). Nursing home entry: Meaning making by relatives. *Australasian Journal on Ageing, 15*(3), 123–126.
Nay, R. (2002). The dignity of risk. *Australian Nursing Journal, 9*(9), 33.
Phillips, J., & Currow, D. (2017). Would reframing aged care facilities as a 'hospice' instead of a 'home' enable older people to get the care they need? *Collegian, 24*(1), 1–2.
Pillemer, K., Burnes, D., Riffin, C., & Lachs, M. S. (2016). Elder abuse: Global situation, risk factors, and prevention strategies. *The Gerontologist, 56*(Suppl_2), S194–S205.
Power, G. A., Swaffer, K., & Thomas, W. H. (2010). *Dementia beyond drugs: Changing the culture of care*. HPP/Health Professions Press.
Ratcliffe, J. (2020). Australians want more funding for higher-quality aged care -- and most are willing to pay extra tax to achieve it. *The Conversation*.
Russell, S. (2017). *Living well in an aged care home* (Research Report, Issue. https://www.aged-carematters.net.au/living-well-in-an-aged-care-home/
Russell Kennedy Lawyers. (2021). *Serious Incident Response Scheme: Second phase of mandatory reporting obligations commences 1 October 2021*. R. K. Lawyers.
Sawyer, D. (2020). CCTV can offer more than protection: Opinion. *Australian Ageing Agenda*.
Smith, C. (2019). *Navigating the maze: An overview of Australia's current aged care system*. Background Paper, Issue. //agedcare.royalcommission.gov.au/news-and-media/navigating-maze-overview-australias-current-aged-care-system.
Westbury, J. (2019). Chemical restraint has no place in aged care, but poorly designed reforms can easily go wrong. *The Conversation*.
World Health Organisation. (2021). *Elder abuse*. https://www.who.int/news-room/fact-sheets/detail/elder-abuse

Index

A
Aboriginal and Torres Strait Islander worker, 18
Aboriginal culture, 18
Aboriginality, 18
Advance care directive (ACD), 13, 52
Advance care planning (ACP), 52–54
Age Care Act 1997, 214
Aged care, 93
  prevailing perception, 55
Aged care assessment team (ACAT), 30
Aged care facilities, 12, 178, 262, 275
Aged Care Funding Instrument (ACFI), 8, 221
Aged care nurses, 17
Aged Care Quality and Safety Commission (ACQSC), 29, 219, 252
Aged Care Quality Standards, 219, 243
Aged-care Royal Commission, 49, 65, 67, 79
Aged care services, 10
Aged care standards, 243
Aged care system, 49, 66, 214
Ageing population, 1, 2
Ageism, 4–6
Alcoholic beverages, 70
Allied health services, 78–79
Alzheimer's disease (AD), 101
Aromatherapy, 78
Artificial intelligence (AI), 145
Australian Aged Care Classification (AN-ACC), 9
Australian Law Reform Commission (ALRC), 255, 269
Australian Medical Association (AMA), 248

B
Benevolent ageism, 5
Benevolent Oppression: Lived Experiences of Nursing Home Life, 272
Bereavement, 188, 189
Best medicine, 88
Bowel and bladder management, 71
Burden *vs.* benefit, 7
Bus trips, 90

C
Cardiopulmonary resuscitation (CPR), 76–77
Care planning, 198
Careful attention, 68
Careful communication, 19
Case conferences, 62
Celebration, 87
Chronic loneliness, 86
Chronic obstructive pulmonary disease (COPD), 157
Clinical care, 55, 57, 64–66, 88, 94
  CMAs (*see* Comprehensive medical assessments (CMAs))
  duty of care, 58–59
  general practitioners (GPs), 59–60
  learning from Nightingale, 57–58
  telehealth, 61
Clinical deterioration, 61
Clinically assisted nutrition and hydration (CANH), 146
Clustered domestic models, 13
Comfort measures, 166

Communication, 86, 141
　aged care facility, 30
　agencies/age groups
　　children presence, 39–41
　　hospitalization/transfers, 42, 43
　　nursing home, 38
　　relatives caring, 38, 39
　component, 33
　definition, 31, 32, 34
　handover conversation
　　nursing home, 44
　　staff reactions, 44–46
　keys and cues, 35
　nursing home, 35
　short-term stay, 36, 37
　waiting list to admission
　　experience, 30
　　foyer/entrance, 25
　　information, 26
　　nursing home, 27–29
　　pre-admission, 26, 27
　　residents/relatives, 25
　　waiting list, 30
Community expectations
　agencies, 271
　alternatives to restraint, 261
　awareness, 267
　CCTV monitoring, 268
　chemical restraint, 259–261
　complaint's process, 257
　consumer outcomes, 243
　design matters, 274, 275
　elder abuse, 263, 264
　families, 247
　home/hospice, 275, 276
　homely community, 271–273
　investigative reporting, 270
　making and managing
　　complaints, 252–255
　managing and responding to risk, 257, 258
　meals, 247, 248
　medical and allied health services, 247
　mission statement, 245
　neglect, 261–263
　negligence, 261–263
　non-clinical factors, 243
　nursing home staff, 255
　physical restraint, 258, 259
　prospective residents, 246
　quality, 249–251
　quality and standards, 247
　reporting elder abuse, 266, 267
　resident records, 248, 249
　resident-resident aggression, 265, 266
　residents'/relatives' committee, 270
　serial complainant, 256
　sexual assault, 269
　size matters, 273, 274
　snacks, 247, 248
　staffing inadequacies, 251, 252
　transparency, 245, 246
　ultimate abuse, 265
　visiting, 247
　Workplace Health and Safety (WHS), 246
Community partnership, 57
Community responses to death and dying
　acknowledging family, 202
　communicating the fact of death, 205
　consequences, 200
　cultural differences, 203
　death as loss of community, 206
　death as part of life, 201
　death denial, 198–199
　good death, 202–203
　humour and death, 201
　notification of dying, 204–205
　planning for death, 206
　reportable deaths, 206
　wish to hasten death, 198–199
Community Visitors Scheme, 51, 85
Compassion fatigue, 159
Compassionate pragmatism, 58
Comprehensive medical assessments
　(CMAs), 61
Comprehensive pain management, 160
Consumer-Directed Care (CDC), 11
Continence management, 71
Conversation, 87
COVID-19, 1, 4, 6, 9, 11, 13, 32, 39, 41, 73, 131, 145, 152, 168, 245
COVID-19 pandemic, 64, 69, 70, 72–74, 82, 86, 123, 191, 202, 225
　advance care planning, 52
　Australian nursing home residents, 64
Creativity, 185
Cultural competence, 17
Cultural recognition, 18
Cultural reform, 2
Culturally and linguistically diverse
　(CALD), 16, 268
Culture, 16
Curative/life-prolonging' stage, 138

D

Dance, 79
Death with dignity, 192
Decision-making style, 8

Index

Dehydration, 71
Dementia
  AD, 101
  behaviour management, 109
  definition, 100
  diagnosis, 100
  DLB, 102
  evidence-based care, 107
  families flourishing/floundering, 125
  misconceptions, 99
  types, 101
  younger onset, 102, 103
Dementia with Lewy bodies (DLB), 102
Depression, 75–76
Digital technology, 71
Dignity, 192–193
Dolls, 90
Duty of care, 58–59
Dying
  dying process, 185–186
  enjoyment and celebration, 184
  naturally, 181–182
  notification of dying, 204–205
  resident, 182–183
Dying process, 149
Dyspnoea, 72

E
Early palliative care, 138
Education, 177
Emotional care, 166
Emotional intelligence (EI), 212
Empathy, 158
End-of-life, 139, 147, 152, 153, 155, 164–167
Enrolled nurse (EN), 65
Enteral feeding, 70
Ethical enquiry, 93
Ethical tone, 6
Ethno-specific nursing homes, 11
Euphemisms, 179, 180, 198
Euthanasia, 192, 193
Evidence-based care
  aversion to water, 119
  bereavement planning, 124
  care at night, 121
  communal lounge, 119
  depression/delirium, 121
  dignity of risk, 114, 115
  end-of-life care, 122, 123
  language/dementia, 110
  LCOs, 124, 125
  leadership/education, 107, 108
  music, 112, 113

nursing home culture, 116
nursing home design, 117
pain/behaviour, 110, 111
psychiatric drug prescribing, 117, 118
qualified carers, 108, 109
residential aged care, 121
restrained usage, 109
segregation, 114
self-care, 120
violent behaviour, 111
Excursions, 90

F
Facility, 12
False comfort, 167
Families flourishing or floundering
  care planning, 131
  communication, 127, 128
  family assessment, 126, 127
  information deficit, 125, 126
  mealtimes, 132, 133
  sexuality, 132
  spiritual care, 130
  touch, 131
  visiting, 129
  wandering, 128, 129
Financial abuse, 263
Frailty, 170
Frontotemporal dementia, 101
Frustration, 183
Funding, 8, 9
Futile care, 193
Futile treatment, 193

G
General practitioners (GPs), 59–60, 140
Geriatricians, 141
Geriatrics, 60
Gerontologists, 60
Gerontology, 60
Gold card, 9
Good death, 178, 202, 203
Grattan Institute Report, 243
Grief, 188–190, 195

H
Hairdressing, 89
Hastened death, 199
Healthcare assistants, 66
Hidden death, 192
Hippocratic Oath, 77

Homelike atmosphere, 13
Human rights, 3–5, 20
Hydration, 145
Hypercognitive cultural story, 105

I
Ill-fitting, 168
Impending death, 191
Independence, 93
Individual, 3
Infection control, 73, 74, 225
Infections, 72
Informal carers
    citizenship, 50
    family, friends and community, 49
    volunteer programme, 50–51
Information technology (IT), 92
Intermittent pain, 160

L
Language, 178
Laughter, 87, 88
Laundering, 89
Leadership
    absent family, 239
    acknowledging night staff, 218, 219
    awareness, 212
    board governance, 213
    career path, 211
    caring for staff, 217, 218
    choosing, right staff, 215–217
    clinical governance, 213, 214
    communication, 212
    community
        accreditation and quality monitoring, 219, 220
        communication, 221
        exit interview, 223, 224
        long service, 222
        orientation, 222
        policies, 220, 221
        procedures, 220, 221
        staff departures, 223
        staff induction, 222
    decision-making, 235
    emotional intelligence (EI), 212
    false comfort, 239
    families' needs, 232–235
    family carer's ageing, 238
    family meeting, 229–232
    family member, 235–237
    innovative education, 225, 226
    nursing home culture, 211
    personal strengths, 211
    policies, 235
    political skills, 211
    procedures, 235
    quality of management, 211
    recognising family suffering, 238, 239
    reforms needed, 214, 215
    resident's family, 227–229
    routine education sessions, 224, 225
    students, 226, 227
    transformational leadership, 211
Lesbian, Gay, Bisexual, Transgender/Gender-Diverse, Intersex, and Queer (LGBTIQ), 19, 20
Liaison system, 151
Limitation of care orders (LCOs), 124
Loneliness, 86
Loss of independence, 93

M
Mealtime experience, 69
Medication advisory committees (MACs), 62
Medication compliance, 62
Medication management, 62–63
Mental health care, 75
Misunderstandings, 179
Mourning bench, 167
Multidisciplinary team, 141
Music, 79

N
National Disability Insurance Scheme (NDIS), 11
National Health and Medical Research Committee's (NHMRC's), 103
Nightingale, 180
Nightingale's wisdom, 180
Not for Resuscitation (NFR), 76–77
Nurse Practitioner (NP), 66–67
Nursing documentation, 59
Nursing home, 49, 54, 56, 211–221, 224, 226, 227, 229, 232–234, 237–239
    aged care placement, 54
    dehydration, 71
    enteral feeding, 70
    learning from Nightingale, 57
    preparing for death (*see* Preparation for death)
    residents' skincare, 67
    RN role, 63–65
    use of alcohol, 70

Index

Nursing home residents, 1, 3, 6, 10–13, 16, 138, 139, 143, 145, 151, 153, 154, 157, 159, 161–163, 168, 170, 182, 185–187, 191, 192, 246, 258, 262, 273, 276
Nutrition, 145
    and hydration, 67–69

O

Optimal pain management, 92
Oral and dental care, 74–75
Osteoarthritis, 161
Oxygen therapy, 72

P

Pain, 159, 160
Pain management, 75
Palliation, 137, 141, 142, 153, 163
Palliative care (PA), 137, 138
    basket, 164
    benefits, 138
    burden, 148
    cheer, 164
    chronic condition, 157
    cloak, 163
    comfort, 166, 167
    compassion, 158, 159
    cover-up, 171
    culture, 165
    death, 148, 149
    dementia, 146
    early, 138
    education, 143, 156, 157
    empathy, 158
    final wishes, 150
    frailty, 170
    GP, 164
    hope, 171
    learning from literature, 162
    learning from patient, 162, 163
    legal, 165, 166
    manager's knowledge, 144
    misunderstanding, 140
    multidisciplinary team, 141, 142
    opinions, 139
    pain, 159–162
    partnership
        business, 152
        families, 154–156
        GP, 153
        liaison system, 151
        nurse, 154
        nursing home, 151
        pharmacists role, 153, 154
    psychiatrist's role, 169
    public perceptions, 140
    reporting, 141
    RN, 145
    sedation, 169
    specialist, 143
    suffer, 167
    sympathy, 158
    terminal stage, 150
Palliative Care Australia (PCA), 152
Palliative care champion, 143
Palliative condition, 146
Parkinson's disease (PD), 102
Partnerships, 21, 52, 53, 60, 65
Perceptions and perspectives of death
    anticipating death, 186
    dignity, euthanasia and futile treatment, 192–194
    dying process, 185–186
    grief among support staff, 189–190
    grief, loss and bereavement, 188–189
    hospital, 190–191
    impending death, 191–192
    solitary death, 192
    volunteer's grief, 190
Personal Care Assistant (PCA), 65–66
Personal insights, 15
Person-centred care (PCC), 3, 105
Persons, 3
Pets, 90
Physical abuse, 263
Physical pain, 161, 168
Podiatry, 88
Poor continence planning, 71
Positive mythology, 5
Post-traumatic stress disorder (PTSD), 218
Premature death, 182
Preparation for death
    conversation, 178–179
    death, 177
    dying, 184–185
    dying naturally, 181–182
    dying resident, 182–183
    education, 177
    euphemisms and misunderstandings, 179–181
    language, 178
    preference and choice, 181
    premature death, 182
    staff preparedness, 183
    sudden death, 183–184
Privacy, 93

Procedural matters, death's aftermath
  care plan, 194
  death notification and certification, 196
  death review, 198–199
  death's effect on other residents, 194–195
  documenting the death, 195
  verification of death, 196
  written record, 194
Psychological abuse, 263
Psychological care, 66, 75
Psychological theory, 5
Psychosocial Care, 84, 85
Psychotropics, 109
Public Service Residential Aged Care Services (PSRACS), 8

Q
Quality continence care, 72
Quality of life, 147

R
Reablement, 78
Registered Nurse (RN), 9, 63–65
Resident, 12
Residential aged care, 2, 16, 19, 20, 66, 79, 90, 152, 159, 169, 190, 253, 260, 262, 272–274
Residential aged care facility (RACF), 13
Residential aged care service management, 63
Residential care, 9
Resident-related risks, 257
Residents, 7
Residents sleeping, 92
Residents' clothing, 89
Respect, 93
Respite care, 15
Right to vote, 50
Robots, 93
Royal Commission, 245
Royal Commission into Aged Care Quality and Safety, 212, 215
Royal Commission's Report, 245

S
Segregation, 114
Self-esteem, 219
Semantic dementia, 101
Sensory loss, 91–92
Serious Incident Response Scheme (SIRS), 255
Sexual abuse, 263
Sexual activity, 84

Sexual expression, 83
Sexuality, 84
Shadow Assistant Minister for Health and Ageing, 251
Shared room, 14
Singing, 79
Single room, 14
Sitting service, 15
Skilled communication, 167
Skincare, 67
Sleep, 92
Social care, 166
Social theory, 5
Solitary death, 192
Specialist palliative care, 143
Spiritual care, 66, 80–82
Spiritual factors, 81
Spirituality, 80, 81
Staff caring, 139
Staff education, 17
Staffing issues, 9, 63
Sudden death, 183, 184
Suffering, 168
Suicide, 186–188, 199
Support staff, 189
Sympathy, 158

T
Telehealth, 61
Telemedicine, 61
Therapeutic, 31
Transport modes, 90

U
Unregistered carers, 66

V
Videoconferencing, 61
Volunteering, 51
Volunteers, 183, 189, 190
  qualified coordinator, 51
  recruitment and supervision, 51

W
Workplace Health and Safety (WHS), 246
World Health Organization (WHO), 4

Y
Young People in Nursing Homes (YPINH), 20
Younger population, 5